A GUIDEBOOK TO MECHANISM IN
ORGANIC CHEMISTRY

A Guidebook to Mechanism in Organic Chemistry

PETER SYKES

M.SC., PH.D., F.R.I.C.

*Fellow of Christ's College and University
Lecturer in Organic Chemistry in the
University of Cambridge*

LONGMAN

LONGMAN GROUP LIMITED
London
*Associated companies, branches and representatives
throughout the world*

FIRST PUBLISHED 1961
SECOND IMPRESSION 1962
THIRD IMPRESSION 1962
FOURTH IMPRESSION 1963
FIFTH IMPRESSION 1963
SECOND EDITION 1965
SECOND IMPRESSION 1966
THIRD IMPRESSION 1967
FOURTH IMPRESSION 1969
THIRD EDITION 1970

TRANSLATIONS INTO
GERMAN, JAPANESE, AND SPANISH, 1964
FRENCH, 1966
ITALIAN, 1967
PORTUGUESE, 1969

SBN 582 44861·1

PRINTED IN GREAT BRITAIN BY
SPOTTISWOODE, BALLANTYNE AND CO. LTD
LONDON AND COLCHESTER

FOR
JOYCE

CONTENTS

PAGE

Foreword *by Professor Lord Todd, F.R.S.* . . . ix

Preface to Third Edition xi

1 Structure, Reactivity, and Mechanism . . . 1

2 Energetics, Kinetics, and the Investigation of Mechanism 35

3 The Strengths of Acids and Bases 52

4 Nucleophilic Substitution at a Saturated Carbon Atom . 73

5 Carbonium Ions, Electron-Deficient Nitrogen and Oxygen
 Atoms, and their Reactions 96

6 Electrophilic and Nucleophilic Substitution in Aromatic
 Systems 119

7 Addition to Carbon–Carbon Double Bonds . . . 156

8 Addition to Carbon–Oxygen Double Bonds . . . 178

9 Elimination Reactions 211

10 Carbanions and their Reactions 232

11 Radicals and their Reactions 256

Select Bibliography 289

Index 291

FOREWORD

THE great development of the theory of organic chemistry or more particularly of our understanding of the mechanism of the reactions of carbon compounds, which has occurred during the past thirty years or so, has wrought a vast change in outlook over the whole of the science. At one time organic chemistry appeared to the student as a vast body of facts, often apparently unconnected, which simply had to be learnt, but the more recent developments in theory have changed all this so that organic chemistry is now a much more ordered body of knowledge in which a logical pattern can be clearly seen. Naturally enough during the long period of development from the initial ideas of Lapworth and Robinson organic chemical theory has undergone continuous modification and it is only in comparatively recent times that it has become of such evident generality (although doubtless still far from finality) that its value and importance to the undergraduate student has become fully realised. As a result the teaching of organic chemistry has been, to some extent, in a state of flux and a variety of experiments have been made and a substantial number of books produced setting out different approaches to it. While it is the writer's opinion that it is unsatisfactory to teach first the main factual part of the subject and subsequently to introduce the theory of reaction mechanism, he is equally convinced that at the present time it is quite impracticable to concentrate almost entirely on theory and virtually to ignore the factual part of the subject. Organic chemical theory has not yet reached a level at which it permits prediction with any certainty of the precise behaviour of many members of the more complex carbon compounds which are of everyday occurrence in the practice of the science. Sound theory is vital to the well-being of organic chemistry; but organic chemistry remains essentially an experimental science.

In Cambridge we are seeking the middle way, endeavouring to build up both aspects of the subject in concert so that there is a minimum of separation between fact and theory. To achieve this the student is introduced at an early stage to the theoretical principles involved and to the essential reaction mechanisms illustrated by a modest number of representative examples. With this approach is coupled a more factual treatment covering the chemistry of the major groups of carbon compounds. Dr. Sykes [who has been intimately associated with this approach] has now written this aptly-named 'Guidebook' to reaction mechanism which sets out in an admirably lucid way what the student requires as a complement to his factual reading. I warmly commend it as a book which will enable students to rationalise many of the facts of organic chemistry, to appreciate the logic of the subject and in so doing to minimise the memory work involved in mastering it.

TODD.

PREFACE TO THIRD EDITION

IN the nine years that have elapsed since this book was first published, detailed knowledge of the pathways by which many organic reactions actually proceed has increased enormously; as, correspondingly, has the degree of sophistication with which we regard the interplay of electronic and steric effects that influence the reactivity and behaviour of molecular species under specified conditions.

The successes that have been achieved in this field are so manifest that one is not any longer in the position of needing to provide any justification of the importance of studying reaction pathways as an *intrinsic part* of organic chemistry even at the elementary level. It does remain true, however, because of the assumptions that have to be made when dealing with large and often complex molecules, that the effective application of electronic and quantum-mechanical considerations to the very bread and butter of organic chemistry still requires an imaginatively educated insight: it is the purpose of this book to seek to provide it.

In preparing this third edition the original arrangement, because it has been found to work in practice, has been largely adhered to. Thus the first chapter still seeks to provide a succinct statement of the basic principles involved, and the rest of the book endeavours to show how these work in explaining the variation of reactivity with structure and reaction conditions, the existence of three main classes of reagent—electrophiles, nucleophiles and radicals—and their involvement in the fundamental processes of organic chemistry—substitution, addition, elimination and rearrangement. A new chapter, 'Energetics, Kinetics, and the Investigation of Mechanism', has been added whose title is self-explanatory, and the rest of the book has been very extensively rewritten with the intention of discussing new material where this is relevant, with providing better examples of the operation of particular effects, and, above all, with providing more satisfactory explanations of why particular things happen in the way they are observed to do.

Preface

I have throughout been particularly conscious of the need to avoid markedly increasing the size (and price!) of the book, or its general level of sophistication: what success the book has enjoyed in the past has stemmed very largely from its being simple in essence, and I am most desirous that this should continue to be the case. The book was initially written primarily for first-year students in universities and technical colleges but has, in practice, been used by a surprisingly large number of sixth form students and I hope this may remain so with this new edition. The main aim, as with earlier editions, is to encourage an *understanding* rather than a mere *knowing* of organic reactions, with all that that implies in terms of the student then being able to put his newly gained understanding to work for him in situations that he has not previously encountered: surely the main aim of studying anything! Apart from being very extensively rewritten, the book has been wholly reset, not only the text matter but also the structural formulae, reaction schemes etc.

I am again most grateful to the many readers who have kindly pointed out errors to me, and also made useful criticisms and suggestions; wherever possible these have been incorporated. I owe a particular debt of gratitude in this respect to Dr. G. M. Clarke, who kindly read the book through recently while the revision was being made, and also to Mr. Hasib Ullah who so generously read the proofs. I should again welcome any further points that readers may care to raise about this new edition.

Cambridge,   PETER SYKES.
September 1969

1 STRUCTURE, REACTIVITY, AND MECHANISM

THE chief advantage of a mechanistic approach to the vast array of disparate information that makes up organic chemistry is the way in which a relatively small number of guiding principles can be used, not only to explain and interrelate existing facts but to forecast the outcome of changing the conditions under which already known reactions are carried out and to foretell the products that may be expected from new ones. It is the business of this chapter to outline some of these guiding principles and to show how they work. As it is the compounds of carbon with which we shall be dealing, something must first be said about the way in which carbon atoms can form bonds with other atoms, especially with other carbon atoms.

ATOMIC ORBITALS

The carbon atom has, outside its nucleus, six electrons which, on the Bohr theory of atomic structure, were believed to be arranged in orbits at increasing distance from the nucleus. These orbits represented gradually increasing levels of energy, that of lowest energy, the 1s, accommodating two electrons, the next, the 2s, also accommodating two electrons, and the remaining two electrons of a carbon atom going into the 2p level, which is actually capable of accommodating a total of six electrons.

The Heisenberg indeterminacy principle and the wave-mechanical view of the electron have made us do away with anything so precisely defined as actual orbits, and instead we can now only quote the relative probabilities of finding an electron at various distances from the nucleus of an atom. The classical orbits have, therefore, been replaced by three-dimensional *atomic orbitals*, which can be said to represent the shape and size of the space around the nucleus in which there is the greatest probability of finding an electron corresponding to a particular, quantised energy level: they are, indeed, rather like three-dimensional electronic contours. The size (energy level) of such atomic orbitals is defined by the principal quantum number, n, their shape and orientation by the subsidiary quantum numbers, l and m.

1

One limitation that theory imposes on such orbitals is that each may accommodate not more than two electrons, these electrons being distinguished from each other by having opposed ('paired') spins. This follows from the Pauli exclusion principle which states that no two electrons in any atom may have exactly the same set of quantum numbers.

It can be shown from wave-mechanical calculations that the $1s$ orbital (corresponding to the classical K shell) is spherically symmetrical about the nucleus of the atom and that the $2s$ orbital is similarly spherically symmetrical but at a greater distance from the nucleus; there is a region between the two latter orbitals where the probability of finding an electron approaches zero (a *spherical nodal surface*):

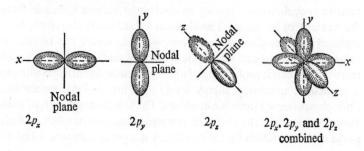

As yet, this marks no radical departure from the classical picture of orbits, but with the $2p$ level (the continuation of the L shell) a difference becomes apparent. Theory now requires the existence of *three $2p$ orbitals*, all of the same shape and energy level (orbitals having the same energy level are described as *degenerate*), arranged mutually at right-angles along notional x, y and z axes and, therefore, designated as $2p_x$, $2p_y$ and $2p_z$, respectively. Further, these three $2p$ orbitals are found to be not spherically symmetrical, like the $1s$ and $2s$, but 'dumb-bell' shaped with a plane, in which there is zero probability of finding an electron (*nodal plane*), passing through the nucleus (at right-angles to the x, y and z axes, respectively) and so separating the two halves of each dumb-bell:

We can thus designate the distribution of the six electrons of the carbon atom, in atomic orbitals, as $1s^2 2s^2 2p_x^1 2p_y^1$. This arises from Hund's rule that electrons will avoid occupying the same orbital so long as there are other energetically equivalent (degenerate, e.g. $2p_x, 2p_y, 2p_z$) atomic orbitals still empty—the $2p_z$ orbital thus remains unoccupied. The 2s orbital takes up its full complement of two electrons before the 2p orbitals begin to be occupied, however, as it is at a slightly lower energy level. This, however, represents the *ground state* of the free carbon atom in which only *two* unpaired electrons (in the $2p_x$ and $2p_y$ orbitals) are available for the formation of bonds with other atoms, i.e. at first sight carbon might appear to be only divalent.

This, however, is contrary to experience for though compounds are known in which carbon is bonded only to two other atoms, e.g. CCl_2 (p. 229), these are often highly unstable; in the enormous majority of its compounds carbon exhibits quadrivalency, e.g. CH_4. This can be achieved by uncoupling the $2s^2$ electron pair and promoting one of them to the vacant $2p_z$ orbital; the carbon atom is then in a higher energy (*excited*) state, $1s^2 2s^1 2p_x^1 2p_y^1 2p_z^1$, and having *four* unpaired electrons it is able to form *four*, rather than only *two*, bonds with other atoms or groups. The large amount of energy produced by forming these two extra bonds considerably outweighs that required (≈ 97 kcal/mole) for the initial $2s^2$ uncoupling and $2s \rightarrow 2p$ promotion.

HYBRIDISATION

A carbon atom combining with four other atoms clearly does not use the 2s and the three 2p atomic orbitals that would now be available, for this would lead to the formation of three bonds mutually at right angles (with the three 2p orbitals) and one different, non-directed bond (with the spherical 2s orbital). Whereas in fact the four C—H bonds in, for example, methane are known to be identical and symmetrically (tetrahedrally) disposed at an angle of 109° 28' to each other. This may be accounted for on the basis of redeploying the 2s and the three 2p atomic orbitals so as to yield four new, identical orbitals which are capable of forming stronger bonds (*cf.* p. 4). These new orbitals are known as sp^3 *hybrid* atomic orbitals and the process by which they are obtained as *hybridisation*:

3

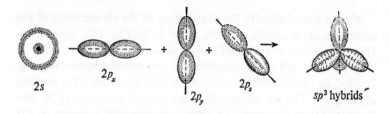

It should, however, be firmly emphasised, despite the diagram above, that hybridisation is a mathematical device in calculation and not a physical reality.

Similar but different redeployment is envisaged when a carbon atom combines with three other atoms, e.g. in ethylene (p. 7): sp^2 hybrid atomic orbitals disposed at 120° to each other in the same plane (*plane trigonal*) are then employed. Finally, when carbon combines with two other atoms, e.g. in acetylene (p. 8): sp^1 hybrid atomic orbitals disposed at 180° to each other (*co-linear*) are employed. In each case the *s* orbital is always involved as it is the one of lowest energy level.

BONDING IN CARBON COMPOUNDS

Bond formation between two atoms is then envisaged as the progressive overlapping of the atomic orbitals of the two participating atoms, the greater the possible overlapping, the stronger the bond so formed. The relative overlapping powers of atomic orbitals have been calculated as follows:

$$s = 1·00$$
$$p = 1·72$$
$$sp^1 = 1·93$$
$$sp^2 = 1·99$$
$$sp^3 = 2·00$$

It will thus be apparent why the use of hybrid orbitals, e.g. sp^3 hybrid orbitals in the combination of one carbon and four hydrogen atoms to form methane, results in the formation of stronger bonds.

When the atoms have come sufficiently close together, it can be shown that their two atomic orbitals are replaced by two *molecular* orbitals, one having less energy and the other more than the energies of the two separate atomic orbitals. These two new molecular orbitals spread over both atoms and either may contain the two electrons:

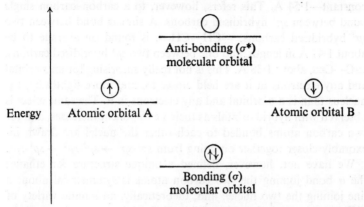

The molecular orbital of reduced energy is called the *bonding orbital* and its occupancy results in the formation of a stable bond between the two atoms; the molecular orbital of increased energy is called the *anti-bonding orbital*, it remains empty in the ground state of the molecule and need not here be further considered in the formation of stable bonds between atoms.

If overlap of the two atomic orbitals has taken place along their major axes, the resultant bonding molecular orbital is referred to as a σ orbital and the bond formed as a σ bond. The σ molecular orbital, and the electrons occupying it, are found to be *localised* symmetrically about the internuclear axis of the atoms that are bonded to each other. Thus on combining with hydrogen, the four hybrid sp^3 atomic orbitals of carbon overlap with the $1s$ atomic orbitals of four hydrogen atoms to form four identical, strong σ bonds, making angles of 109° 28′ with each other (the regular tetrahedral angle), in methane. A similar, exactly regular, tetrahedral structure will result with, for example, CCl_4 but with, say, CH_2Cl_2, though the arrangement will remain tetrahedral, it will depart very slightly from exact symmetry; the two large chlorine atoms will take up more room than hydrogen so that the H—C—H and Cl—C—Cl bond angles will differ slightly from 109° 28′ and from each other.

(i) Carbon–carbon single bonds

The combination of two carbon atoms, for example in ethane, results from the axial overlap of two sp^3 atomic orbitals, one from each

5

carbon atom, to form a strong σ bond between them. The carbon–carbon bond length in saturated compounds is found to be pretty constant—1·54 Å. This refers, however, to a carbon–carbon single bond between sp^3 hybridised carbons. A similar bond between two sp^2 hybridised carbons, $=CH—CH=$, is found on average to be about 1·47 Å in length, and one between two sp^1 hybridised carbons, $\equiv C—C\equiv$, about 1·38 Å. This is not really surprising for an s orbital and any electrons in it are held closer to, and more tightly by, the nucleus than is a p orbital and any electrons in it. The same effect is observed with hybrid orbitals as their s component increases, and for two carbon atoms bonded to each other the nuclei are drawn inexorably closer together on going from sp^3-$sp^3 \to sp^2$-$sp^2 \to sp^1$-sp^1.

We have not, however, defined a unique structure for ethane; the σ bond joining the two carbon atoms is symmetrical about a line joining the two nuclei, and, theoretically, an infinite variety of different structures is still possible, defined by the position of the hydrogens on one carbon atom relative to the position of those on the other. The two extremes of the possible species are known as the *eclipsed* and *staggered* forms; they and the infinite variety of structures lying between them are known as *conformations* of the ethane molecule. Conformations are defined as different arrangements of the same group of atoms that can be converted into one another without the breaking of any bonds.

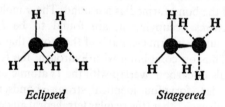

Eclipsed   *Staggered*

The staggered conformation is likely to be the more stable of the two for the hydrogen atoms are as far apart as they can get (3·1 Å) and any so-called 'non-bonded' interaction between them is thus at a minimum, whereas in the eclipsed conformation they are suffering the maximum of crowding (2·3 Å, slightly less than the sum of their van der Waals radii). The long cherished principle of free rotation about a carbon–carbon single bond is not contravened, however, as it has been shown that the eclipsed and staggered conformations differ by only ≈ 3 kcal/mole in energy content and this is small enough to allow their ready interconversion through the agency of ordinary thermal motions at room temperature, the rotation frequency at 25° being $\approx 10^{12}$ per sec. That such crowding *can* lead to a real restriction of rotation about a carbon–carbon single bond has been confirmed by the isolation of two forms of $CHBr_2CHBr_2$, though admittedly only at low temperatures where collisions between molecules do not provide enough energy to effect the interconversion.

(ii) Carbon–carbon double bonds

In ethylene each carbon atom is bonded to only *three* other atoms, two hydrogens and one carbon. Strong σ bonds are formed with these three atoms by the use of *three* orbitals derived by hybridising the $2s$ and, this time, *two* only of the carbon atom's $2p$ atomic orbitals— an atom will normally only mobilise as many hybrid orbitals as it has atoms or groups to form strong σ bonds with. The resultant sp^2 hybrid orbitals all lie in the same plane and are inclined at 120° to each other (*plane trigonal orbitals*). In forming the molecule of ethylene, two of the sp^2 orbitals of each carbon atom are seen as overlapping with the $1s$ orbitals of two hydrogen atoms to form two strong σ C—H bonds, while the third sp^2 orbital of each carbon atom overlap axially to form a strong σ C—C bond between them. It is found experimentally that the H—C—H and H—C—C bond angles are in fact 116·7° and 121·6°, respectively. The departure from 120° is hardly surprising seeing that different trios of atoms are involved.

This then leaves, on each carbon atom, *one unhybridised* $2p$ atomic orbital at right angles to the plane containing the carbon and hydrogen atoms. These two $2p$ atomic orbitals are parallel to each other and can themselves overlap, resulting in the formation of a bonding molecular orbital spreading over both carbon atoms and situated above and below the plane (i.e. it has a node in the plane of the molecule) containing the two carbon and four hydrogen atoms ($\diagdown$ indicates bonds to atoms lying *behind* the plane of the paper and $\blacktriangledown$ bonds to those lying in *front* of it):

This new bonding molecular orbital is known as a π orbital, and the electrons that occupy it as π electrons. The new π bond that is thus formed has the effect of drawing the carbon atoms closer together thus the C$=$C distance in ethylene is 1·33 Å compared with a C—C distance of 1·54 Å in ethane. The *lateral* overlap of the p atomic orbitals that occurs in forming a π bond is less effective than the axial overlap that occurs in forming a σ bond and the former is thus weaker than the latter. This is reflected in the fact that the energy of a carbon–carbon double bond, though more than that of a single bond is, indeed, less than twice as much. Thus the C—C bond energy in ethane is 83 kcal/mole, while that of C$=$C in ethylene is only 143 kcal/mole.

The lateral overlap of the two $2p$ atomic orbitals, and hence the strength of the π bond, will clearly be at a maximum when the two carbon and four hydrogen atoms are exactly coplanar, for it is only in this position that the p atomic orbitals are exactly parallel to each other and thus capable of the maximum overlapping. Any disturbance of this coplanar state by twisting about the σ bond joining the two carbon atoms would lead to reduction in π overlapping and hence a decrease in the strength of the π bond: it will thus be resisted. A theoretical justification is thus provided for the long observed resistance to rotation about a carbon–carbon double bond. The distribution of the π electrons in two layers, above and below the plane of the molecule, and extending beyond the carbon–carbon bond axis means that a region of negative charge is effectively waiting there to welcome any electron-seeking reagents (e.g. oxidising agents), so that it comes as no surprise to realise that the characteristic reactions of a carbon–carbon double bond are predominantly with such reagents (*cf.* p. 156). Here the classical picture of a double bond has been replaced by an alternative in which the two bonds joining the carbon atoms, far from being identical, are considered to be different in nature, strength and position.

(iii) Carbon–carbon triple bonds

In acetylene each carbon atom is bonded to only *two* other atoms, one hydrogen and one carbon. Strong σ bonds are formed with these two atoms by the use of *two* hybrid orbitals derived by hybridising the $2s$ and, this time, *one* only of the carbon atom's $2p$ atomic orbitals. The resultant sp^1 hybrid orbitals are co-linear. Thus, in forming the molecule of acetylene, these hybrid orbitals are used to form strong σ bonds between each carbon atom and one hydrogen atom and between the two carbon atoms themselves, resulting in a linear molecule having *two* unhybridised $2p$ atomic orbitals, at right angles to each other, on each of the two carbon atoms. The atomic orbitals on one carbon atom are parallel to those on the other and can thus overlap with each other resulting in the formation of *two* π bonds in planes at right angles to each other:

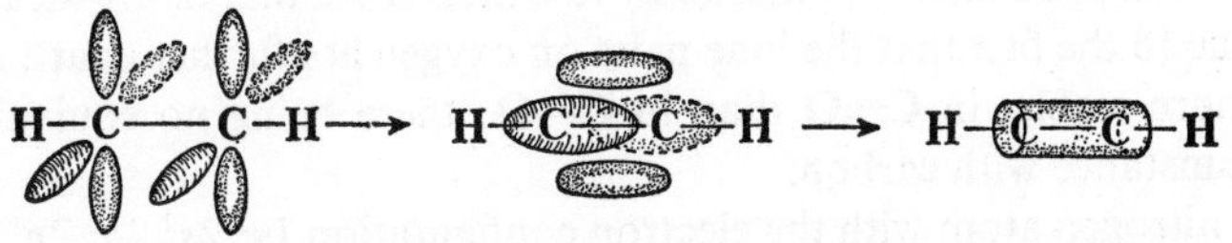

The acetylene molecule is thus effectively sheathed in a cylinder of negative charge. The $C\equiv C$ bond energy is 194 kcal/mole, so that the increment due to the third bond is less than that occurring on going from a single to a double bond. The $C\equiv C$ bond distance is 1·20 Å so that the carbon atoms have been drawn still further together, but here again the decrement on going $C=C \rightarrow C\equiv C$ is smaller than that on going $C-C \rightarrow C=C$.

(iv) Carbon–oxygen and carbon–nitrogen bonds

An oxygen atom has the electron configuration $1s^2 2s^2 2p_x^2 2p_y^1 2p_z^1$ and it too, on combining with other atoms, can be looked upon as utilising hybrid orbitals so as to form the strongest possible bonds. Thus on combining with the carbon atoms of two methyl groups to form dimethyl ether, CH_3-O-CH_3, the oxygen atom can use four sp^3 hybrid orbitals: two to form σ bonds by overlap with an sp^3 orbital of each of the two carbon atoms and the other two to accommodate its two lone pairs of electrons. The $C-O-C$ bond angle is found to be 110°, the $C-O$ bond length, 1·42 Å, and the bond energy, 86 kcal/mole.

An oxygen atom can also form a double bond to carbon, thus in

$$\overset{\textstyle O}{\underset{\textstyle \|}{}}$$

acetone, CH_3-C-CH_3, oxygen uses three sp^2 hybrid orbitals: one to form a σ bond by overlap with an sp^2 orbital of the carbon atom and the other two to accommodate its two lone pairs of electrons. This leaves an unhybridised p orbital on both oxygen and carbon that can overlap with each other laterally (*cf.* $C=C$, p. 7) to form a π bond:

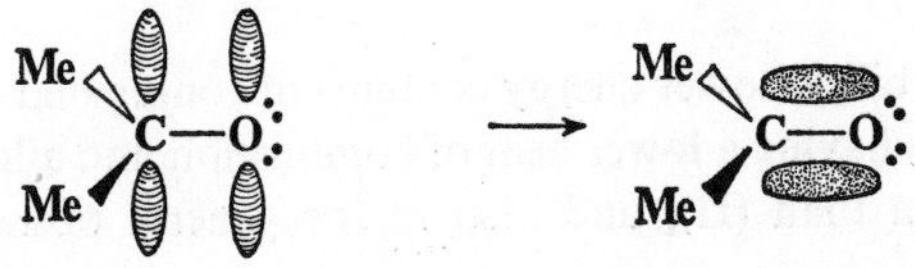

9

The C—C—O bond angle is found to be $\approx 120°$, the C=O bond length, 1·22 Å, and the bond energy, 179 kcal/mole. The fact that this is very slightly greater than twice the C—O bond energy, whereas the C=C bond energy is markedly less than twice that of C—C, may be due to the fact that the lone pairs on oxygen are further apart, and so more stable, in C=O than in C—O; there being no equivalent circumstance with carbon.

A nitrogen atom with the electron configuration $1s^2\, 2s^2\, 2p_x^1\, 2p_y^1\, 2p_z^1$ can also be looked upon as using hybrid orbitals in forming single, $\geq$C—N$\lessdot$, double, $\geq$C=N$\cdot$, and triple, –C≡N:, bonds with carbon. In each case one such orbital is used to accommodate the nitrogen lone pair of electrons and in double and triple bond formation one and two π bonds, respectively, are also formed by lateral overlap of the unhybridised p orbitals on nitrogen and carbon. Average bond lengths and bond energies are single, 1·47 Å and 73 kcal/mole, double, 1·27 Å and 147 kcal/mol, and triple, 1·16 Å and 213 kcal/mole.

(v) Conjugation

When we come to consider molecules that contain more than one multiple bond, e.g. dienes with two C=C bonds, it is found that compounds in which the bonds are *conjugated* (alternating multiple and single; I) are slightly more stable than those in which they are *isolated* (II):

$$\text{Me—CH—CH—CH—CH}_2 \longrightarrow \text{Me—CH—CH—CH—CH}_2$$

(I)

$$\text{CH}_2\text{—CH—CH}_2\text{—CH—CH}_2 \longrightarrow \text{CH}_2\text{—CH—CH}_2\text{—CH—CH}_2$$

(II)

This extra stability (lower energy content) of conjugated molecules is revealed in (I) having a lower heat of combustion and a lower heat of hydrogenation than (II), and also in the general observation that isolated double bonds can often be made to migrate quite readily so that they become conjugated:

$$MeCH{=}CH{-}CH_2{-}\underset{\underset{Me}{|}}{C}{=}O \xrightarrow[\text{catalyst}]{\text{Base}} MeCH_2{-}CH{=}CH{-}\underset{\underset{Me}{|}}{C}{=}O$$

Conjugation is not of course confined to carbon–carbon multiple bonds.

With both (I) and (II) above, overlap of the *p* atomic orbitals on adjacent carbon atoms could lead to the formation of two localised π bonds as shown, and the compounds would thus be expected to resemble ethylene, only twice as it were! This is indeed found to be the case with (II), but (I) does exhibit differences in terms of the slightly greater stability referred to above and also in spectroscopic behaviour (see below). On looking more closely at (I), however, it is seen that overlap could also take place between the *p* atomic orbitals of the two centre carbon atoms of the conjugated system, as well as between these and the *p* orbitals on the terminal carbon atoms. We could thus envisage a bonding molecular orbital involving all four carbon atoms (III)

Me$-$CH$-$CH$-$CH$-$CH$_2$ CH$_2{=}$CH$-$CH$=$CH$_2$

(III)   (IV)

CH$_2$$-CH-CH-CH_2$

(V)

in which the two electrons that occupy it are said to be *delocalised* as they are now spread over, and held in common by, the whole of the conjugated system; a further molecular orbital would, of course, be needed to accommodate the remaining two electrons, and this is found to be of a somewhat higher energy level.

The effect of any such delocalisation that actually takes place would clearly be to impose considerable restriction on rotation about the central C$-$C bond in (III) and (V)—all four *p* atomic orbitals would have to be parallel—which is indeed observed as highly preferred conformations of the compounds. The C$_2$$-C_3$ bond in butadiene (IV) is found to be slightly shorter (1·47 Å) than a normal C$-$C single

11

bond (1·54 Å), but not as short as might have been expected if delocalisation were making a really significant contribution: the overlap with each other of the *p* orbitals on C_2 and C_3 is apparently less effective than their overlap with the *p* orbitals on C_1 and C_4, respectively, i.e. the real structure of butadiene in fact departs little from the classical formulation (IV). The observed shortening of the C_2—C_3 bond may in any case be due simply to the state of hybridisation (*sp²*) of the two carbon atoms involved (*cf* p. 6). It thus seems likely that the observed stabilisation of a conjugated diene (compared with an isomeric isolated diene) may not be ascribable simply to delocalisation either, but that the state of hybridisation of the carbon atoms involved and the differing strength of the bonds formed between them also play a part.

Delocalisation is, however, much involved in stabilising the excited states of dienes, and of polyenes in general, i.e. in lowering the energy level of their excited states. The effect of this is to reduce the energy gap between ground and excited states of conjugated molecules, as compared with those containing isolated double bonds, and this energy gap is progressively lessened as the extent of conjugation increases. This means that the amount of energy required to effect the promotion from ground to excited state decreases with increasing conjugation, i.e. the wavelength at which the necessary radiation is absorbed increases. Simple dienes absorb in the ultra-violet region but as the extent of conjugation increases the absorption gradually moves towards the visible range, i.e. the compound becomes coloured. This is illustrated by the series of αω-diphenylpolyenes below:

$C_6H_5(CH=CH)_nC_6H_5$	*Colour*
$n = 1$	colourless
$n = 2-4$	yellow
$n = 5$	orange
$n = 8$	red

(vi) Benzene and aromaticity

One of the major problems of elementary organic chemistry is the detailed structure of benzene. The known planar structure of the molecule implies *sp²* hybridisation with *p* atomic orbitals, at right angles to the plane of the nucleus, on each of the six carbon atoms (VI):

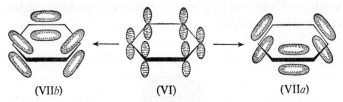

(VII*b*)   (VI)   (VII*a*)

Overlapping could, of course, take place 1,2; 3,4; 5,6; or 1,6; 5,4; 3,2 leading to formulations corresponding to the Kekulé structures (VII*a* and VII*b*) but any one *p* orbital could also overlap in both directions leading to delocalised molecular orbitals (*cf.* butadiene). The six *p* atomic orbitals can give rise to six molecular orbitals (*n* atomic orbitals always give rise to *n* molecular orbitals) whose energy levels can be represented as:

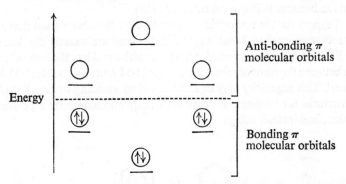

The delocalised bonding orbital of lowest energy is cyclic and embraces all the six carbon atoms of the ring (VIII*a*): two electrons are thus accommodated. The two further delocalised bonding orbitals, of equal energy (VIII*b* and VIII*c*), also encompass all six carbon atoms but each has a nodal plane (*cf.* p. 7) perpendicular to the plane of the ring, in addition to that coinciding with the plane of the ring; each of these molecular orbitals accommodates two electrons:

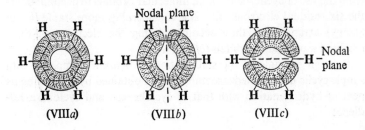

(VIII*a*)   (VIII*b*)   (VIII*c*)

13

The net result is an annular cloud of negative charge above and below the plane of the ring (IX):

(IX)

The influence of this charge cloud on the type of reagents that will attack benzene is discussed below (p. 119).

Support for the above view is provided by the observation that all the carbon–carbon bond lengths in benzene are exactly the same, 1·39 Å, i.e. benzene is a regular hexagon with bond lengths somewhere in between the normal values for a single (1·54 Å) and a double (1·33 Å) bond. This regularity may be emphasised by avoiding writing Kekulé structures for benzene, as these are clearly an inadequate representation, and instead using:

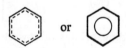

There remains, however, the question of the much remarked stability of benzene. Part of this no doubt arises from the disposition of the three plane trigonal σ bonds about each carbon at their optimum angle of 120° (the regular hexagonal angle), but a larger part stems from the use of cyclic, delocalised molecular orbitals to accommodate the six residual electrons; this is a considerably more stable (lower energy) arrangement than accommodating the electrons in three localised π molecular orbitals (*cf*. p. 7).

A rough estimate of the stabilisation of benzene compared with simple cyclic unsaturated structures can be obtained by comparing its heat of hydrogenation with that of cyclohexene and cyclohexa-1,3-diene:

$$\text{cyclohexene} + H_2 \longrightarrow \text{cyclohexane} \qquad \Delta H = -28\cdot6 \text{ kcal/mole}$$

$$\text{cyclohexadiene} + 2H_2 \longrightarrow \text{cyclohexane} \qquad \Delta H = -55\cdot6 \text{ kcal/mole}$$

$$\text{benzene} + 3H_2 \longrightarrow \text{cyclohexane} \qquad \Delta H = -49\cdot8 \text{ kcal/mole}$$

The heat of hydrogenation of the diene is very nearly twice that of cyclohexene and the heat of hydrogenation of the three double bonds in a Kekulé structure should thus be of the order of $-28\cdot6 \times 3 = -85\cdot8$ kcal/mole; but when benzene is actually hydrogenated only 49·8 kcal/mole are evolved. Thus interaction of the π electrons in benzene may be said to result in the molecule being stabler by 36 kcal/mole than if no such interaction took place (the stabilisation arising from such interaction in conjugated dienes is only ≈ 6 kcal/mole). This amount by which benzene is stabilised compared with a theoretical cyclohexatriene should properly be called its *stabilisation energy*, it is, however, often called its *delocalisation energy*, which immediately begs the question as to how much of the stabilisation is actually due to delocalisation. The term *resonance energy*, though still widely used, is highly unsatisfactory on semantic grounds as it immediately conjures up visions of rapid oscillation between one structure and another, e.g. the Kekulé structures, thus entirely misrepresenting the real state of affairs.

The stability conferred on benzene by cyclic delocalisation etc. also explains why the characteristic reactions of simple aromatic systems are substitutions rather than the addition reactions that might, from the classical Kekulé structures, be expected and which are indeed realised with non-cyclic conjugated trienes. For addition would lead to a product in which delocalisation, though still possible, could now involve only four carbon atoms and would have lost its characteristic cyclic character (XI; *cf.* butadiene), whereas substitution results in the retention of delocalisation essentially similar to that in benzene with all that it implies (X):

Addition   Substitution
(XI)   (X)

In other words, substitution can take place with overall retention of aromaticity, addition cannot (*cf.* p. 120).

(vii) Conditions necessary for delocalisation

The difficulty in finding a satisfactory representation for the carbon–carbon bonding in benzene brings home to us the fact that our normal way of writing bonds between atoms as single, double or triple, involving two, four and six electrons, respectively, is clearly inadequate: some bonds involve other, even fractional, numbers of electrons. This is seen very clearly in the acetate ion (XII)

$$CH_3-C\overset{O}{\underset{O^\ominus}{\diagdown}}$$

(XII)

where, in flat contradiction of the above formula, X-ray crystallography shows that the two oxygen atoms are indistinguishable from each other, the two carbon–oxygen bond distances being the same and involving the same number of electrons.

These difficulties have led to the convention of representing molecules that cannot adequately be written as a single classical structure by a combination of two or more classical structures, the so-called *canonical structures*, linked by a double-headed arrow. The way in which one of these structures can be related to another often being indicated by curved arrows, the *tail* of the curved arrow indicating where an electron pair moves *from* and the *head* of the arrow where it moves to:

$$CH_3C\overset{O}{\underset{O^\ominus}{\diagdown}} \quad \leftrightarrow \quad CH_3C\overset{O^\ominus}{\underset{O}{\diagdown}} \quad \equiv \quad \left[CH_3C\overset{O}{\underset{O}{\diagdown}}\right]^\ominus$$

(XIIIa)   (XIIIb)   (XIV)

It cannot be too firmly emphasised, however, that the acetate ion does *not* have two possible, and alternative, structures which are rapidly interconvertible, but a single, real structure (XIV)—sometimes referred to as a hybrid—for which the classical (canonical) structures (XIIIa) and (XIIIb) are less exact, limiting approximations.

A certain number of limitations must be borne in mind, however, when considering delocalisation and its representation through two or more classical structures as above. Broadly speaking, the more canonical structures that can be written for a compound, the greater the delocalisation of electrons and the more stable the compound will be. These structures must not vary too widely from each other in energy content, however, or those of higher energy will contribute so little to the hybrid as to make their contribution virtually irrelevant. The stabilising effect is particularly marked when the structures have the same energy content, as with (XIIIa) and (XIIIb) above. Structures involving separation of charge (*cf.* p. 22) may be written but, other things being equal, these are usually of higher energy content than those in which such separation has not taken place, and hence contribute correspondingly less to the hybrid. The structures written must all contain the same number of paired electrons and the constituent atoms must all occupy essentially the same positions relative to each other in each canonical structure. If delocalisation is to be significant, all atoms attached to unsaturated centres must lie in the same plane or nearly so; examples where delocalisation, with consequent stabilisation, is actually prevented by steric factors are discussed subsequently (p. 26).

The delocalisation that is so effective in promoting the stability of aromatic compounds results when there are no *partially* occupied orbitals of the same energy. The complete filling of such orbitals can be shown to occur in cyclic systems with $2 + 4n$ π electrons, and 6π electrons ($n = 1$) is the arrangement that occurs by far the most commonly in aromatic compounds. 10π electrons ($n = 2$) are present in naphthalene (delocalisation energy, 61 kcal/mole) and 14π electrons ($n = 3$) in anthracene and phenanthrene (delocalisation energies, 84 and 91 kcal/mole, respectively) and though these substances are not monocyclic like benzene, the introduction of the trans-annular bonds that makes them bi- and tri-cyclic, respectively, seems to cause relatively little perturbation so far as delocalisation of the π electrons over the cyclic group of ten or fourteen carbon atoms is concerned.

Quasi aromatic structures are also known in which the stabilised cyclic species is an ion, e.g. the tropylium (cycloheptatrienyl) cation (p. 100), the cyclopentadienyl anion (p. 235), in both of which $n = 1$, and even more surprisingly cyclopropenyl cations (p. 100) for which $n = 0$. Further, the ring structure need not be purely carbocyclic and pyridine (p. 146), for example, with a nitrogen atom in the ring, is at least as highly stabilised as benzene.

THE BREAKING AND FORMING OF BONDS

A covalent bond between two atoms can essentially be broken in the following ways:

$$R:X \quad \rightleftharpoons \quad \begin{array}{l} R\cdot + \cdot X \\ R:^{\ominus} + X^{\oplus} \\ R^{\oplus} + :X^{\ominus} \end{array}$$

In the first case each atom separates with one electron leading to the formation of highly reactive entities called free radicals, owing their reactivity to their unpaired electron; this is referred to as *homolytic fission* of the bond. Alternatively, one atom may hold on to both electrons, leaving none for the other, the result in the above case being a negative and a positive ion, respectively. Where R and X are not identical, the fission can, of course, take place in either of two ways, as shown above, depending on whether R or X retains the electron pair. Either of these processes is referred to as *heterolytic fission*, the result being the formation of an *ion pair*. Formation of a covalent bond can, of course, take place by the reversal of any of these processes.

Such free radicals or ion pairs are formed transiently as reactive intermediates in a very wide variety of organic reactions as will be shown below. Reactions involving radicals tend to occur in the gas phase and in solution in non-polar solvents and to be catalysed by light and by the addition of other radicals (p. 256). Reactions involving ionic intermediates take place more readily in solution in polar solvents. Many of these ionic intermediates can be considered as carrying their charge on a carbon atom, though the ion is often stabilised by delocalisation of the charge, to a greater or lesser extent, over other carbon atoms or atoms of different elements:

$$CH_2{=}CH{-}CH_2{-}\overset{\bullet\bullet}{\underset{}{O}}H$$

$\Big\downarrow H^{\oplus}$

$$CH_2{=}CH{-}CH_2{-}\overset{\oplus}{\underset{H}{O}}H \xrightarrow{-H_2O} [CH_2{=}CH{-}\overset{\oplus}{C}H_2 \leftrightarrow \overset{\oplus}{C}H_2{-}CH{=}CH_2]$$

$$CH_3{-}\overset{\overset{O}{\|}}{C}{-}CH_3 \xrightarrow{\ominus OH} [CH_3{-}\overset{O}{C}\overset{\ominus}{C}H_2 \leftrightarrow CH_3{-}\overset{\overset{O^{\ominus}}{|}}{C}{=}CH_2] + H_2O$$

When a positive charge is carried on carbon the entity is known as a *carbonium ion* and when a negative charge, a *carbanion*. Though such ions may be formed only transiently and be present only in minute concentration, they are nevertheless often of paramount importance in controlling the reactions in which they participate.

FACTORS AFFECTING ELECTRON-AVAILABILITY IN BONDS AND AT INDIVIDUAL ATOMS

In the light of what has been said above, any factors that influence the relative availability of electrons (the *electron density*) in particular bonds or at particular atoms in a compound might be expected to affect very considerably its reactivity towards a particular reagent; for a position of high electron availability will be attacked with difficulty if at all by, for example, $^{\ominus}$OH, whereas a position of low electron availability is likely to be attacked with ease, and vice versa with a positively charged reagent. A number of such factors have been recognised.

(i) Inductive effect

In a covalent single bond between unlike atoms the electron pair forming the σ bond is never shared absolutely equally between the two atoms; it tends to be attracted a little more towards the more electronegative atom of the two. Thus in an alkyl halide

$$\underset{(XV)}{{>}C{-}Cl} \qquad \underset{(XVIa)}{\overset{\delta+\ \ \delta-}{{>}C{-}Cl}} \qquad \underset{(XVIb)}{{>}C{\rightarrow}Cl}$$

19

the electron density tends to be greater nearer chlorine than carbon (XV) as the former is the more electronegative; this is generally represented in classical formulae by (XVI*a*) or (XVI*b*). If the carbon atom bonded to chlorine is itself attached to further carbon atoms, the effect can be transmitted further:

$$\text{C--C--C} \rightarrow \text{C} \twoheadrightarrow \text{Cl}$$
$$4 \quad 3 \quad 2 \quad 1$$

The effect of the chlorine atom's partial appropriation of the electrons of the carbon–chlorine bond is to leave C_1 slightly electron-deficient; this it seeks to rectify by, in turn, appropriating slightly more than its share of the electrons of the σ bond joining it to C_2, and so on down the chain. The effect of C_1 on C_2 is less than the effect of Cl on C_1, however, and the transmission quickly dies away in a saturated chain, usually being too small to be noticeable beyond C_2.

Most atoms and groups attached to carbon exert such *inductive effects* in the same direction as chlorine, i.e. they are electron-withdrawing, owing to their being more electronegative than carbon, the major exception being alkyl groups which are electron-donating.* Though the effect is quantitatively rather small, it is responsible for the increase in basicity that results when one of the hydrogen atoms of ammonia is replaced by an alkyl group (p. 64), and, in part at any rate, for the readier substitution of the aromatic nucleus in toluene than in benzene. Several suggestions have been made to account for the electron-donating abilities of CH_3, CH_2R, CHR_2 and CR_3, none of which is wholly convincing and the matter can be said to be unsettled.

All inductive effects are permanent polarisations in the ground state of the molecule and are therefore manifested in its physical properties, for example, its dipole moment.

(ii) Mesomeric or conjugative effect

This is essentially a further statement of the electron redistribution that can take place in unsaturated and especially in conjugated systems via their π orbitals. An example is the carbonyl group (p. 178) whose properties are not entirely satisfactorily represented by the

* The metal atoms in, for example, lithium alkyls and Grignard reagents, both of which compounds are largely covalent, are also electron-donating, leading to negatively polarised carbon atoms in each case: $R \twoheadleftarrow Li$ and $R \twoheadleftarrow MgHal$ (*cf*. p. 192).

20

classical formulation (XVII), nor by the extreme polar structure (XVIII), that may be derived from it by an electron shift as shown:

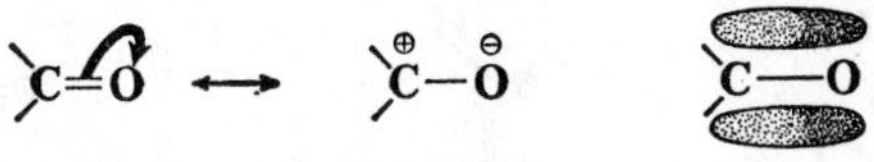

(XVII)   (XVIII)   (XIX)

The actual structure is somewhere in between, i.e. a hybrid of which the above are the canonical forms, perhaps best represented by (XIX) in which the readily polarisable π electrons are drawn preferentially towards the more electronegative oxygen atom. If the carbonyl group is conjugated with a $\diagup$C$=$C$\diagdown$ bond, the above polarisation can be transmitted further via the π electrons:

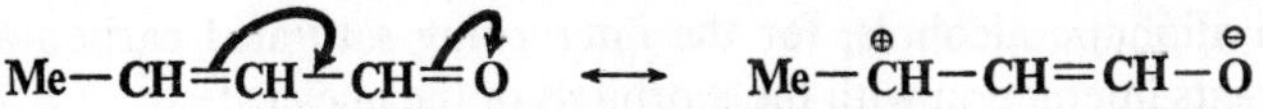

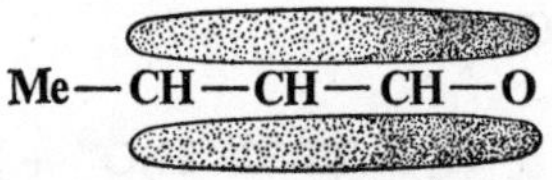

(XX)

Delocalisation takes place (XX), so that an electron-deficient atom results at C_3, as well as at C_1 as in a simple carbonyl compound. The difference between this transmission via a conjugated system and the inductive effect in a saturated system is that here the effect suffers much less diminution by its transmission, and the polarity alternates.

The stabilisation that can result by delocalisation of a positive or negative charge in an ion via its π orbitals can be a potent feature in making the formation of the ion possible in the first place (*cf.* p. 54). It is, for instance, the stabilisation of the phenoxide ion (XXI) by delocalisation of its charge via the delocalised π orbitals of the nucleus that is largely responsible for the acidity of phenol (*cf.* p. 55):

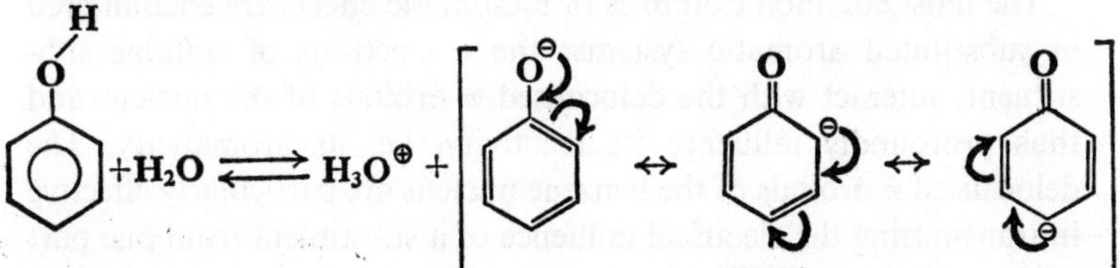

(XXI)

An apparently similar delocalisation can take place in undissociated phenol itself involving an unshared electron pair on the oxygen atom

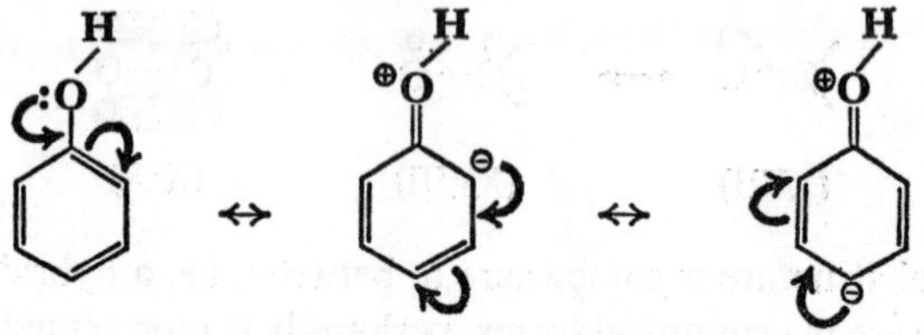

but this involves separation of charge and will thus be correspondingly less effective than the stabilisation of the phenoxide ion which does not.

Similar stabilisation of the anion with respect to the neutral molecule cannot occur with benzyl alcohol, which is thus no more acidic than aliphatic alcohols, for the intervening saturated carbon atom prevents interaction with the π orbitals of the nucleus:

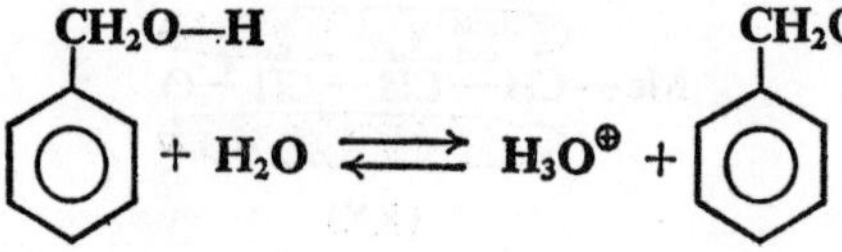

(XXII)

In other words, the phenoxide ion (XXI) is stabilised with respect to the phenol molecule whereas the benzyl oxide ion (XXII) is not so stabilised with respect to the benzyl alcohol molecule. The even greater acidity of RCO_2H is due largely to the particularly effective stabilisation by delocalisation of the carboxylate anion (*cf.* p. 16).

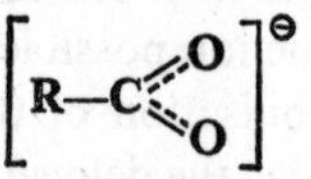

(XXIII)

The most common examples of mesomeric effects are encountered in substituted aromatic systems: the π electrons of suitable substituents interact with the delocalised π orbitals of the nucleus and thus profoundly influence its reactivity, i.e. its aromaticity. The delocalised π orbitals of the benzene nucleus are particularly effective in transmitting the electrical influence of a substituent from one part of the molecule to another:

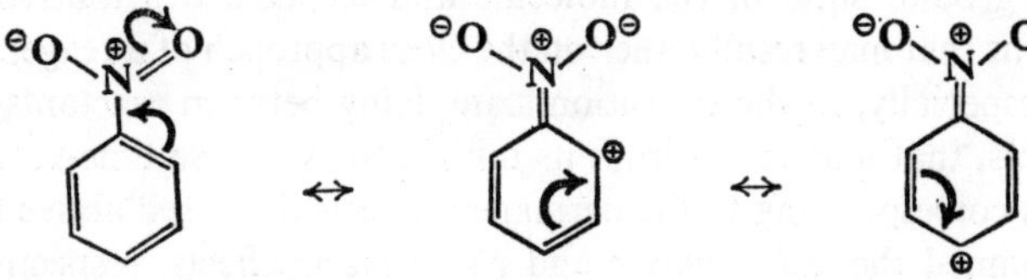

Thus the nitro group in nitrobenzene lowers the density of negative charge over the nucleus, as compared with benzene itself: it is an *electron-withdrawing group*, in contrast to the negatively charged oxygen atom in the phenoxide ion (XXI), which is an *electron-donating group*. Because of the presence of an electron-withdrawing group, nitrobenzene will be less readily attacked than benzene itself by positive ions or electron-deficient reagents (oxidising agents such as **KMnO₄**, for example) which, as will be seen below (p. 119), are exactly the type of reagents involved in normal aromatic substitution reactions.

Mesomeric, like inductive, effects are permanent polarisations in the ground state of the molecule and are therefore manifested in the physical properties of the compounds in which they occur. The essential difference between inductive and mesomeric effects is that the former occur essentially in saturated groups or compounds, the latter in unsaturated and, especially, conjugated compounds. The former involve the electrons in σ bonds, the latter those in π bonds and orbitals. Inductive effects are transmitted over only quite short distances in saturated chains before dying away, whereas mesomeric effects may be transmitted from one end to the other of quite large molecules provided that conjugation (i.e. delocalised π orbitals) is present through which they can proceed. Either effect influences the behaviour of compounds in both essentially static and dynamic situations: in both the position of equilibria and rates of reaction, in the strength of acids and bases as much as in the reactivity of alkyl halides or the relative ease of substitution of different aromatic species.

(iii) Time-variable effects

Some workers have sought to distinguish between effects such as the two considered above which are permanent polarisations manifested in the ground state of the molecule and changed distributions of electrons that may result either on the close approach of a reagent or, more especially, in the transition state, lying between reactants and products, that may result from its initial attack. These time-variable factors corresponding to the permanent effects discussed above have been named the *inductomeric* and *electromeric* effects, respectively. Any such effects can be looked upon as *polarisabilities* rather than as polarisations, for the distribution of electrons reverts to that of the ground state of the molecule attacked if either of the reactants is removed without reaction being allowed to take place or, if a transition state is actually formed, it decomposes to yield the starting materials again.

Such time-variable effects, being only temporary, will not, of course, be reflected in the physical properties of the compounds concerned. It has often proved impossible to distinguish experimentally between permanent and time-variable effects, but it cannot be too greatly emphasised that despite the difficulties in distinguishing what proportions of a given effect are due to permanent and to time-variable factors, the actual close approach of a reagent may have a profound effect in enhancing reactivity in a reactant molecule and so in promoting reaction.

(iv) Hyperconjugation

The inductive effect of alkyl groups is normally found to be in the order

$$CH_3 \rightarrow \; < \; Me \rightarrow CH_2 \rightarrow \; < \; \begin{matrix} Me \searrow \\ \\ Me \nearrow \end{matrix} CH \rightarrow \; < \; \begin{matrix} Me \searrow \\ Me \rightarrow C \rightarrow \\ Me \nearrow \end{matrix}$$

as would be expected. When, however, the alkyl groups are attached to an unsaturated system, e.g. a double bond or a benzene nucleus, this order is found to be disturbed and in the case of some conjugated systems actually reversed. It thus appears that alkyl groups are capable, in these circumstances, of giving rise to electron release by a mechanism different from the inductive effect, and of which methyl is the most successful exponent. This has been explained as proceeding by an extension of the conjugative or mesomeric effect, delocalisation taking place in the following way:

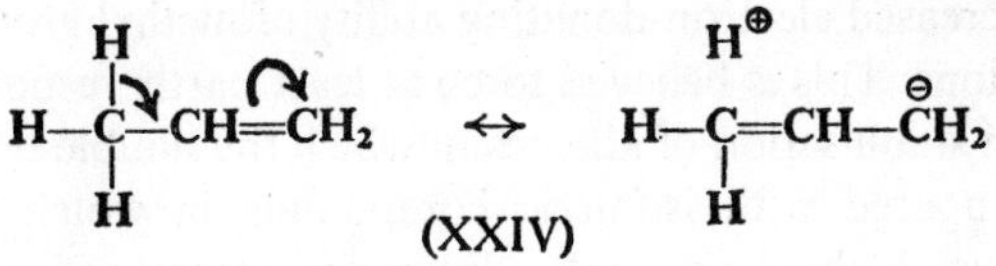

$$H-\overset{\displaystyle H}{\underset{\displaystyle H}{C}}CH=CH_2 \quad \leftrightarrow \quad H-\overset{\displaystyle H^{\oplus}}{\underset{\displaystyle H}{C}}=CH-\overset{\ominus}{C}H_2$$

(XXIV)

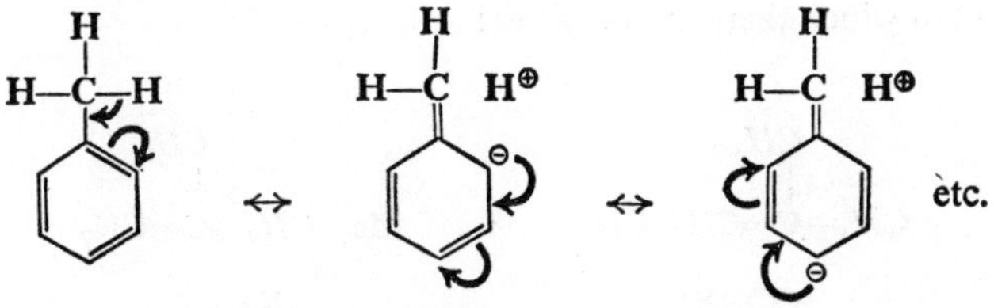

(XXV)

This effect has been called *hyperconjugation* and has been used successfully to explain a number of otherwise unconnected phenomena. It should be emphasised that it is not suggested that a proton actually becomes free in (XXIV) or (XXV), for if it moved from its original position one of the conditions necessary for delocalisation to occur would be controverted (p. 17).

The reason for reversal of the expected electron-donating ability in the series $CH_3 \rightarrow MeCH_2 \rightarrow Me_2CH \rightarrow Me_3C$ is that hyperconjugation depends for its operation on hydrogen attached to carbon atoms α- to the unsaturated system. This is clearly at a maximum with CH_3 (XXIV) and nonexistent with Me_3C (XXVIII), provided it is assumed that no similar effect of comparable magnitude occurs in C—C bonds,

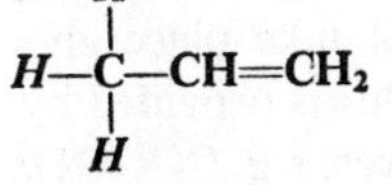

$$H-\overset{\displaystyle H}{\underset{\displaystyle H}{C}}-CH=CH_2 \qquad Me-\overset{\displaystyle H}{\underset{\displaystyle H}{C}}-CH=CH_2 \qquad Me-\overset{\displaystyle Me}{\underset{\displaystyle H}{C}}-CH=CH_2$$

(XXIV) 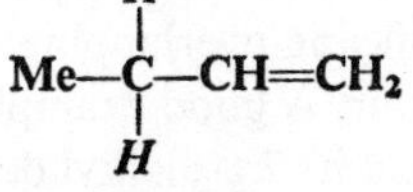   (XXVI) 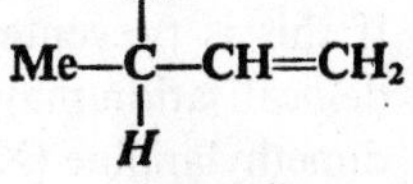   (XXVII)

$$Me-\overset{\displaystyle Me}{\underset{\displaystyle Me}{C}}-CH=CH_2$$

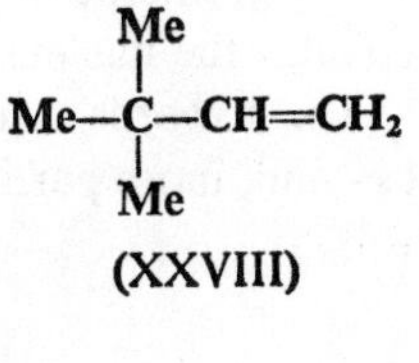

(XXVIII)

hence the increased electron-donating ability of methyl groups under these conditions. This is believed to be at least partly responsible for the increased stabilisation of alkenes in which the double bond is not terminal compared with isomeric compounds in which it is, i.e. (XXIX) in which there are nine α-hydrogen atoms compared with (XXX) in which there are only five:

$$CH_3-\overset{\overset{\displaystyle CH_3}{|}}{C}=CH-CH_3 \qquad\qquad Me-CH_2-\overset{\overset{\displaystyle CH_3}{|}}{C}=CH_2$$

(XXIX)   (XXX)

This leads to their preferential formation in reactions which could lead to either compound on introduction of the double bond (p. 219), and even to the fairly ready isomerisation of the less into the more stable compound.

Although hyperconjugation has proved useful on a number of occasions, its validity is by no means universally accepted and much further work needs to be done on its theoretical justification.

STERIC EFFECTS

We have to date been discussing factors that may influence the relative availability of electrons in bonds or at particular atoms in a compound, and hence influence that compound's reactivity. The working or influence of these factors may, however, be modified or even nullified by the operation of steric factors; thus effective delocalisation via π orbitals can only take place if the p or π orbitals on the atoms involved in the delocalisation can become parallel or fairly nearly so. If this is prevented, significant overlapping cannot take place and delocalisation may not occur. A good example of this is provided by dimethylaniline (XXXI) and its 2,6-dialkyl derivatives, e.g. (XXXII). The **NMe₂** group in (XXXI), being electron-donating (due to the unshared electron pair on nitrogen interacting with the delocalised π orbitals of the nucleus), activates the nucleus towards attack by the diazonium cation **PhN₂$^{\oplus}$**, i.e. towards azo-coupling, leading to preferential substitution at *o*- and, more particularly, *p*-positions (*cf.* p. 138):

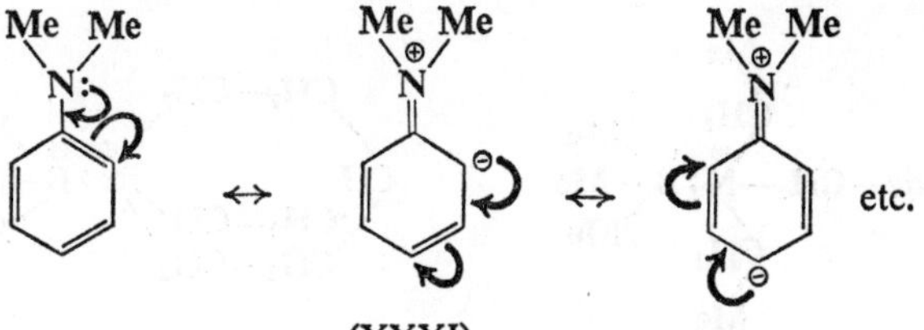

(XXXI)

The 2,6-dimethyl derivative (XXXII) does not couple under these conditions, however, despite the fact that the methyl groups that have been introduced are much too far away for their bulk to interfere directly with attack at the *p*-position. The failure to couple at this position is, in fact, due to the two methyl groups in the *o*-positions to the NMe_2 interfering sterically with the two methyl groups attached to nitrogen and so preventing these lying in the same plane as the benzene nucleus. This means that the *p* orbitals of nitrogen and the ring carbon atom to which it is attached are prevented from becoming parallel to each other and their overlapping is thus inhibited. Electronic interaction with the nucleus is thus largely prevented and transfer of charge to the *p*-position, with consequent activation to attack by $PhN_2^{\oplus}$ as in (XXXI), does not now take place:

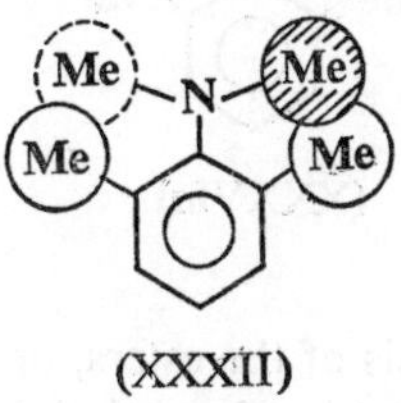

(XXXII)

The most common steric effect, however, is the classical 'steric hindrance' in which it is apparently the sheer bulk of groups that is influencing the reactivity of a site in a compound directly and not by promoting or inhibiting electron-availability. This has been investigated closely in connection with the stability of the complexes formed by trimethylboron with a wide variety of amines. Thus the complex (XXXIII) formed with triethylamine dissociates extremely readily whereas the complex (XXXIV) with quinuclidine, which can be looked upon as having three ethyl groups on nitrogen that are 'held back' from interfering sterically with attack on the nitrogen atom, is very stable:

Me
/
CH$_2$
\$\oplus$ $\ominus$ Me
Me—CH$_2$—N : B—Me
/ \
CH$_2$ Me
\
Me

CH$_2$—CH$_2$
/ \$\oplus$ $\ominus$ Me
CH N : B—Me
\`·CH$_2$···CH$_2$·/ \
CH$_2$—CH$_2$ Me

(XXXIII)   (XXXIV)

That this difference is not due to differing electron availability at the nitrogen atom in the two cases is confirmed by the fact that the two amines differ very little in their strengths as bases (*cf.* p. 71): the uptake of a proton constituting very much less of a steric obstacle than the uptake of the relatively bulky BMe$_3$.

More familiar examples of steric inhibition, however, are probably the difficulties met with in esterifying tertiary acids (XXXV) and 2,6-disubstituted benzoic acids (XXXVI*a*, *cf.* p. 207)

R$_3$CCO$_2$H

CO$_2$H
Me⤹⤸Me

(XXXV)   (XXXVI*a*)

CO$_2$H
CH$_2$
Me⤹⤸Me

(XXXVI*b*)

and then in the hydrolysis of the esters, or other derivatives such as amides, once made. That this effect is indeed steric is suggested by its being much greater in magnitude than can be accounted for by any influence the alkyl substituents might be expected to have on electron availability and also by its non-occurrence in the aromatic species if these substituents are in the *m*- or *p*-positions. Further, if the carboxyl group is moved away from the nucleus by the introduction of a CH$_2$ group, the new acid (XXXVI*b*) may now be esterified as readily as the unsubstituted species: the functional group is now beyond the steric range of the methyl substituents.

It should be emphasised that such steric inhibition is only an extreme case and any factors which disturb or inhibit a particular orientation of the reactants with respect to each other, short of

28

preventing their close approach, can also profoundly affect the rate of reactions: a state of affairs that is often encountered in reactions in biological systems.

CLASSIFICATION OF REAGENTS

Reference has already been made to electron-donating and electron-withdrawing groups, their effect being to render a site in a molecule electron-rich or electron-deficient, respectively. This will clearly nfluence the type of reagent with which the compound will most readily react. An electron-rich compound, such as phenoxide ion, (XXXVII)

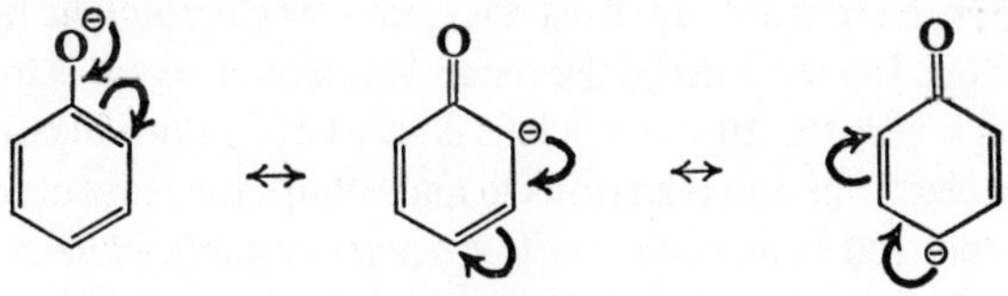

(XXXVII)

will tend to be most readily attacked by positively charged ions such as $PhN_2^{\oplus}$, the diazonium cation, or by other species which, though not actually ions themselves, possess an atom or centre which is electron-deficient, for example the sulphur atom in sulphur trioxide:

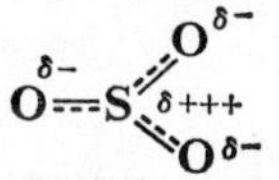

Diazo-coupling (p. 131) or sulphonation (p. 127) takes place on a carbon atom of the nucleus rather than on oxygen because of the charge-transfer from oxygen to carbon that can take place as shown above and because of the greater stability of the carbon rather than the oxygen-substituted products.

Conversely, an electron-deficient centre, such as the carbon atom in methyl chloride (XXXVIII)

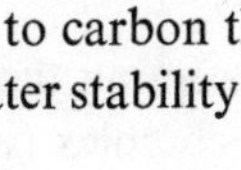

(XXXVIII)

29

will tend to be most readily attacked by negatively charged ions such as $^\ominus$OH, $^\ominus$CN, etc., or by other species which, though not actually ions themselves, possess an atom or centre which is electron-rich, for example the nitrogen atom in ammonia or amines, $H_3N:$ or $R_3N:$. It must be emphasised that only a *slightly* unsymmetrical distribution of electrons is required for a reaction's course to be dominated: the presence of a full-blown charge on a reactant certainly helps matters along but is far from being essential. Indeed the requisite unsymmetrical charge distribution may be induced by the mutual polarisation of reagent and substrate on their close approach as when bromine adds to ethylene (p. 156).

In reactions of the first type the reagent is looking for a position in the substrate to be attacked where electrons are especially readily available; such reagents are thus referred to as *electrophilic reagents* or *electrophiles*. In reactions of the second type the reagent is looking for a position where the atomic nucleus is short of its normal complement of orbital electrons and is anxious to make it up: the reagents employed are thus referred to as *nucleophilic reagents* or *nucleophiles*.

This differentiation can be looked upon as a special case of the acid/base idea. The classical definition of acids and bases is that the former are proton-donors and the latter proton-acceptors. This was made more general by Lewis who defined acids as compounds prepared to accept electron pairs and bases as substances that could provide such pairs. This would include a number of compounds not previously thought of as acids and bases, e.g. boron trifluoride (XXXIX)

$$
\begin{array}{ccc}
\underset{\text{(XXXIX)}}{\overset{F}{\underset{F}{\overset{|}{F}}}\!\!-\!\!B + :N\!\!\underset{Me}{\overset{Me}{-}}\!\!Me} & \rightleftharpoons & \underset{\text{(XL)}}{\overset{F}{\underset{F}{\overset{|}{F}}}\!\!\blacktriangleright\!\!B\!:\!\overset{\oplus}{N}\!\!\underset{Me}{\overset{Me}{\triangleleft}}\!\!Me}
\end{array}
$$

which acts as an acid by accepting the electron pair on nitrogen in trimethylamine to form the complex (XL), and is therefore referred to as a *Lewis acid*. Electrophiles and nucleophiles in organic reactions can be looked upon essentially as acceptors and donors, respectively, of electron pairs from and to other atoms, most frequently *carbon*. Electrophiles and nucleophiles also, of course, bear a relationship to oxidising and reducing agents for the former can be looked upon as electron-acceptors and the latter as electron-donors. A number of the more common electrophiles and nucleophiles are listed below.

Electrophiles

$H^{\oplus}$, $H_3O^{\oplus}$, HNO_3, H_2SO_4, HNO_2 (i.e. $^{\oplus}NO_2$, SO_3 and $^{\oplus}NO$ respectively), $PhN_2^{\oplus}$

BF_3, $AlCl_3$, $ZnCl_2$, $FeCl_3$, Br_2, $\overset{*}{I}$—Cl, $\overset{*}{NO}$—Cl, $\overset{*}{CN}$—Cl, H_2O_2, O_3

$$R_3\overset{\oplus}{\underset{*}{C}}, \quad \underset{*}{>}C{=}O, \quad R{-}\overset{\overset{O}{\|}}{\underset{*}{C}}{-}Cl, \quad R{-}\overset{\overset{O}{\|}}{\underset{*}{C}}{-}O{-}\overset{\overset{O}{\|}}{C}{-}R, \quad \underset{*}{CO_2}$$

Nucleophiles

$H^{\ominus}$, $H_2\overset{\ominus}{N}$, $HO^{\ominus}$, $RO^{\ominus}$, $RS^{\ominus}$, $RCO_2^{\ominus}$, $Hal^{\ominus}$, $HSO_3^{\ominus}$, $^{\ominus}CN$, $RC{\equiv}C^{\ominus}$,

$^{\ominus}CH(CO_2Et)_2$

(XLI)

$>\!O\!:$, $\quad \overset{\rightharpoonup}{N}\!:$, $\quad >\!S\!:$

$R\underset{*}{MgBr}$, $\underset{*}{RLi}$, $Li\overset{*}{AlH_4}$, $*\!\langle\bigcirc\rangle\!-\!\overset{*}{O}^{\ominus}$

Where a reagent is starred, the star indicates the atom that accepts electrons from, or donates electrons to, the substrate as the case may be. It rapidly becomes apparent that no clear distinction can be made between what constitutes a reagent and what a substrate, for though HNO_3, $^{\ominus}OH$, etc., are normally thought of as reagents, the diethyl malonate carbanion (XLI) could, at will, be either reagent or substrate, when reacted with, for example, an alkyl halide. The reaction of the former on the latter is a nucleophilic attack, while that of the latter on the former would be looked upon as an electrophilic attack; but from the standpoint of whichever reactant a reaction itself is viewed, its essential nature is not for a moment in doubt.

It should not be forgotten, however, that reactions involving free radicals as the reactive entities are also known. These are much less susceptible to variations in electron density in the substrate than are reactions involving polar intermediates, but they are greatly affected by the addition of small traces of substances that either liberate or remove radicals. They are considered in detail below (p. 266).

TYPES OF REACTION

Within this classification there are essentially four kinds of reaction which organic compounds can undergo:

(*a*) Displacement (or substitution) reactions.

(*b*) Addition reactions.
(*c*) Elimination reactions.
(*d*) Rearrangements.

In (*a*) it is displacement from carbon that is normally referred to but the atom displaced can be either hydrogen or another atom or group. In electrophilic substitution reactions it is often hydrogen that is displaced, classical aromatic substitution (p. 119) being a good example:

In nucleophilic substitution reactions, it is often an atom other than hydrogen that is displaced (pp. 73, 150):

$$NC^{\ominus} + R\!-\!Br \longrightarrow NC\!-\!R + Br^{\ominus}$$

but nucleophilic displacement of hydrogen is also known (p. 148)

though hydride ion is not actually liberated as such as will be seen subsequently (p. 149). Radical-induced displacement reactions are also known, for example the halogenation of alkanes, though it should be emphasised that these are not *direct* displacements on carbon (*cf.* p. 276).

Addition reactions, too, can be electrophilic, nucleophilic or radical-induced depending on the type of species that initiates the process. Addition to simple carbon–carbon double bonds is normally either an electrophilic or radical reaction: an example is the addition of **HBr**

$$>C=C< \xrightarrow{\text{HBr}} >\overset{}{\underset{\text{H}}{C}}-\overset{\text{Br}}{\underset{}{C}}<$$

which can be initiated by the attack of either $H^\oplus$ (p. 160) or $Br\cdot$ (p. 271) on the double bond. By contrast, the addition reactions exhibited by the carbon–oxygen double bond in simple aldehydes and ketones are usually nucleophilic in character (p. 178). An example is the base-catalysed formation of cyanhydrins in liquid **HCN**:

$$>\overset{\delta+}{C}\overset{\delta-}{=}O \underset{\text{slow}}{\rightleftharpoons} >C\overset{O^\ominus}{\underset{CN}{}} \underset{\text{fast}}{\overset{\text{HCN}}{\rightleftharpoons}} >C\overset{OH}{\underset{CN}{}} + {}^\ominus CN$$

Elimination reactions are, of course, essentially the reversal of addition reactions; the most common is the loss of atoms or groups from adjacent carbon atoms to yield alkenes:

$$>\overset{}{\underset{Br}{C}}-\overset{H}{\underset{}{C}}< \xrightarrow{-HBr}$$
$$>C=C<$$
$$>\overset{}{\underset{OH}{C}}-\overset{H}{\underset{}{C}}< \xrightarrow{-H_2O}$$

Rearrangements may also proceed via intermediates that are essentially cations, anions, or radicals, though those involving carbonium ions or other electron-deficient species are by far the most common. They can involve either the mere migration of a functional group (p. 103) as in the allylic system

33

$$\underset{\text{CH=CH}_2}{\overset{\text{OH}}{>\!\!C<}} \quad \overset{\text{H}^{\oplus}}{\longrightarrow} \quad \left[\underset{\text{CH=CH}_2}{\overset{\oplus}{>\!\!C}} \quad \leftrightarrow \quad \underset{\text{CH--}\overset{\oplus}{\text{CH}}_2}{>\!\!C} \right] + \text{H}_2\text{O}$$

$$\Big\downarrow \text{H}_2\text{O}$$

$$\underset{\text{CH--CH}_2\text{--OH}}{>\!\!C} + \text{H}^{\oplus}$$

or the actual rearrangement of the carbon skeleton of a compound as in the pinacol (XLII) → pinacolone (XLIII) change (p. 107):

$$\underset{\substack{| \quad | \\ \text{HO} \quad \text{OH}}}{\text{Me}_2\text{C--CMe}_2} \quad \overset{\text{H}^{\oplus}}{\longrightarrow} \quad \text{Me}_3\text{CCOMe}$$

$$\text{(XLII)} \qquad\qquad\qquad \text{(XLIII)}$$

The actual rearrangement step is often followed by a displacement, addition or elimination reaction before a final, stable product is obtained.

2 ENERGETICS, KINETICS, AND THE INVESTI- GATION OF MECHANISM

WE have now listed a number of electronic and steric factors that can influence the reactivity of a compound in a given situation, and also the types of reagent that might be expected to attack particular centres in such a compound especially readily. We have as yet, however, had little to say directly about how these electronic and steric factors, varying from one structure to another, actually operate in energetic and kinetic terms to influence the course and rate of a reaction. These considerations are of major importance, not least for the light they might be expected to throw on the detailed pathway by which a reaction proceeds.

ENERGETICS OF REACTION

In considering that conversion of starting materials into products which constitutes an organic reaction, one of the things that we particularly want to know is '*how far* will the reaction go over towards products?' Systems tend to move towards their most stable state, so we might expect that the more stable the products are compared with the starting materials, the further over in the former's favour any equilibrium between them might be expected to lie, i.e. the larger $\Delta_{stability}$ is in the diagram below, the greater the expected conversion into products:

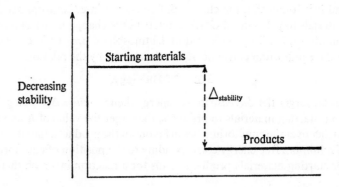

However, it quickly becomes apparent that the simple energy change that occurs on going from starting materials to products and that may readily be measured as the heat of reaction, ΔH, (H is a measure of the heat content, or *enthalpy*, of a compound and ΔH is preceded by a minus sign if the products have a lower heat content than the starting materials; when there is such a decrease in enthalpy the reaction is exothermic) is not an adequate measure of the difference in stability between them, for there is often found to be no correlation between ΔH and the equilibrium constant for the reaction, K. Highly exothermic reactions are known with only small equilibrium constants (little conversion of starting materials into products) and some reactions with large equilibrium constants are known that are actually endothermic (enthalpy of products higher than that of starting materials): clearly some factor in addition to enthalpy must be concerned in the relative stability of chemical species.

That this should be so is a corollary of the Second Law of Thermodynamics which is concerned essentially with probabilities and with the tendency for ordered systems to become disordered: a measure of the degree of disorder of a system being provided by its *entropy*, S. In seeking their most stable condition, systems tend towards *minimum* energy (actually enthalpy, H) and *maximum* entropy (disorder or randomness), a measure of their relative stability must thus embrace a compromise between H and S and is provided by the *Gibb's free energy*, G, which is defined by

$$G = H - TS$$

where T is the absolute temperature. The free energy change during a reaction, at a particular temperature, is thus given by

$$\Delta G = \Delta H - T\Delta S$$

and it is found that the change in free energy in going from starting materials to products, $\Delta G°$ ($\Delta G°$ refers to the change under standard conditions, i.e. for 1 mole and at 1 atmosphere pressure), is related to the equilibrium constant, K, for the change by the relation

$$-\Delta G° = 2 \cdot 303 RT \log K$$

i.e. the larger the *decrease* in free energy (hence, *minus* $\Delta G°$) on going from starting materials to products, the larger the value of K and the further over the equilibrium lies in favour of the products, the situation of minimum free energy corresponding to the position of equilibrium for starting materials/products. Thus for a reaction in which there is

36

no free energy change ($\Delta G° = 0$), $K = 1$ corresponding to 50% conversion of starting materials into products. Increasing *positive* values of $\Delta G°$ imply rapidly *decreasing* fractional values of K (the relationship is a logarithmic one) corresponding to extremely little conversion into products, while increasing *negative* values of $\Delta G°$ imply correspondingly rapidly *increasing* values of K; thus a $\Delta G°$ of -10 kcal/mole corresponds to an equilibrium constant of $\approx 10^7$ and essentially complete conversion into products. A knowledge of the standard free energies of our starting materials and of our products, which have been measured for a large number of organic compounds, thus enables us to predict the expected extent of the conversion of the former into the latter.

The ΔH factor for the change can be equated with the difference in energies of the bonds in the starting materials and in the products, and an approximate value of ΔH for a reaction can often be predicted from tables of standard bond energies: which is hardly unexpected, as it is from ΔH data that the average bond energies have been compiled in the first place! The entropy factor cannot be explained quite so readily, but effectively it relates to the number of possible ways in which their total, aggregate energy may be shared out among an assembly of molecules, and also to the number of ways in which an individual molecule's quanta of energy may be shared out for translational, rotational, and vibrational purposes, of which the translational is likely to be by far the largest in magnitude. Thus for a reaction in which there is an increase in the number of molecular species on going from starting materials to products

$$A \rightleftharpoons B + C$$

there is likely to be a sizeable increase in entropy because of the gain in translational freedom. The $-T\Delta S$ term may then be large enough to outweigh the $+\Delta H$ term of an endothermic reaction, thus leading to a negative value for ΔG and an equilibrium that lies well over in favour of products. If the reaction is exothermic anyway (ΔH negative), ΔG will of course be even more negative and the equilibrium constant, K, correspondingly larger still. Where the number of participating species decreases on going from starting materials to products there is likely to be a decrease in entropy (ΔS negative), hence

$$\Delta G = \Delta H - (-)T\Delta S$$

37

and unless the reaction is sufficiently exothermic (ΔH negative and large enough) to counterbalance this, ΔG will be positive and the equilibrium thus well over in favour of starting materials. Cyclisation reactions may also be attended by a decrease in entropy (though less than when the number of participating species decreases on forming products),

$$CH_3(CH_2)_3CH{=}CH_2 \; \rightleftharpoons \; \begin{array}{c} H_2 \\ C \\ H_2C \quad CH_2 \\ H_2C \quad CH_2 \\ C \\ H_2 \end{array}$$

for though there is no significant difference in translational entropy, a constraint is imposed by the change on rotation about the carbon–carbon single bonds, which is essentially free in the open chain starting material but is greatly restricted in the cyclic product.

It should not be overlooked that the entropy term involves temperature ($T\Delta S$) while the enthalpy (ΔH) term does not and their relative contributions to the free energy change may be markedly different for the same reaction carried out at widely different temperatures.

KINETICS OF REACTION

Though a negative $\Delta G°$ is a necessary condition if a change is to take place at all under a given set of conditions, the free energy criterion is not of itself sufficient for it tells us nothing about *how fast* the starting materials are converted into products. Thus for the oxidation of cellulose

$$(C_6H_{10}O_5)_n + 6nO_2 \; \rightleftharpoons \; 6nCO_2 + 5nH_2O$$

$\Delta G°$ is negative and large in magnitude so that the equilibrium lies essentially completely over in favour of CO_2 and H_2O, but a newspaper (very largely cellulose) can be read in the air (or even in oxygen!) at room temperature for long periods of time without it noticeably fading away to gaseous products: the *rate* of the conversion is extremely slow under these conditions despite the very large $-\Delta G°$, though it is, of course, speeded up at higher temperatures. The conversion of starting materials into products, despite a negative $\Delta G°$, is rarely if ever a mere run down-hill (I), there is generally a barrier to be overcome *en route* (II):

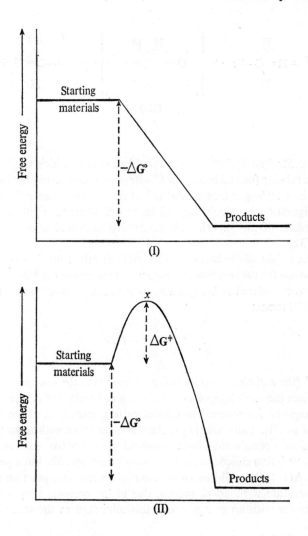

(i) **Reaction rate and free energy of activation**

The position x in the *energy profile* (II) corresponds to the least stable configuration through which the starting materials pass in their conversion to products and is generally referred to as the *activated complex* or *transition state*. It should be emphasised that this is merely a highly unstable state that is passed through in a dynamic process, and *not* an intermediate that can actually be isolated. An example is (III)

$$\text{HO}^{\ominus} + \text{H} \overset{\text{H}}{\underset{\text{H}}{\rightarrow}} \text{C} - \text{Br} \quad \rightarrow \quad \left[\text{HO} \overset{\delta-}{\cdots\cdots} \overset{\text{H}}{\underset{\text{H}}{\text{C}}} \overset{\text{H}}{\cdots\cdots} \overset{\delta-}{\text{Br}} \right] \quad \rightarrow \quad \text{HO} - \overset{\text{H}}{\underset{\text{H}}{\text{C}}} \blacktriangleleft \text{H} + \text{Br}^{\ominus}$$

(III)

in the alkaline hydrolysis of methyl bromide in which the **HO—C** bond is being formed before the **C—Br** bond is completely broken and the three hydrogen atoms attached to carbon are passing through a configuration in which they all lie in one plane (at right-angles to the plane of the paper). This reaction is discussed in detail below (p. 73).

The height of the barrier in (II), $\Delta G^{\ddagger}$ is called the *free energy of activation* for the reaction (the higher it is the slower the reaction) and can be considered as being made up of enthalpy ($\Delta H^{\ddagger}$) and entropy ($T\Delta S^{\ddagger}$) terms:

$$\Delta G^{\ddagger} = \Delta H^{\ddagger} - T\Delta S^{\ddagger}$$

$\Delta H^{\ddagger}$ (the *enthalpy of activation*) corresponds to the energy necessary to effect the stretching or even breaking of bonds that is an essential prerequisite for reaction to take place (e.g. stretching of the **C—Br** bond in III). Thus reacting molecules must bring with them to any collision a certain minimum threshold of energy for reaction to be possible (often called simply the *activation energy*, ΔE, but equatable with $\Delta H^{\ddagger}$); the well-known increase in the rate of a reaction as the temperature is raised is, indeed, due to the growing proportion of molecules with an energy above this minimum as the temperature rises.

The $\Delta S^{\ddagger}$ term (the *entropy of activation*) again relates to randomness. If formation of the transition state requires the imposition of a high degree of organisation in the way the reactant molecules must approach each other, and also of the concentration of their energy in particular linkages so as to allow of their ultimate breakage, then the attainment of the transition state is attended by a sizeable decrease in entropy (randomness), and the possibility of its formation is correspondingly low.

(ii) Kinetics and the rate-limiting step

Experimentally, the measurement of reaction rates consists in investigating the rate at which starting materials disappear and/or products appear at a particular (constant) temperature and seeking to relate this to the concentration of one or all of the reactants. The reaction may be followed by a variety of methods, e.g. titrimetric, spectroscopic, colorimetric, etc., but the crucial step normally involves testing the crude kinetic data against various possible functions of concentration until a reasonable fit is obtained. Thus for the reaction

$$CH_3Br + {}^{\ominus}OH \rightarrow CH_3OH + Br^{\ominus}$$

it comes as no surprise to find a rate equation

$$Rate = k[CH_3Br][{}^{\ominus}OH]$$

where k is known as the *rate constant* for the reaction, which is said to be of *second order* overall, *first order* with respect to CH_3Br and *first order* with respect to ${}^{\ominus}OH$.

That the stoichiometry is often an inadequate guide, however, is shown for reactions such as the base-catalysed bromination of acetone

$$CH_3COCH_3 + Br_2 \rightarrow CH_3COCH_2Br + HBr$$

which is found to follow the rate equation

$$Rate = k[CH_3COCH_3][{}^{\ominus}OH]$$

i.e. bromine is apparently not involved but the alkalinity of the solution is (*cf.* p. 249). Clearly bromine must be involved in some stage of the overall reaction as it is incorporated into the final product, but it is not involved in the process whose rate we are actually measuring. In fact few, if any, organic reactions are simple one-shot processes as depicted in (II) and the stoichiometry of a reaction as conveyed by its equation is essentially 'chemical book-keeping', which may well give no guide whatever to the actual pathway by which the reaction takes place. This is obvious enough in an extreme example such as the formation of hexamine

$$6CH_2O + 4NH_3 \rightarrow C_6H_6N_4 + 6H_2O$$

where the chance of the simultaneous collision of *six* molecules of CH_2O and *four* of NH_3 in a *ten*-body collision are effectively non-existent. But even where the stoichiometry is less extreme reactions are

41

normally composite consisting of a number of successive steps (each a two-body collision) of which we are actually measuring the slowest, and thus *rate-limiting*, one—the kinetic 'bottleneck' on the production line converting starting materials into products:

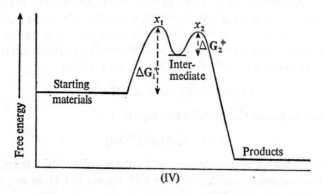

(IV)

In (IV) starting materials are being converted via transition state x_1 into an intermediate, which then decomposes into products via a second transition state x_2. As depicted above the formation of the intermediate via x_1, is the more energy-demanding ($\Delta G_1^{\neq} > \Delta G_2^{\neq}$) of the two steps and hence will be the slower, i.e. the stage whose rate our kinetic experiments will actually be measuring. It is followed by a fast (less energy demanding), non-rate-limiting conversion of the intermediate into products. The above bromination of acetone, can, under certain conditions, be said to follow an idealised pattern corresponding to (IV), in which slow, rate determining removal of proton by base results in the formation of the carbanion intermediate (V), which then undergoes rapid, non-rate-limiting attack by Br_2 to

$$HO^{\ominus} + H\!-\!CH_2COCH_3 \xrightarrow[\text{Slow}]{} H_2O + {}^{\ominus}CH_2COCH_3$$

$$(V)$$

$$\xrightarrow[\text{fast}]{Br_2} BrCH_2COCH_3 + Br^{\ominus}$$

yield bromoacetone and bromide ion as the products. It should be emphasised that though this explanation is a reasonable deduction from the experimentally established rate equation, the latter cannot be said actually to *prove* the former. Our experimentally determined rate equation will give us information about the species that are involved up to and including the rate-limiting stage of a reaction but only inferentially about any intermediates, and not at all, except by default, about the species that are involved in rapid non-rate-limiting processes beyond this stage.

In considering the effect that a change of conditions, e.g. of solvent or in the structure of the starting material, might be expected to have on the rate of a reaction, we need to know what effect such changes will have on the stability (free energy level) of the transition state: any factors which serve to stabilise it will lead to its more rapid formation, and the opposite will also apply. It is seldom if ever possible to obtain such detailed information about the high-energy transition states and the best we can commonly do is to take the relevant intermediates as models for them, seeing what effect one might expect such changes to have on these models. Such a model is not unreasonable; the transiently formed intermediate in (IV) closely resembles, in terms of free energy level, the transition state that precedes it and might be expected to resemble it in structure as well. Certainly such an intermediate is normally likely to be a better model for the transition state than the starting material would be. Thus σ complexes in aromatic electrophilic substitutions are used as models for the transition states that are their immediate precursors (p. 138).

The effect of catalysts in increasing the rate at which a reaction will take place is to provide an alternative path of less energetic demand, often through the formation of a new and more stable (lower energy) intermediate:

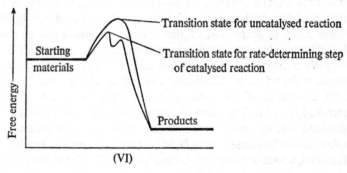

(VI)

Thus the rate of hydration of an alkene directly with water is often extremely slow

$$>C=C< \ + \ H_2O \ \longrightarrow \ \overset{\displaystyle HO}{\underset{\displaystyle H}{>C-C<}}$$

but it can be greatly speeded up by the presence of an acid catalyst which effects initial protonation of the alkene to a carbonium ion intermediate. This is then followed by easy and rapid attack on the positively charged ion by a water molecule acting as a nucelophile, and finally by liberation of a proton which is able to function again as a catalyst (p. 162):

(iii) Kinetic *v.* thermodynamic control

Where a starting material may be converted into two or more alternative products, e.g. in electrophilic attack on an aromatic species that already carries a substituent (p. 138), the proportions in which the alternative products are formed is often determined by their relative rates of formation: the faster a product is formed the more of it there will be in the final product mixture; this is known as *kinetic control*. This is not always what is observed however, for if one or more of the alternative reactions is reversible, or if the products are readily interconvertible directly under the conditions of the reaction, the composition of the final product mixture may be dictated not by the relative rates of formation of the different products but by their relative thermodynamic stabilities in the reaction system: we are then seeing *thermodynamic* or *equilibrium control*. Thus the nitration of toluene is found to be kinetically controlled whereas the Friedel-Crafts alkylation of the same species is often thermodynamically controlled (p. 144). The form of control that operates may also be influenced by the reaction condition, thus the sulphonation of naphthalene with concentrated H_2SO_4 at 80° is essentially kinetically controlled, whereas at 160° it is thermodynamically controlled (p. 146).

INVESTIGATION OF REACTION MECHANISMS

It is seldom, if ever, possible to provide complete and entire information, structural, energetic, and stereochemical, about the pathway that is traversed by any chemical reaction: too much is involved.

Sufficient data can nevertheless often be gathered to show that one theoretically possible mechanism is just not compatible with the experimental results, or to demonstrate that of several alternatives one is a good deal more likely than the others.

(i) Kinetics

The largest body of information has come—and still does come—from kinetic studies, but the interpretation of kinetic data in mechanistic terms is not always quite as simple as might at first sight be supposed. Thus the effective reacting species, whose concentration really determines the reaction rate, may well differ considerably from what we put into the reaction mixture to start with and whose changing concentration we are actually seeking to measure. Thus in aromatic nitration the effective attacking species is usually $^{\oplus}NO_2$ (p. 122), but it is HNO_3 that we put into the reaction mixture and whose changing concentration we are measuring; the relationship between the two may well be complex and so, therefore, may be the relation between the rate of reaction and $[HNO_3]$. Despite the fact that the essential reaction is a simple one, it may not be easy to deduce this from the quantities that we can readily measure.

Then again, if the aqueous hydrolysis of the alkyl halide, **RHal**, is found to follow the rate equation

$$\text{Rate} \propto [\textbf{RHal}]$$

it is not necessarily safe to conclude that the rate-determining step does *not* involve the participation of water simply on the grounds that $[H_2O]$ does not appear in the rate equation, for if water is being used as the solvent its concentration would remain virtually unchanged whether or not it actually participated in the rate-limiting stage. The point could perhaps be settled by carrying out the hydrolysis in another solvent, e.g. HCO_2H, and by using a much smaller concentration of water as the nucleophile; the hydrolysis may then be found to follow the rate equation

$$\text{Rate} \propto [\textbf{RHal}][H_2O]$$

but the actual mechanism of hydrolysis could well have changed on altering the solvent, so that we are not, of necessity, any the wiser about what really went on in the aqueous solution to begin with!

The vast majority of organic reactions are carried out in solution and quite small changes in the solvent used can have the profoundest effects on reaction rates and mechanisms. Particularly is this so when polar intermediates, for example carbonium ions or carbanions as constituents of ion pairs, are involved for such species normally carry an envelope of solvent molecules about with them, which greatly affects their stability (and hence their ease of formation), and which is strongly influenced by the composition and nature of the solvent employed, particularly its polarity and ion-solvating capabilities. By contrast reactions that involve radicals (p. 256) are much less influenced by the nature of the solvent (unless this is itself capable of reacting with radicals), but are greatly influenced by the addition of radical sources (e.g. peroxides) or radical absorbers (e.g. quinones), or by light which may initiate reaction through the production of radicals by photochemical activation, e.g. $Br_2 \xrightarrow{h\nu} Br\cdot \ \cdot Br$.

(ii) Isotope effects

Simple kinetic data will often not tell us whether a particular bond is broken in the rate-limiting stage of a reaction or not and further refinements have then to be resorted to. If, for example, the bond concerned is C—H, the question may be settled by comparing the rates of reaction, under the same conditions, of the compound in which we are interested and the corresponding compound containing the heavier isotope, deuterium, in a C—D bond; the latter bond is harder to break and its reactions correspondingly slower, ≈ 7-fold at 25°. Thus in the oxidation

$$Ph_2CHOH \rightarrow Ph_2C{=}O$$

which may be effected by alkaline permanganate, Ph_2CHOH is found to be oxidised 6·7 times as rapidly as Ph_2CDOH; the reaction is said to exhibit a *primary kinetic isotope effect* and breaking of the C—H bond must clearly be involved in the rate-limiting stage of the reaction. By contrast benzene, C_6H_6, and hexadeuterobenzene, C_6D_6, are found to nitrate at essentially the same rate and the C—H bond-breaking that must occur at some stage in the overall process

thus cannot be involved in the rate-limiting stage (*cf.* p. 123).

46

Isotopes can also be used to solve mechanistic problems that are non-kinetic. Thus the aqueous hydrolysis of esters to yield an acid and an alcohol could, in theory, proceed by cleavage at (*a*) alkyl/oxygen fission, or (*b*) acyl/oxygen fission:

$$R-\overset{\overset{\displaystyle O}{\|}}{C}\overset{a}{\vdots}O\overset{}{\vdots}R' \quad\xrightarrow[\substack{(a)\\ H_2{}^{18}O}]{} \quad R-\overset{\overset{\displaystyle O}{\|}}{C}-O-H + HO^{18}-R'$$

$$\xrightarrow[]{(b)} \quad R-\overset{\overset{\displaystyle O}{\|}}{C}-{}^{18}OH + H-OR'$$

If the reaction is carried out in water enriched in the heavier oxygen isotope ^{18}O, (*a*) will lead to an alcohol which is ^{18}O enriched and an acid which is not, while (*b*) will lead to an ^{18}O enriched acid but a normal alcohol. Most esters are in fact found to yield an ^{18}O enriched acid indicating that hydrolysis, under these conditions, proceeds via (*b*) acyl/oxygen fission (p. 205). It should of course be emphasised that these results are only valid provided that neither acid nor alcohol, once formed, can itself exchange its oxygen with water enriched in ^{18}O, as has indeed been shown to be the case.

Heavy water, D_2O, has often been used in a rather similar way. Thus in the Cannizzaro reaction of benzaldehyde (p. 188)

$$Ph-\overset{\overset{\displaystyle O}{\|}}{C}-H + Ph-\overset{\overset{\displaystyle O}{\|}}{C}-H \xrightarrow[H_2O]{^{\ominus}OH} Ph-\overset{\overset{\displaystyle O}{\|}}{C}-O^{\ominus} + Ph-\overset{\overset{\displaystyle OH}{|}}{\underset{\underset{\displaystyle H}{|}}{C}}-H$$

the question arises of whether the second hydrogen atom that becomes attached to carbon, in the molecule of benzyl alcohol that is formed, comes from the solvent (H_2O) or from a second molecule of benzaldehyde. Carrying out the reaction in D_2O is found to lead to the formation of *no* **PhCHDOH**, thus demonstrating that the second hydrogen atom could not have come from water and must, therefore, have been provided by direct transfer from a second molecule of benzaldehyde.

(iii) Intermediates

Perhaps the most concrete evidence that can be obtained about the mechanism of a reaction is provided by the actual isolation from the reaction mixture of one or more intermediate species in the conversion of starting materials into products. Thus in the Hofmann reaction (p. 111) by which amides are converted into amines

$$R-\overset{\overset{\displaystyle O}{\|}}{C}-NH_2 \xrightarrow[\ominus OH]{Br_2} RNH_2$$

it is, with care, possible to isolate the N-bromoamide, **RCONHBr**, its anion, **RCONBr$^{\ominus}$**, and an isocyanate, **RNCO**, thus going some considerable way to eludicate the overall mechanism of the reaction. It is of course necessary to establish beyond all doubt that any species isolated really *is* an intermediate—and not merely an alternative product—by showing that it may be converted, under the normal reaction conditions, into the usual reaction products at a rate at least as fast as the overall reaction under the same conditions.

It is much more usual not to be able to isolate any intermediates at all, but this does not necessarily mean that none are formed, merely that they may be too labile to permit of their isolation. Their occurrence may then often be inferred from physical, particularly spectroscopic, measurements made on the system. Thus in the formation of oximes from a number of carbonyl compounds by reaction with hydroxylamine (p. 185),

$$\overset{R \qquad R}{\underset{\underset{\displaystyle O}{\|}}{C}} \xrightarrow{NH_2OH} \overset{R \qquad R}{\underset{\underset{\displaystyle \overset{..}{N}}{\|}}{C}} + H_2O$$
$$\qquad\qquad\qquad\qquad \overset{}{\underset{OH}{}}$$

the infra-red absorption band characteristic of $>C=O$ in the starting material disappears rapidly and may have gone completely before the band characteristic of $>C=N^{\frown}$ in the product even begins to appear. Clearly an intermediate must be formed, and further evidence suggests that it is

$$\overset{R \qquad R}{\underset{HO \qquad NHOH}{C}}$$

which forms rapidly and then breaks down only slowly to yield the products, the oxime and water.

Where we have reason to suspect the involvement of a particular species as a labile intermediate in the course of a reaction it may be possible to confirm our suspicions by introducing into the reaction mixture, with malice aforethought, a reactive species which we should expect our postulated intermediate to react with particularly readily, and thereby 'trap' it. Thus in the hydrolysis of chloroform with strong bases, the highly electron-deficient dichlorocarbene, CCl_2, which has been suggested as a labile intermediate (p. 229), could be trapped by introducing into the reaction mixture the electron-rich *cis* but-2-ene (VII) and then isolating the resultant stable cyclopropane derivative (VIII), whose formation can hardly be accounted for in any other way:

$$(VII) \quad \overset{\text{Me} \quad \text{Me}}{\diagdown\!\!=\!\!\diagup} \quad \underset{CCl_2}{} \quad \longrightarrow \quad \overset{\text{Me} \quad \text{Me}}{\underset{Cl}{\bigtriangledown}}\overset{Cl}{} \quad (VIII)$$

(iv) Stereochemical criteria

Information about the stereochemical course followed by a particular reaction can also provide useful insight into its mechanism, and may well introduce more stringent criteria than any suggested mechanistic scheme will have to meet. Thus the fact that the base-catalysed bromination of an optically active ketone such as (IX)

$$\underset{\substack{(+) \\ (IX)}}{Ph CO\overset{*}{C}HMeEt} \xrightarrow[\ominus OH]{Br_2} \underset{(\pm)}{Ph CO\overset{*}{C}BrMeEt}$$

leads to an optically inactive racemic product (p. 250) indicates that the reaction must proceed through a planar intermediate which can undergo attack equally well from either side leading to equal amounts of the two mirror-image forms of the product. Equally, the fact that cyclopentene (X) adds on bromine under polar conditions to yield the *trans* dibromide (XI), only, indicates that the mechanism of the

$$(X) \quad \varhexagon \xrightarrow{Br_2} \quad \text{(XI)}$$

49

reaction cannot simply be direct addition of the bromine molecule to the double bond, for this must lead to the *cis* dibromide (XII),

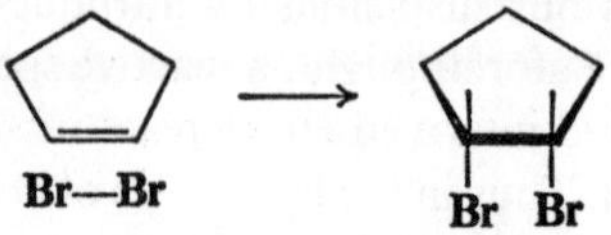

and must be at least a two stage process (p. 157). Many elimination reactions also take place more readily with the *trans* member of a pair of *cis*/*trans* isomers (p. 216), as seen in the conversion of *syn* and *anti* aldoxime acetates into a common nitrile:

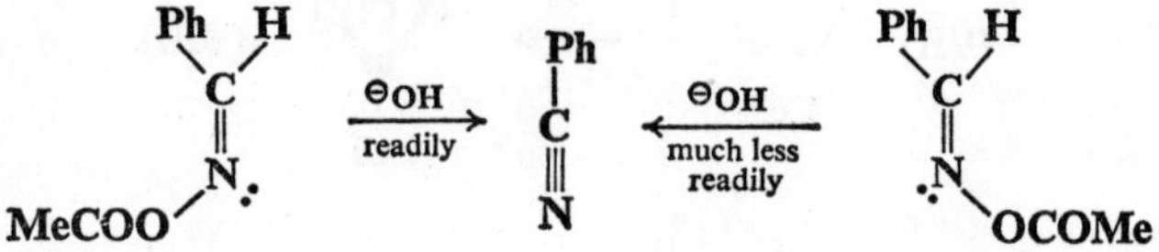

This clearly sets limitations to which any mechanism advanced for the reaction will have to conform and gives the lie to that prime tenet of 'lasso chemistry' by which groups are eliminated most readily when closest together:

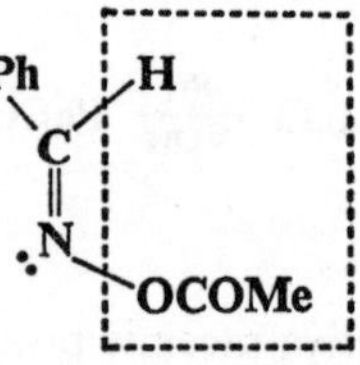

(v) The unexpected

Perhaps the most fundamental information about a reaction is provided by establishing the structure of the product or products that are formed and relating their structure to that of the starting material; this information can be particularly informative when an entirely unexpected product is obtained. Thus in the reaction of *p*-chlorotoluene (XIII) with amide ion, $^{\ominus}NH_2$, in liquid ammonia (p. 153) we obtain not only the expected *p*-toluidine (XIV) but also the entirely unexpected *m*-toluidine (XV) which is in fact the major product. (XV) clearly cannot be obtained from (XIII) by a simple substitution process and is formed either by an entirely different route than is (XIV), or if the two products do arise from some common intermediate then clearly neither is obtained by simple, direct substitution.

A reaction that is found to proceed unexpectedly faster or slower than those of compounds of related structure under the same conditions may also point to a mechanism different from the one that might otherwise have been assumed. Thus the observed rates of hydrolysis of the chloromethanes with strong bases, under comparable conditions, are found to vary as follows:

$$CH_3Cl \gg CH_2Cl_2 \ll CHCl_3 \gg CCl_4$$

clearly suggesting that chloroform undergoes attack in a different manner from the other compounds, as has indeed already been suggested above (p. 49).

The degree of success with which a suggested mechanism can be said to delineate the course of a particular reaction is not determined solely by its ability to account for the known facts; the acid test is how successful it is at forecasting a change in rate, or even in the nature of the products formed, when the conditions under which the reaction is carried out or the structure of the starting material are changed. Some of the suggested mechanisms we shall encounter measure up to these criteria better than do others, but the overall success of a mechanistic approach to organic reactions is demonstrated by the way in which the application of a few relatively simple guiding principles can bring light and order to bear on a vast mass of disparate information about equilibria, reaction rates, and the relative reactivity of organic compounds. We shall now go on to consider some simple examples of this.

3 THE STRENGTHS OF ACIDS AND BASES

MODERN electronic theories of organic chemistry have been highly successful in a wide variety of fields in relating behaviour with structure, but nowhere has this been more marked than in accounting for the relative strengths of organic acids and bases. According to the definition of Arrhenius, acids are compounds that yield hydrogen ions, $H^{\oplus}$, in solution while bases yield hydroxide ions, $^{\ominus}OH$. Such definitions are reasonably adequate if reactions in water only are to be considered, but the acid/base relationship has proved so useful in practice that the concepts of both acids and bases have become considerably more generalised. Thus Brønsted defined acids as substances that would give up protons, i.e. *proton donors*, while bases were *proton acceptors*. The first ionisation of sulphuric acid in aqueous solution is then looked upon as:

$$H_2SO_4 + H_2O \rightleftarrows H_3O^{\oplus} + HSO_4^{\ominus}$$

Acid	*Base*	*Con-*	*Con-*
		jugate	*jugate*
		acid	*base*

Here water is acting as a base by accepting a proton and is thereby converted into its so-called *conjugate acid*, $H_3O^{\oplus}$, while the acid, H_2SO_4, by donating a proton is converted into its *conjugate base*, $HSO_4^{\ominus}$.

The more generalised picture provided by Lewis who defined acids as molecules or ions capable of coordinating with unshared electron pairs, and bases as molecules or ions which have such unshared electron pairs available for coordination, has already been referred to (p. 30). Lewis acids include such species as boron trifluoride (I) which reacts with trimethylamine to form a solid salt (m.p. 128°):

$$\text{Me} \diagdown \atop \text{Me} \rightarrow \text{N:} \quad \text{B} \diagdown \atop \diagdown \text{F} \rightleftarrows \text{Me} \diagdown \atop \text{Me} \rightarrow \overset{\oplus}{\text{N}}:\overset{\ominus}{\text{B}} \diagdown \atop \diagdown \text{F}$$

(I)

Other common examples are aluminium chloride, stannic chloride, zinc chloride, etc. We shall, at this point, be concerned essentially with proton acids, and the effect of structure on the strength of a number of organic acids and bases will now be considered in turn. Compounds in which it is a **C—H** bond that is ionised will be considered subsequently (p. 232), however.

ACIDS

(i) pK_a

The strength of an acid, **HX**, in water, i.e. the extent to which it is dissociated, may be determined by considering the equilibrium:

$$H_2O: + HX \rightleftharpoons H_3O^{\oplus} + X^{\ominus}$$

Then the equilibrium constant is given by

$$K_a \approx \frac{[H_3O^{\oplus}][X^{\ominus}]}{[HX]}$$

the concentration of water being taken as constant as it is present in such large excess. It should be emphasised that K_a, the *acidity constant* of the acid in water, is only approximate if concentrations instead of activities have been used. The constant is influenced by the composition of the solvent in which the acid is dissolved (see below) and by other factors but it does, nevertheless, serve as a useful guide to acid strength. In order to avoid writing negative powers of 10, K_a is generally converted into pK_a ($pK_a = -\log_{10} K_a$); thus while K_a for acetic acid in water at 25° is 1.79×10^{-5}, $pK_a = 4.76$. The *smaller* the numerical value of pK_a, the *stronger* is the acid to which it refers.

(ii) Effect of solvent

The influence of the solvent on the dissociation of acids (and of bases) can be profound; thus hydrogen chloride which is a strong acid in water is not ionised in benzene. Water is a most effective ionising solvent on account (*a*) of its high dielectric constant, and (*b*) of its ion-solvating ability. The higher the dielectric constant of a solvent the smaller the electrostatic energy of any ions present in it; hence the more stable such ions are in solution, and the less ready are they to recombine with each other.

Ions in solution strongly polarise solvent molecules near them and the greater the extent to which this can take place, the greater the stability of the ion, which is in effect stabilising itself by spreading its charge. Water is extremely readily polarised and ions stabilise themselves when dissolved in it by collecting a solvation envelope of water molecules around them. A major virtue of water as an ionising solvent is its facility in solvating *both* cations and anions.

Water also has the advantage of itself being able to function as an acid or a base with equal facility, which further increases its usefulness and versatility as an ionising solvent. It does however have the disadvantage as an ionising solvent for organic compounds that some of them are insufficiently soluble in the unionised form to dissolve in it in the first place.

(iii) The origin of acidity in organic compounds

Acidity in an organic compound, YH, may be influenced by
  (a) The strength of the Y—H bond,
  (b) The electronegativity of Y,
  (c) Factors stabilizing $Y^{\ominus}$ compared with YH,
but of these (a) is not normally found to be a limiting factor. The effect of (b) is reflected in the fact that the pK_a of methanol, CH_3O—H, is ≈ 16 while that of methane, H_3C—H, is ≈ 43, oxygen being considerably more electronegative than carbon. By contrast, the pK_a of formic acid,

$$H-\overset{\overset{\textstyle O}{\|}}{C}-O-H$$

is 3·77. This is in part due to the electron-withdrawing carbonyl group enhancing the electron affinity of the oxygen atom to which the incipient proton is attached but much more important is the stabilisation possible in the resultant formate anion compared with the undissociated formic acid molecule:

There is extremely effective delocalisation, with consequent stabilisation, in the formate anion involving as it does two canonical structures of identical energy and though delocalisation can take place in the formic acid molecule also, this involves separation of charge and will consequently be much less effective as a stabilising influence (*cf.* p. 17). The effect of this differential stabilisation is somewhat to discourage the recombination of proton with the formate anion, the equilibrium is to this extent displaced to the right, and formic acid is, by organic standards, a moderately strong acid.

With alcohols there is no such factor stabilising the alkoxide ion, $RO^{\ominus}$, relative to the alcohol itself and alcohols are thus very much less acidic than carboxylic acids. With phenols, however, there is again the possibility of relative stabilisation of the anion (II) by delocalisation of its negative charge through interaction with the π orbitals of the aromatic nucleus:

(II)

Delocalisation also occurs in the undissociated phenol molecule (*cf.* p. 22) but, involving charge separation, is less effective than in the anion (II), thus leading to some reluctance on the part of the latter to recombine with a proton. Phenols are indeed found to be stronger acids than alcohols (the pK_a of phenol itself is 9·95) but considerably weaker than carboxylic acids. This is due to the fact that delocalisation of the negative charge in the carboxylate anion involves structures of identical energy content (see above), and of the centres involved two are highly electronegative oxygen atoms; whereas in the phenoxide ion (II) the structures involving negative charge on the nuclear carbon atoms are likely to be of higher energy content than the one in which it is on oxygen and, in addition, of the centres here involved only one is a highly electronegative oxygen atom. The relative stabilisation of the anion with respect to the undissociated molecule is thus likely to be less effective with a phenol than with a carboxylic acid leading to the lower relative acidity of the former.

(iv) Simple aliphatic acids

The replacement of the non-hydroxylic hydrogen atom of formic acid by an alkyl group might be expected to produce a weaker acid as the electron-donating inductive effect of the alkyl group would reduce the residual electron affinity of the oxygen atom carrying the incipient proton and so reduce the strength of the acid. In the anion the increased electron availability on oxygen would serve to promote its recombination with proton as compared with the formate/formic acid system:

$$\left[Me{\rightarrow}C{\underset{\textstyle O}{\overset{\textstyle O}{\diagup}}} \right]^{\ominus} \qquad \left[H{-}C{\underset{\textstyle O}{\overset{\textstyle O}{\diagup}}} \right]^{\ominus}$$

We should thus expect the equilibrium to be shifted to the left compared with formic acid/formate and it is in fact found that the pK_a of acetic acid is 4·76, compared with 3·77 for formic acid. However, the overall change in structure effected in so small a molecule as formic acid on replacement of H by CH_3 makes it doubtful whether so simple an argument is really valid; it could well be that the relative solvation possibilities in the two cases are markedly affected by the considerably different shapes of the two small molecules. It is important to remember that the value of the acidity constant, K_a, of an acid is related to the standard free energy change for the ionisation, $\Delta G°$, by the relation

$$-\Delta G° = 2·303 RT \log K_a$$

and that $\Delta G°$ includes both enthalpy and entropy terms:

$$\Delta G° = \Delta H° - T\Delta S°$$

Thus it is found for the ionisation of acetic acid in water at 25° ($K_a = 1·79 \times 10^{-5}$) that $\Delta G° = 6·5$ kcal, $\Delta H° = -0·13$ kcal, and $\Delta S° = -22$ cal/degree; while for formic acid the corresponding figures are: $\Delta G° = 5·1$ kcal, $\Delta H° = -0·07$ kcal, and $\Delta S° = -17·7$ cal/degree. At first sight it is perhaps a little surprising that the $\Delta H°$ values are so small, this stems from the energy of solvation of the ions formed essentially cancelling out the dissociation energy of the $O—H$ bond in the carboxyl group of the undissociated acid.

The difference in $\Delta G°$ in the two cases, and hence the difference in

K_a for the two acids, thus arises from a difference in the entropy terms. There are two species on each side of the equilibrium and differences in translational entropy on dissociation will thus be small. However, the two species are neutral molecules on one side of the equilibrium and ions on the other; the main feature that contributes to $\Delta S°$ is thus the solvation sheaths of water molecules that surround $RCO_2{}^{\ominus}$ and $H_3O^{\oplus}$ and the consequent restriction, in terms of increased orderliness, that is thereby imposed on the solvent water molecules; the increase in orderliness not being quite so great as might have been expected as there is already a good deal of orderliness in liquid water itself. The difference in strength between formic and acetic acids thus does indeed relate to the differential solvation of their anions as was suggested above.

Further substitution of alkyl groups in acetic acid has much less effect than this first introduction and, being now essentially a second-order effect, the influence on acid strength is not always regular, steric and other influences playing a part; pK_a values are observed as follows:

$$Me_2CHCO_2H \quad Me_3CCO_2H$$
$$4·86 \qquad\qquad 5·05$$

$$CH_3CO_2H \quad MeCH_2CO_2H$$
$$4·76 \qquad\quad 4·88$$

$$Me(CH_2)_2CO_2H \quad Me(CH_2)_3CO_2H$$
$$4·82 \qquad\qquad\quad 4·86$$

If there is a doubly bonded carbon atom adjacent to the carboxyl group the acid strength is increased. Thus acrylic acid, $CH_2{=}CHCO_2H$, has a pK_a of 4·25 compared with 4·88 for the saturated analogue, propionic acid. This is due to the fact that the unsaturated α-carbon atom is sp^2 hybridised, which means that electrons are drawn closer to the carbon nucleus than in a saturated, sp^3 hybridised atom due to the rather larger s contribution in the sp^2 hybrid. The result is that sp^2 hybridised carbon atoms are less electron donating than saturated sp^3 hybridised ones, and so acrylic acid though still weaker than formic acid is stronger than propionic. The effect is much more marked with the sp^1 hybridised carbon atom of a triple bond, thus the pK_a of propiolic acid, $HC{\equiv}CCO_2H$, is 1·84. An analogous situation occurs with the hydrogen atoms of ethylene and acetylene; those of the former are little more acidic than the hydrogens in ethane, whereas those of acetylene are sufficiently acidic to be readily replaceable by a number of metals.

(v) Substituted aliphatic acids

The effect of introducing electron-withdrawing substituents into fatty acids is more marked. Thus halogen, with an inductive effect acting in the opposite direction to alkyl, might be expected to increase the strength of an acid so substituted, and this is indeed observed as pK_a values show:

$$CH_3 \rightarrow CO_2H$$
4·76

$$F \leftarrow CH_2 \leftarrow CO_2H$$
2·66

$$\underset{Cl}{\overset{Cl}{\diagdown}}CH \leftarrow CO_2H$$
1·29

$$Cl \leftarrow CH_2 \leftarrow CO_2H$$
2·86

$$Br \leftarrow CH_2 \leftarrow CO_2H$$
2·90

$$Cl \leftarrow \overset{Cl}{\underset{Cl}{C}} \leftarrow CO_2H$$
0·65

$$I \leftarrow CH_2 \leftarrow CO_2H$$
3·16

The relative effect of the different halogens is in the expected order, fluorine being the most electronegative (electron-withdrawing) and producing a hundredfold increase in strength of fluoracetic acid as compared with acetic acid itself. The effect is very much greater than that produced, in the opposite direction, by the introduction of an alkyl group, and the introduction of further halogens still produces large increases in acid strength: trichloracetic is thus a very strong acid.

Here again it is important to remember that K_a (and hence pK_a) is related to $\Delta G°$ for the ionisation, and that $\Delta G°$ includes both $\Delta H°$ and $\Delta S°$ terms. In this series of halogen-substituted acetic acids $\Delta H°$ is found to differ little from one compound to another, the observed change in $\Delta G°$ along the series being due largely to variation in $\Delta S°$. This arises from the substituent halogen atom effecting delocalisation of the negative charge over the whole of the anion,

$$\left[F \leftarrow CH_2 \leftarrow C \underset{O}{\overset{O}{\diagdown}} \right]^{\ominus} \quad \left[CH_3 \rightarrow C \underset{O}{\overset{O}{\diagdown}} \right]^{\ominus}$$

the latter thus imposes correspondingly less powerful restriction on the water molecules surrounding it than does the unsubstituted acetate anion whose charge is largely concentrated, being confined substantially to $-CO_2^{\ominus}$. There is therefore a smaller decrease in entropy on ionisation of the halogen-substituted acetic acids than with acetic acid itself. This effect is particularly pronounced with CF_3CO_2H for whose ionisation $\Delta G° = 0.3$ kcal, compared with 6·5 kcal for CH_3CO_2H, while $\Delta H°$ is essentially the same for the ionisation of both acids.

The introduction of a halogen further away than in the α-position to the carboxyl group has much less effect, its inductive effect quickly dying in a saturated chain as the following pK_a values show:

$$MeCH_2CH_2CO_2H \qquad MeCH_2CHClCO_2H$$
$$4·82 \qquad\qquad\qquad 2·84$$

$$MeCHClCH_2CO_2H \qquad CH_2ClCH_2CH_2CO_2H$$
$$4·06 \qquad\qquad\qquad 4·52$$

Other electron-withdrawing groups, e.g. $R_3\overset{\oplus}{N}-$, $-CN$, $-NO_2$, $>CO$, $-CO_2R$ increase the strength of fatty acids, as also do hydroxyl and methoxyl groups; the unshared electrons on the oxygen atoms of the last two groups are not able to exert a mesomeric effect in the opposite direction to their inductive effect owing to the intervening saturated carbon atom. These effects are seen in the pK_a values:

$$O_2N \leftarrow CH_2 \leftarrow CO_2H \qquad\qquad EtO_2C \leftarrow CH_2 \leftarrow CO_2H$$
$$1·68 \qquad\qquad\qquad\qquad 3·35$$

$$\overset{\oplus}{Me_3N} \leftarrow CH_2 \leftarrow CO_2H \qquad\qquad MeCO \leftarrow CH_2 \leftarrow CO_2H$$
$$1·83 \qquad\qquad\qquad\qquad 3·58$$

$$NC \leftarrow CH_2 \leftarrow CO_2H \qquad\qquad Me\ddot{O} \leftarrow CH_2 \leftarrow CO_2H$$
$$2·47 \qquad\qquad\qquad\qquad 3·53$$

$$H\ddot{O} \leftarrow CH_2 \leftarrow CO_2H$$
$$3·83$$

(vi) Phenols

Analogous effects can be observed with substituted phenols, the presence of electron-withdrawing groups in the nucleus increasing their acidity. In the case of a nitro-group, the inductive effect will fall off with distance as we go o- → m- → p-nitrophenol but there will be an electron-withdrawing mesomeric effect when the nitro group is in the *o*- or *p*-, but not in the *m*-position, and this, too, will promote ionisation by stabilisation of the resultant anion. We might therefore expect *o*- and *p*-nitrophenols to be more acidic than the *m*-compound which is, in fact, found to be the case:

	pK_a
C_6H_5OH	9·9
o-$O_2NC_6H_4OH$	7·2
m-$O_2NC_6H_4OH$	8·35
p-$O_2NC_6H_4OH$	7·14
2,4-$(O_2N)_2C_6H_3OH$	4·01
2,4,6-$(O_2N)_3C_6H_2OH$	1·02

Thus picric acid is a very strong acid. The effect of introducing electron-donating alkyl groups is small:

	pK_a
C_6H_5OH	9·95
o-MeC_6H_4OH	10·28
m-MeC_6H_4OH	10·08
p-MeC_6H_4OH	10·19

The resulting substituted phenols are very slightly weaker acids, but the effect is marginal and irregular, indicating that the effect of such substituents in destabilising the phenoxide ion by disturbing the interaction of its negative charge with the delocalised π orbitals of the aromatic nucleus is small, as might have been expected.

Interactions between the anions and solvent water molecules can obviously play a part here also in influencing ΔG°, and differential effects depending on the pattern of substitution are to be expected.

(vii) Aromatic carboxylic acids

Benzoic acid, with a pK_a of 4·20, is a stronger acid than its saturated analogue cyclohexane carboxylic acid ($pK_a \approx 4·87$) suggesting that a phenyl group, like a double bond, is here exerting an electron-withdrawing effect—compared with a saturated carbon atom—on the carboxyl group, due to the sp^2 hybridised carbon atom to which the carboxyl group is attached (*cf*. p. 57). The introduction of alkyl groups has very little effect on the strength of benzoic acid (*cf*. similar introduction in aliphatic acids)

	pK_a
$C_6H_5CO_2H$	4·20
$m\text{-}MeC_6H_4CO_2H$	4·24
$p\text{-}MeC_6H_4CO_2H$	4·34

but electron-withdrawing groups increase its strength, the effect, as with the phenols, being most pronounced when they are in the *o*- and *p*-positions:

	pK_a
$C_6H_5CO_2H$	4·20
$o\text{-}O_2NC_6H_4CO_2H$	2·17
$m\text{-}O_2NC_6H_4CO_2H$	3·45
$p\text{-}O_2NC_6H_4CO_2H$	3·43
$3,5\text{-}(O_2N)_2C_6H_3CO_2H$	2·83

The particularly marked effect with $o\text{-}NO_2$ is probably due to the very short distance over which the powerful inductive effect is operating.

The presence of groups such as **OH**, **OMe** or halogen having an electron-withdrawing inductive effect but an electron-donating mesomeric effect when in the *o*- and *p*-positions may, however, cause the *p*-substituted acids to be weaker than the *m*- and, on occasion, weaker even than the unsubstituted acid itself, e.g. *p*-hydroxybenzoic acid:

pK_a of $XC_6H_4CO_2H$

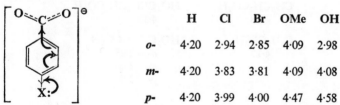

	H	Cl	Br	OMe	OH
o-	4·20	2·94	2·85	4·09	2·98
m-	4·20	3·83	3·81	4·09	4·08
p-	4·20	3·99	4·00	4·47	4·58

It will be noticed that this compensating effect becomes more pronounced in going $Cl \approx Br \rightarrow OH$, i.e. in increasing order of readiness with which the atom attached to the nucleus will part with its electron pairs. The behaviour of *o*-substituted acids is, as seen above, often anomalous. Their strength is sometimes considerably

61

greater than expected due to direct interaction of the adjacent groups, e.g. intramolecular hydrogen bonding stabilises the anion (IV) from salicylic acid (III) by delocalising its charge, an advantage not shared by the *m*- and *p*-isomers nor by *o*-methoxybenzoic acid:

(III)   (IV)

The effect is even more pronounced where hydrogen bonding can occur with a hydroxyl group in both *o*-positions and 2,6-dihydroxy-benzoic acid has a pK_a of 1·30.

(viii) Dicarboxylic acids

As the carboxyl group itself has an electron-withdrawing inductive effect, the presence of a second such group in an acid would be expected to make it stronger, as shown by the following pK_a values:

HCO_2H	HO_2CCO_2H
3·77	1·23
CH_3CO_2H	$HO_2CCH_2CO_2H$
4·76	2·83
$CH_3CH_2CO_2H$	$HO_2CCH_2CH_2CO_2H$
4·88	4·19
$C_6H_5CO_2H$	$HO_2CC_6H_4CO_2H$
4·17	*o*- 2·98
	m- 3·46
	p- 3·51

The effect is very pronounced but falls off sharply as soon as the carboxyl groups are separated by more than one saturated carbon atom. Maleic acid (V) is a much stronger acid than fumaric (VI) (pK_a^1 is 1·92 compared with 3·02) due to the hydrogen bonding that can take place with the former, but not the latter, stabilising the anion (*cf.* salicylic acid):

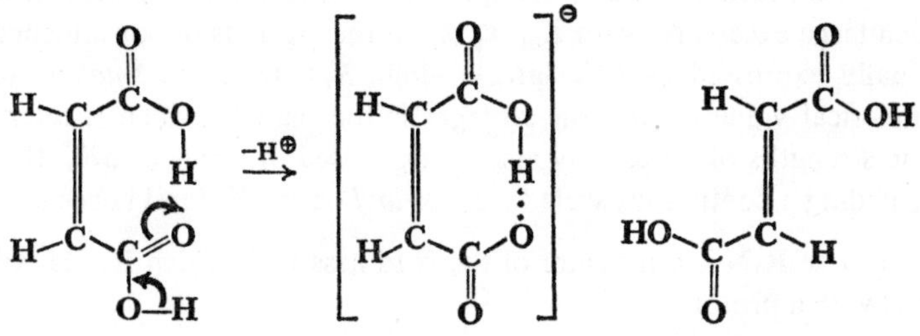

(V)   (VI)

The second dissociation of fumaric occurs more readily than that of maleic acid, however (pK_a^2 is 4·38 compared with 6·23) because of the greater difficulty in removing the proton from the negatively charged cyclic system of the latter. Oxalic, malonic and succinic acids are each weaker in their second dissociations than formic, acetic and propionic acids, respectively, because the second proton has to be removed from a negatively charged species containing an electron-donating substituent, i.e. $^{\ominus}O_2C$, which might be expected to de-stabilise the anion with respect to the undissociated acid as compared with the unsubstituted system:

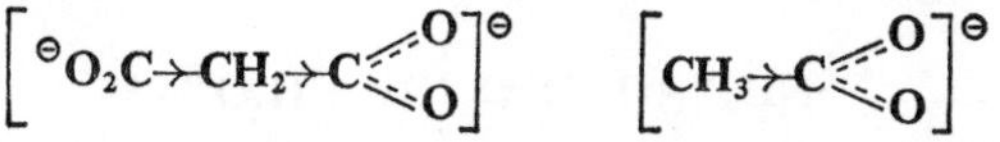

$$\left[{}^{\ominus}O_2C \rightarrow CH_2 \rightarrow C {\overset{O}{\underset{O}{\diagup}}}\right]^{\ominus} \qquad \left[CH_3 \rightarrow C {\overset{O}{\underset{O}{\diagup}}}\right]^{\ominus}$$

There is also likely to be a pronounced difference in ΔS° in the two cases, for the anion bearing two separated negative charges will impose considerably greater restraint on the surrounding solvent water molecules than will the singly charged anion.

BASES

(i) pK_b

The strength of a base, R_3N, in water, may be determined by considering the equilibrium:

$$R_3N: + HOH \rightleftharpoons \overset{\oplus}{R_3N}:H + {}^{\ominus}OH$$

Then the equilibrium constant, in water, is given by

$$K_b \approx \frac{[\overset{\oplus}{R_3NH}][{}^{\ominus}OH]}{[R_3N]}$$

63

the concentration of water being taken as constant as it is present in such large excess. As with K_a, K_b is, on the grounds of convenience, usually expressed as pK_b ($pK_b = -\log_{10} K_b$), then the *smaller* the numerical value of pK_b the *stronger* is the base to which it refers. The strengths of bases may also be expressed in terms of pK_a, thus providing a continuous scale for acids *and* bases. K_a (and hence pK_a) for a base R_3N is a measure of the readiness with which $R_3\overset{\oplus}{N}H$ will part with a proton

$$R_3\overset{\oplus}{N}H + H_2O \;\rightleftarrows\; R_3N + H_3O^{\oplus}$$

i.e. of the acidity of $R_3\overset{\oplus}{N}H$, and is defined by:

$$K_a \approx \frac{[R_3N][H_3O^{\oplus}]}{[R_3\overset{\oplus}{N}H]}$$

pK_b values in water may be converted into pK_a by use of the relation:

$$pK_a + pK_b = 14 \cdot 00 \text{ (at } 25°)$$

Considering the acidity of $NH_4^{\oplus}$

$$NH_4^{\oplus} + H_2O \;\rightleftarrows\; NH_3 + H_3O^{\oplus}$$

it is found that $\Delta G° = 12 \cdot 6$ kcal, $\Delta H° = 12 \cdot 4$ kcal, and $\Delta S° = -0 \cdot 7$ cal/degree at 25°. Thus the position of equilibrium is effectively determined by $\Delta H°$, the effect of $\Delta S°$ being all but negligible: a result that is in marked contrast to the behaviour of many acids as we have seen above (p. 56). The reason for the small effect of $\Delta S°$ is that here there is one charged species (a positive ion) on each side of the equilibrium, and these ions have closely comparable effects in restricting the solvent water molecules that surround them so that their entropies of solvation tend to cancel each other out.

(ii) Aliphatic bases

As increasing strength in nitrogenous bases is related to the readiness with which they are prepared to take up protons and, therefore, to the availability of the unshared electron pair on nitrogen, we might expect to see an increase in basic strength as we go: $NH_3 \rightarrow RNH_2 \rightarrow R_2NH \rightarrow R_3N$, due to the increasing inductive effect of successive alkyl groups making the nitrogen atom more negative. An actual series of amines have pK_b values as follows, however:

$$Me \rightarrow NH_2$$
3·36

$$\underset{Me}{\overset{Me}{\diagdown}} NH$$
3·23

$$\underset{Me}{\overset{Me}{\diagdown}} N - Me$$
4·20

$$NH_3$$
4·75

$$Et \rightarrow NH_2$$
3·33

$$\underset{Et}{\overset{Et}{\diagdown}} NH$$
3·07

$$\underset{Et}{\overset{Et}{\diagdown}} N - Et$$
3·12

It will be seen that the introduction of an alkyl group into ammonia increases the basic strength markedly as expected, ethyl having a very slightly greater effect than methyl. The introduction of a second alkyl group further increases the basic strength but the net effect of introducing the second alkyl group is much less marked than with the first. The introduction of a third alkyl group to yield a tertiary amine, however, *decreases* the basic strength in both the series quoted.

This is due to the fact that the basic strength of an amine in water is determined not only by electron-availability on the nitrogen atom, but also by the extent to which the cation, formed by uptake of a proton, can undergo solvation and so become stabilised. The more hydrogen atoms attached to nitrogen in the cation, the greater the possibilities of solvation via hydrogen bonding between these and water:

$$R_2N \overset{\oplus}{\underset{H}{\diagup}} \underset{H}{\overset{H}{\cdots}} \underset{O-H}{\cdots} \quad > \quad R_3\overset{\oplus}{N}H \leftarrow :O \overset{H}{\underset{H}{\diagdown}}$$

Thus as we go along the series $NH_3 \rightarrow RNH_2 \rightarrow R_2NH \rightarrow R_3N$, though the inductive effect would tend to be *increase* the basicity, progressively less stabilisation of the cation by hydration will occur, which will tend to *decrease* the basicity. The *net* effect of introducing successive alkyl groups thus becomes progressively smaller, and an actual changeover takes place on going from a secondary to a tertiary amine. If this is the real explanation, no such changeover should occur

65

if measurements of basicity are made in a solvent in which hydrogen-bonding cannot take place; it has, indeed, been found that in chloro-benzene the order of basicity of the butylamines is

$$BuNH_2 < Bu_2NH < Bu_3N$$

though their pK_b values in water are 3·39, 2·72 and 4·13, respectively.

Quaternary alkylammonium salts, e.g. $R_4N^{\oplus}I^{\ominus}$, are known, on treatment with moist silver oxide—'**AgOH**'—to yield basic solutions comparable in strength with the mineral alkalis. This is readily understandable for the base so obtained, $R_4N^{\oplus \ominus}OH$, is bound to be completely ionised as there is no possibility, as with tertiary amines etc., of reverting to an unionised form:

$$R_3\overset{\oplus}{N}H + {}^{\ominus}OH \rightarrow R_3N: + H_2O$$

The effect of introducing electron-withdrawing groups, e.g. **Cl**, **NO₂**, close to a basic centre is, naturally, to decrease the basicity, due to their electron-withdrawing inductive effect (*cf.* substituted anilines below, p. 69); thus

$$\begin{matrix} F_3C \\ F_3C \rightarrow N: \\ F_3C \end{matrix}$$

is virtually non-basic due to the three powerfully electron-withdrawing **CF₃** groups.

The change is particularly pronounced with groups such as $>C=O$, for not only is the nitrogen atom, with its electron pair, then bonded to an sp^2 hybridised carbon (*cf.* p. 57) but, much more significantly, a mesomeric effect can also operate:

$$R-\overset{\overset{\displaystyle O}{\|}}{C}-\ddot{N}H_2 \quad \leftrightarrow \quad R-\overset{\overset{\displaystyle O^{\ominus}}{|}}{C}=\overset{\oplus}{N}H_2$$

Thus amides are only very weakly basic in water (pK_b for acetamide = 14·5) and if two $>C=O$ groups are present, the resultant imides, far from being basic, are often sufficiently acidic to form alkali metal salts, e.g. phthalimide:

The effect of delocalisation in *increasing* the basic strength of an amine is seen in guanidine, $HN=C(NH_2)_2$, which, with the exception of the quaternary alkylammonium hydroxides above, is among the strongest organic bases known, having too small a pK_b in water for it to be accurately measured. Both the neutral molecule and the cation, $H_2\overset{\oplus}{N}=C(NH_2)_2$, resulting from its protonation, are stabilised by delocalisation

but in the cation the positive charge is spread symmetrically by the contribution to the hybrid of three exactly equivalent structures of equal energy. No comparably effective delocalisation occurs in the neutral molecule (in which two of the contributing structures involve separation of charge) with the result that the cation is greatly stabilised with respect to it, thus making protonation 'energetically profitable' and guanidine an extremely strong base.

(iii) Aromatic bases

The exact reverse of the above is seen with aniline which is a very weak base, having a pK_b of 9·38 compared to 4·75 for ammonia and 3·46 for methylamine. In aniline the nitrogen atom is again bonded to an sp^2 hybridised carbon atom but, more significantly, the unshared electron pair on nitrogen can interact with the delocalised π orbitals of the nucleus:

If the aniline is protonated, any such interaction, with resultant stabilisation, is prohibited, as the electron pair on nitrogen is no longer unshared:

(VII)

Here the aniline molecule is stabilised with respect to the anilinium cation (VII) and it is, therefore, 'energetically *un*profitable' for aniline to take up a proton; it thus functions as a base with the utmost reluctance which is reflected in its pK_b of 9·38, compared with that of cyclohexylamine, 3·32. The effect is naturally more pronounced when further phenyl groups are introduced on nitrogen; thus diphenyl-amine, Ph_2NH is an extremely weak base ($pK_b = 13·2$) while triphenyl-amine, Ph_3N is, by ordinary standards, not basic at all.

The overriding importance of this mesomeric destabilising of the anilinium cation (VII) with respect to the aniline molecule in determining the basic strength of aniline is confirmed by the relatively small and irregular effects produced in pK_b when methyl groups are introduced on the nitrogen atom or in the ring; for these groups would not be expected to influence markedly the interaction of the nitrogen's unshared pair with the delocalised π orbitals of the benzene nucleus (*cf.* the small effect produced by introducing alkyl groups into the nucleus of phenol, p. 60). Thus the substituted anilines have pK_b values:

$PhNH_2$	$PhNHMe$	$PhNMe_2$	$MeC_6H_4NH_2$	
9·38	9·60	9·62	*o-*	9·62
			m-	9·33
			p-	9·00

The small, base-strengthening inductive effect they usually exert is not large enough to influence the destabilisation of the cation to any significant extent, and may in any case be modified by steric and solvation considerations. A group with a more powerful inductive effect, e.g. NO_2, has rather more influence. This is intensified when the nitro group is in the *o*- or *p*-position for the interaction of the unshared electron pair of nitrogen with the delocalised π orbitals of the benzene nucleus is then enhanced and the cation even further destabilised with respect to the neutral molecule resulting in further weakening of the base. Thus the nitro-anilines have pK_b values:

PhNH₂	O₂NC₆H₄NH₂	
9·38	*o*-	14·28
	m-	11·55
	p-	13·02

The extra base-weakening effect when the substituent is in the *o*-position is due in part to the short distance over which its inductive effect is operating and also to direct interaction, both steric and by hydrogen bonding, with the NH_2 group (*cf.* the case of *o*-substituted benzoic acids, p. 61). *o*-Nitroaniline is such a weak base that its salts are largely hydrolysed in aqueous solution, while 2,4-dinitroaniline is insoluble in aqueous acids, and 2,4,6,-trinitroaniline resembles an amide; it is indeed called picramide and readily undergoes hydrolysis to picric acid.

With substituents such as **OH** and **OMe** having unshared electron pairs, an electron-donating, i.e. base-strengthening, mesomeric effect can be exerted from the *o*- and *p*-, but not from the *m*-position with the result that the *p*-substituted aniline is a stronger base than the corresponding *m*-compound. The *m*-compound is a weaker base than aniline itself due to the electron-withdrawing inductive effect exerted by the oxygen atom in each case. As so often, the effect of the *o*-substituent remains somewhat anomalous due to direct interaction with the NH_2 group by both steric and polar effects. The substituted anilines have pK_b values as follows:

PhNH₂	HOC₆H₄NH₂		MeOC₆H₄NH₂	
9·38	*o*-	9·28	*o*-	9·51
	m-	9·83	*m*-	9·80
	p-	8·50	*p*-	8·71

An interesting case is provided by 2,4,6-trinitroaniline (VIII) and 2,4,6-trinitrodimethylaniline (IX), the latter being about forty thousand times as strong as the former while aniline itself is only about twice as strong as dimethylaniline. This is due to the fact that the influence of nitro groups in the substituted dimethylaniline (IX) is essentially confined to their inductive effects. The dimethylamino group is sufficiently large to interfere sterically with the very large nitro groups in both *o*-positions, and the *p* orbitals on the nitrogen atoms of both NMe_2 and NO_2 groups are thus prevented from becoming parallel to the orbitals of the nuclear carbon atoms. As a consequence, mesomeric shift of unshared electrons from the amino-nitrogen atom to the oxygen atom of the nitro groups via the delocalised orbitals of the nucleus (*cf.* p. 69) is also prevented, and base-weakening by mesomeric electron-withdrawal does not, therefore, occur.

In trinitro-aniline (VIII), however, the NH_2 group is sufficiently small for no such limitation to be imposed and hydrogen-bonding between the *o*-NO_2 groups and the hydrogens of the NH_2 may help to hold these groups in the required orientation: the *p* orbitals may thus become parallel and interact strongly, the base then being enormously weakened by the very powerful electron-withdrawing mesomeric effects of the three nitro groups:

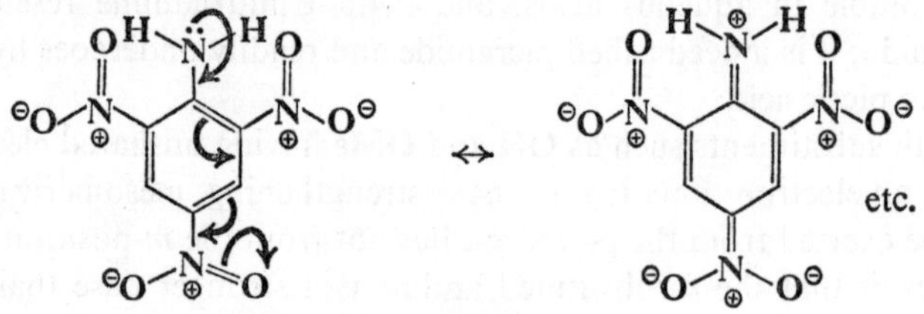

(VIII)

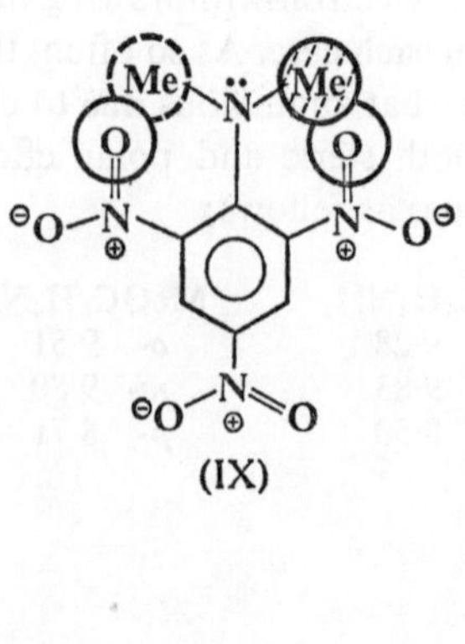

(IX)

(iv) Heterocyclic bases

Pyridine, , has pK_b 8·96 and is a very much weaker base than

the aliphatic tertiary amines (e.g. pK_b of $Et_3N = 3·12$) and this weakness is found to be characteristic of bases in which the nitrogen atom is multiply bonded. This is due to the fact that as the nitrogen atom becomes progressively more multiply bonded its lone-pair of electrons is accommodated in an orbital that has progressively more *s* character; they are thus drawn closer to the nitrogen nucleus and held more tightly by it, thereby becoming less available for forming a bond with proton, with a consequent decline in the basicity of the compound (*cf.* p. 57). As we go $\overset{\diagdown}{\diagup}N: \rightarrow \overset{\diagdown}{\diagup}N: \rightarrow \equiv N:$ in, for example, $R_3N: \rightarrow$ $C_5H_5N: \rightarrow RC\equiv N:$, the unshared pairs are in sp^3, sp^2 and sp^1 orbitals, respectively, and the declining basicity is reflected in the two pK_b values quoted above and the fact that the basicity of alkyl cyanides is too small to measure.

With quinuclidine, however,

the unshared electron pair is again in an sp^3 orbital and its pK_b (3·42) is almost identical with that of triethylamine (3·12).

Pyrrole (X) is found to exhibit some aromatic character (though this is not so pronounced as with benzene or pyridine) and does not behave like a conjugated diene as might otherwise have been expected:

$$\underset{\underset{\text{(X)}}{\overset{..}{\underset{H}{N}}}{\diagup}$$

For such aromaticity to be achieved, six π electrons of the ring atoms must occupy stable delocalised orbitals. This necessitates the contribution of *two* electrons by the nitrogen atom and, though the resultant electron cloud will be deformed towards nitrogen because of the more electronegative nature of that atom as compared with carbon, nitrogen's electron pair will not be readily available for taking up a proton:

(X)

The situation resembles that already encountered with aniline (p. 68) in that the cation (XI), obtained if protonation is forced upon pyrrole (protonation is shown as taking place on nitrogen, but it may occur on the α-carbon atom as happens with C-alkylated pyrroles),

(XI)

is destabilised with respect to the neutral molecule (X); but the effect is here more pronounced, for to function as a base pyrrole has to lose *all* its aromatic character and consequent stability. This is reflected in its pK_b of ≈ 13.6 compared with 9.38 for aniline; it is thus only a very weak base and functions as an acid, albeit a very weak one, in that the hydrogen atom attached to nitrogen may be removed by a strong base, e.g. $^{\ominus}NH_2$, the resultant anion being stabilised by delocalisation.

No such considerations can, of course, apply to the fully-reduced pyrrole, pyrrolidine

which has a pK_b of 2.73, closely similar to that of diethylamine, 3.07.

4 NUCLEOPHILIC SUBSTITUTION AT A SATURATED CARBON ATOM

A TYPE of reaction that has probably received more study than any other—largely due to the monumental work of Ingold and his school—is nucleophilic substitution at a saturated carbon atom: the classical displacement reaction exemplified by the conversion of an alkyl halide to an alcohol by the action of alkali:

$$HO^{\ominus} + R\!-\!Hal \rightarrow HO\!-\!R + Hal^{\ominus}$$

Investigation of the kinetics of such reactions has shown that there are essentially two extreme types, one in which

$$\text{Rate} \propto [RHal][^{\ominus}OH] \tag{1}$$

and another in which

$$\text{Rate} \propto [RHal] \tag{2}$$

i.e. is independent of $[^{\ominus}OH]$. In many examples the kinetics are mixed, showing both types of rate law simultaneously, or are otherwise complicated, but cases are known which do exemplify the simple relations shown above.

RELATION OF KINETICS TO MECHANISM

The hydrolysis of methyl bromide in aqueous alkali has been shown to proceed according to equation (1) and this is interpreted as involving the participation of both halide and hydroxyl ion in the rate-determining (i.e. slowest) step of the reaction. Ingold has suggested a transition state in which the attacking hydroxyl ion becomes partially bonded to the reacting carbon atom before the incipient bromide ion has become wholly detached from it; thus part of the energy necessary to effect the breaking of the C—Br bond is then supplied by that produced in forming the HO—C bond. Calculation shows that an approach by the hydroxyl ion along the line of centres of the carbon and bromine atoms is that of lowest energy requirement. This can be represented:

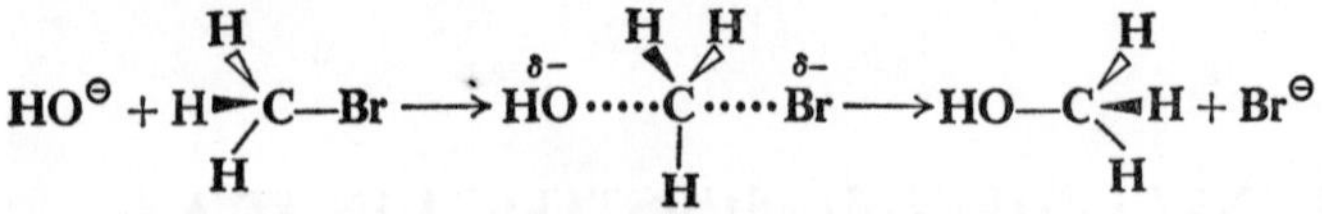

$$HO^{\ominus} + H \blacktriangleright \overset{\displaystyle H}{\underset{\displaystyle H}{C}}\!\!-Br \longrightarrow \overset{\delta-}{HO}\cdots\cdots \overset{\displaystyle H \quad H}{\underset{\displaystyle H}{C}}\cdots\cdots \overset{\delta-}{Br} \longrightarrow HO\!-\!\overset{\displaystyle H}{\underset{\displaystyle H}{C}}\!\blacktriangleleft H + Br^{\ominus}$$

Transition state

The negative charge is spread in the transition state in the course of being transferred from hydroxyl to bromine, and the hydrogen atoms attached to the carbon atom attacked pass through a position in which they all lie in one plane (at right angles to the plane of the paper as drawn here). This type of mechanism has been named by Ingold S_N2, standing for Substitution Nucleophilic bimolecular.

A certain element of confusion is to be met with both in text-books and in the literature over the use of the term bimolecular, particularly in its confusion with second order as applied to reactions. It is probably simplest to reserve the latter purely for a description of the type of kinetic equation that a reaction follows: thus equation (1) represents a second order reaction, being first order in each reactant and equation (2) represents a first order reaction; while molecularity is reserved for a description of the mechanism proposed, being used in the sense of specifying the number of species that are actually undergoing covalency changes in the rate-determining stage. Thus, hydrolysis of methyl bromide under the conditions specified, not only exhibits second order kinetics but, as represented mechanistically, is clearly a bimolecular reaction.

By contrast, the hydrolysis of *t*-butyl chloride in alkali is found kinetically to follow equation (2), i.e. as the rate is independent of [$^{\ominus}$OH], this can play no part in the rate-determining step. This has been interpreted as indicating that the halide undergoes slow ionisation (in fact, completion of the $R \rightarrowtail Hal$ polarisation that has already been shown to be present in such a molecule) as the rate-determining step, followed by rapid, non rate-determining attack by $^{\ominus}$OH or, if that is suitable, by solvent, the latter often predominating:

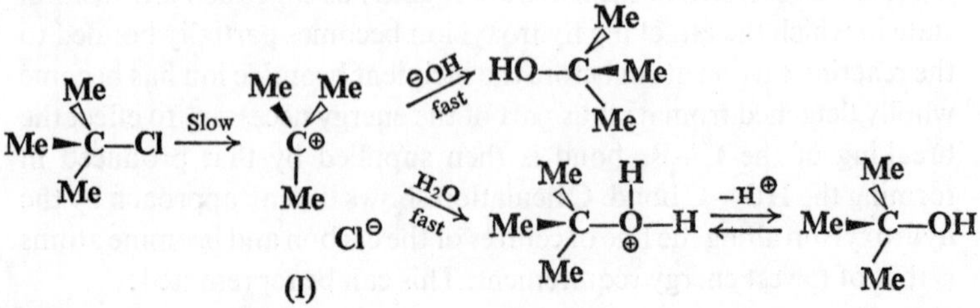

$$Me\blacktriangleright \overset{\displaystyle Me}{\underset{\displaystyle Me}{C}}\!\!-Cl \xrightarrow{\text{Slow}} \overset{\displaystyle Me \quad Me}{\underset{\displaystyle Me}{C^{\oplus}}} \begin{array}{c} \xrightarrow[\text{fast}]{\ominus OH} HO\!-\!\overset{\displaystyle Me}{\underset{\displaystyle Me}{C}}\!\blacktriangleleft Me \\ \\ \xrightarrow[\text{fast}]{H_2O} Me\blacktriangleright\overset{\displaystyle Me \quad H}{\underset{\displaystyle Me}{C}}\!-\!\overset{\oplus}{O}\!-\!H \underset{}{\overset{-H^{\oplus}}{\rightleftharpoons}} Me\blacktriangleright\overset{\displaystyle Me}{\underset{\displaystyle Me}{C}}\!-\!OH \end{array}$$

$$Cl^{\ominus}$$

(I)

74

This type of mechanism has been named S_N1, i.e. Substitution Nucleophilic unimolecular. The energy necessary to effect the initial ionisation is largely recovered in the energy of solvation of the ions so formed. The cation (I) in which the carbon atom carries a positive charge is a carbonium ion and during its formation the initially tetrahedral carbon atom collapses to a more stable planar state in which the three methyl groups are as far apart as they can get; attack by $^\ominus OH$ or solvent can then take place from either side. If this assumption of a planar state is inhibited by steric or other factors (*cf.* p. 81), the carbonium ion will be formed only with difficulty if at all, i.e. ionisation, and hence reaction by the S_N1 mechanism, may then not take place.

Kinetics alone can, in some cases, be an insufficient guide as to which mechanism is being followed, unless the reaction is investigated under more than one set of conditions. Thus where the solvent can act as a nucleophilic reagent, e.g. H_2O under S_N2 conditions:

$$\text{Rate} \propto [\text{RHal}][\text{H}_2\text{O}]$$

But as $[H_2O]$ is effectively constant, the rate becomes proportional to [RHal] and study of the kinetics in water alone would erroneously suggest that the reaction was of the S_N1 type. Such attack by the solvent, in this case H_2O, is known as *solvolysis*.

EFFECT OF SOLVENT

The solvent in which a reaction is carried out may exert a profound effect on the mechanism by which such reaction takes place. So far as the hydrolysis or solvolysis of a given halide is concerned, the more polar the solvent employed the more likely is the reaction to proceed via the S_N1 rather than the S_N2 mode, and such changeovers, as the solvent is varied, are well known. This change in mechanism is in part due to a solvent of high dielectric constant promoting ionisation but also to the fact that ions so produced will become highly solvated in suitable solvents, e.g. water. This solvation process is attended by the liberation of considerable amounts of energy which may go a long way towards providing the energy necessary for ionisation, which is thus further promoted. That such solvation effects are of great importance is confirmed by the fact that though S_N1 reactions are not unknown in the vapour phase, where solvation of ions is naturally impossible, they are very much less common than those in solution.

Thus ΔH for $CH_3Cl \rightarrow CH_3 \cdot \cdot Cl$ in the gas phase is $+80$ kcal/mole whereas the calculated value of ΔH for $CH_3Cl \rightarrow CH_3^{\oplus} \ Cl^{\ominus}$, also in the gas phase, is $+227$ kcal/mole; the calculated value of ΔH for the ionisation of CH_3Cl in aqueous solution is only $+63$ kcal/mole, however. This is still too high an energetic demand and so ionisation of methyl chloride does not take place in aqueous solution, the energetic demand with, for example, Me_3CBr is considerably lower and ionisation in aqueous solution becomes a reasonable proposition (*cf.* p. 74), though in equilibrium terms its extent is still only very small.

For a halide that undergoes hydrolysis by the S_N1 mode in a number of different solvents, the rate of hydrolysis is observed to increase as the solvent becomes more polar and/or a better medium for solvating ions: thus the rate of hydrolysis (S_N1) of *t*-butyl chloride is 30,000 times as fast in 50 % aqueous ethanol as in ethanol alone. The effect of change of solvent on a halide whose hydrolysis proceeds by the S_N2 mechanism throughout is much less marked and the direction of the effect on the rate of reaction depends on the relative disposition of charge in the reactants and in the transition state. Thus in the commonest example (the reaction of an anion with a neutral molecule)

$$Y^{\ominus} + R\!-\!Br \rightarrow \overset{\delta-}{Y}\!\cdots\!R\!\cdots\!\overset{\delta-}{Br} \rightarrow Y\!-\!R + Br^{\ominus}$$

the charge that is originally concentrated on $Y^{\ominus}$ becomes more widely dispersed in the transition state. Increasing polarity of the solvent will tend to slow down such a reaction for it will stabilise $Y^{\ominus}$ more effectively than it will stabilise the transition state.

The different effects on reaction rate that are induced in the two cases may assist in determining whether a particular displacement is proceeding by the S_N1 or S_N2 mode.

EFFECT OF STRUCTURE

An interesting sequence is provided by hydrolysis of the series of halides:

$$CH_3\!-\!Br \qquad Me\!-\!CH_2\!-\!Br \qquad \underset{Me}{\overset{Me}{>}}CH\!-\!Br \qquad \underset{Me}{\overset{Me}{>}}C\!-\!Br$$

The first and last members are described in the literature as undergoing ready hydrolysis, the two intermediate members being more resistant. Measurement of rates of hydrolysis with dilute alkali in aqueous ethanol gives the plot

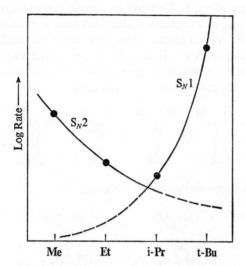

Based on Ingold, *Structure and Mechanism in Organic Chemistry*, by permission of Cornell University Press.

and further kinetic investigation shows a change in order of reaction, and hence presumably of mechanism, as the series is traversed. Thus methyl and ethyl bromide show second order kinetics, isopropyl bromide shows a mixture of second and first order, with the former predominating—the total rate here being a minimum for the series— while *t*-butyl bromide exhibits first order kinetics.

In seeking an explanation for the implied changeover in mechanistic pathway we need to consider, in each case, the effect on the transition state of both electronic and steric factors. For S_N2 attack, the enhanced inductive effect of an increasing number of methyl groups as we go across the series might be expected to make the carbon atom that bears the bromine progressively less positive and hence less readily attacked by $^{\ominus}OH$. This effect is probably small and steric factors are of more significance, thus $^{\ominus}OH$ will find it progressively more difficult to attack the bromine-carrying carbon as the latter becomes more heavily substituted. Further, the resultant S_N2 transition state has *five* other groups around this carbon (compared with only *four* in

77

the initial halide), so that the larger the three original substituents other than **Br** are the more crowded the transition state will be and the more loath therefore to form in the first place.

The nature of the S_N1 transition state is not so clear cut, but a carbonium ion, as a constituent of an ion pair, is usually taken as a model for it. As the above series is traversed the carbonium ion becomes increasingly stabilised and hence, so far as it is a model for the transition state that precedes it, more readily formed. This stabilisation arises from the operation of inductive effects

and also by hyperconjugation (p. 24), e.g.

via the hydrogen atoms attached to the α-carbons, the above series of ions having 0, 3, 6 and 9 such hydrogen atoms, respectively. Support for such an interaction of the **H—C** bonds with the carbon atom carrying the positive charge is provided by substituting **H** by **D** in the original halide, the rate of formation of the ion is then found to be slowed down by $\approx 10\%$ per deuterium atom incorporated: a result compatible only with the **H—C** bonds being involved in the ionisation. The relative contributions of hyperconjugation and inductive effects to the stabilisation of carbonium ions is open to debate, but it is significant that a number of carbonium ions will only form at all if they can take up a planar arrangement, the state in which hyperconjugation will operate most effectively.

The planar state is of course the most stable sterically, for the *three* substituents (*cf. four* in the initial halide, and *five* in the S_N2 transition state) are as far apart from each other as they can get, this advantage becoming the more pronounced the bigger they are. Thus both electronic and steric effects *slow* S_N2 and *accelerate* S_N1 attack as the series is traversed, i.e. as substitution on the bromine-carrying carbon atom increases, and the reason for the mechanistic changeover thus becomes apparent.

A similar change of mechanism is observed, but considerably sooner, in traversing the series:

$$CH_3—Cl \quad PhCH_2—Cl \quad Ph_2CH—Cl \quad Ph_3C—Cl$$

S_N1 hydrolysis is here observed at the second member and with $Ph_3C—Cl$ the ionisation is so pronounced that the compound shows electrical conductivity when dissolved in liquid SO_2. The reason for the greater promotion of ionisation, with consequent more rapid changeover to the S_N1 mechanism, is the considerable stabilisation of the carbonium ion that is here possible by delocalisation of the positive charge

i.e. a classical example of an ion stabilised by charge delocalisation via the agency of the delocalised π orbitals of the benzene nucleus (*cf.* the negatively charged phenoxide ion, p. 21). In terms of overall reactivity, benzyl chloride is rather similar to *t*-butyl chloride; the effect will become progressively more pronounced, and S_N1 attack further facilitated, with Ph_2CHCl and Ph_3CCl as the possibilities of delocalising the positive charge are increased in the carbonium ions obtainable from these halides.

S_N2 attack on the CH_2 in benzyl chloride is also considerably faster than on the CH_2 in, for example, $MeCH_2Cl$ (≈ 100 times as fast) for the sp^2 hybridised carbon in the S_N2 transition state can use its unhybridised p orbital to interact not only with the entering $^\ominus OH$ and the leaving $Cl^\ominus$ but also with the π orbital system of the phenyl substituent thereby stabilising itself.

Similar carbonium ion stabilisation can occur with allyl halides:

$$CH_2{=}CH—CH_2Cl \longrightarrow [CH_2{=}CH{-}CH_2 \leftrightarrow CH_2—CH{=}CH_2] + Cl^\ominus$$

S_N1 attack is thus promoted and allyl, like benzyl, halides are normally extremely reactive as compared with e.g. $CH_3CH_2CH_2Cl$ and $PhCH_2CH_2CH_2Cl$ respectively where such carbonium ion stabilisation cannot take place. S_N2 attack is also greatly speeded up by interaction, in the transition state, of the carbon atom attacked and the π orbital of the double bond, thus resembling benzyl chloride above.

By contrast, vinyl halides such as $CH_2=CH-Cl$ and halogenobenzenes are very unreactive towards nucleophiles. This stems from the fact that the halogen atom is now bonded to an sp^2 hybridised carbon with the result that the electron pair of the C—Cl bond is drawn closer to carbon (p. 4) than in the bond to an sp^3 hybridised carbon. The C—Cl is found to be stronger, and thus less easily broken, than in, for example, CH_3CH_2Cl, and the C—Cl dipole is smaller; there is thus less tendency to ionisation (S_N1) and a less positive carbon for $^\ominus OH$ to attack (S_N2); the π electrons of the double bond also inhibit the close approach of an attacking nucleophile. The double bond would not help to stabilise either the S_N2 transition state or the carbonium ion involved in the S_N1 route. Very much the same considerations apply to phenyl halides with their sp^2 hybridised carbons and the π orbital system of the benzene nucleus; their reactions, which though often bimolecular are not in fact S_N2 in nature, are discussed further below (p. 150).

The influence of steric factors on mechanism is particularly observed when substitution takes place at the β-position. Thus in the series

$$CH_3-CH_2-Hal \qquad Me-CH_2-CH_2-Hal \qquad \begin{array}{c} Me \\ \diagdown \\ CH-CH_2-Hal \\ \diagup \\ Me \end{array}$$

$$\begin{array}{c} Me \\ \diagdown \\ Me-C-CH_2-Hal \\ \diagup \\ Me \end{array}$$

it is found that the S_N2 reaction rate falls as we pass along the series, the drop being particularly marked as we go from the isobutyl to the neopentyl halide, in which attack 'from the back' by e.g. $^\ominus OH$ on the α-carbon atom along the line of centres might be expected to be very highly hindered. The main factor inhibiting S_N2 attack is however the highly crowded transition state

$$\begin{array}{c} Me \\ | \\ Me-C-Me \\ | \\ HO\cdots\overset{\delta-}{\underset{\diagup\ \diagdown}{C}}\cdots\overset{\delta-}{Cl} \\ H\ \ H \end{array}$$

that would have to be formed. The drop in rate probably owes little to the inductive effect of the increasing number of substituent methyl groups owing to the interposition of a saturated carbon atom (the β-carbon) between them and the carbon atom (α-) to be attacked.

A very interesting example of the effect of structure on the reactivity of a halide is provided by 1-bromotriptycene (II):

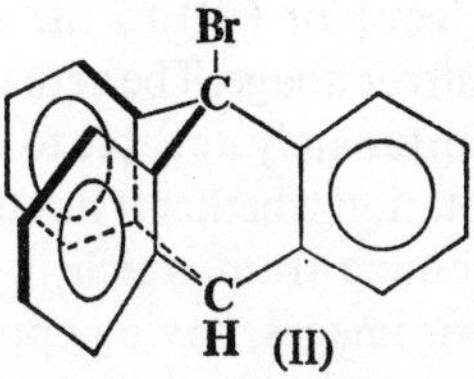

The bromine atom in this compound is virtually inert to nucleophiles. S_N2 'attack from the back' is inhibited sterically by the cage-like structure, and the formation of a transition state, in which the three groups attached to the carbon atom attacked must pass through a co-planar arrangement, is prevented as this atom is held rigidly in position by the substituents attached to it. S_N1 attack is also inhibited because the carbonium ion that would be formed by ionisation is unable to stabilise itself for, being unable to achieve coplanarity with its substituents, charge-delocalisation with them (p. 101) cannot take place. This is in marked contrast to Ph_3CBr in which the C—Br bond is in an exactly analogous local environment but in which there is no such bar to ionisation: the two halides are found to differ in their rate of reaction under the same conditions by a factor of $\approx 10^{23}$!

STEREOCHEMICAL IMPLICATIONS OF MECHANISM

The hydrolysis of an optically active halide presents some interesting stereochemical features. Thus considering both mechanisms in turn:

(i) S_N2 mechanism

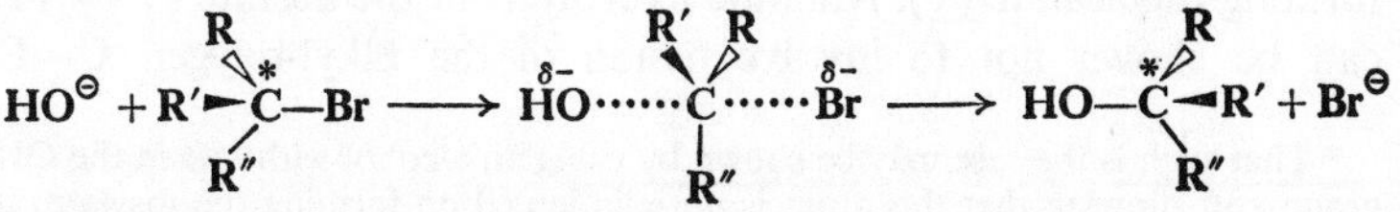

Dextro, i.e. (+)   ?

It will be seen that the arrangement or pattern of the three groups attached to the carbon atom attacked has been effectively turned inside out. The carbon atom is said to have undergone reversal or, as more usually expressed, *inversion* of its *configuration* (the pattern or arrangement in space of the groups attached to it). Indeed, if the product could be the bromide instead of the corresponding alcohol it would be found to rotate the plane of polarisation of plane polarised light in the opposite direction (i.e. laevo or (−)) to the starting material for it would, of course, be its mirror image. The actual product is the alcohol, however, and we are unfortunately not able to tell merely by observing its direction of optical rotation whether it has the same or the opposite configuration to the bromide from which it was derived; for compounds, other than mirror images, having opposite configurations do not necessarily exhibit opposite directions of optical rotation, any more than do compounds having the same configuration necessarily exhibit the same direction of optical rotation. Thus in order to confirm that the above S_N2 reaction is, in practice, actually attended by an inversion of configuration, as the theory requires, it is necessary to have an independent method for relating the configuration of, e.g., a halide and the corresponding alcohol.

(ii) Determination of relative configuration

This turns essentially on the fact that if an asymmetric compound undergoes a reaction in which a bond joining one of the groups to the asymmetric centre is broken, then the centre may—though it need not of necessity—undergo inversion of configuration; while if the compound undergoes reaction in which no such bond is broken then the centre will preserve its configuration intact. Thus in the series of reactions on the asymmetric alcohol (III) formation of an ester with toluene *p*-sulphonyl chloride is known not to break the **C—O** bond of the alcohol,* hence the tosylate (IV) must have the same configuration as the original alcohol. Reaction of this ester with acetate ion is known to be a replacement in which $p\text{-MeC}_6\text{H}_4\text{SO}_2\text{O}^{\ominus}$ is expelled and $\text{MeCO}_2^{\ominus}$ introduced,* hence the **C—O** bond *is* broken in this reaction and inversion of configuration can take place in forming the acetate (V). Alkaline hydrolysis of the acetate (V → VI) can be shown not to involve fission of the alkyl-oxygen **C—O**

---

* That such is the case may be shown by using an alcohol with ^{18}O in the **OH** group and showing that this atom is not eliminated on forming the tosylate; it is however, eliminated when the tosylate is reacted with $\text{MeCO}_2^{\ominus}$.

$$
\underset{\text{(III)}}{\underset{R''}{\overset{R}{R'\!\!-\!\!C\!\!-\!\!OH}}} \xrightarrow{p\text{-MeC}_6\text{H}_4\text{SO}_2\text{Cl}} \underset{\text{(IV)}}{\underset{R''}{\overset{R}{R'\!\!-\!\!C\!\!-\!\!OSO}_2\text{C}_6\text{H}_4\text{Me-}p}} \xrightarrow{\text{Cl}^\ominus} \underset{\text{(VII)}}{\underset{R''}{\overset{R}{\text{Cl}\!\!-\!\!C\!\!-\!\!R'}}}
$$

$$\Big\downarrow \text{MeCO}_2^\ominus$$

$$
\underset{\text{(VI) } (-)}{\underset{R''}{\overset{R}{\text{HO}\!\!-\!\!C\!\!-\!\!R'}}} \xleftarrow{\ominus\text{OH}} \underset{\text{(V)}}{\underset{R''}{\overset{R}{\text{MeCOO}\!\!-\!\!C\!\!-\!\!R'}}}
$$

linkage*, so the alcohol (VI) must have the same configuration as the acetate (V). As (VI) is found to be the mirror image of the starting material, an inversion of configuration must have taken place during the series of reactions and it can only have taken place during the reaction of acetate ion with the tosylate (IV). A number of further reactions with the tosylate show that an inversion of configuration takes place with a variety of anions and hence it may be concluded with a considerable degree of confidence that it takes place on reaction with chloride ion, so that the chloride (VII), like the acetate (V), has opposite configuration to the original alcohol (III). Now that it is thus possible to show that S_N2 reactions are normally attended by inversion of configuration, independent demonstration that a particular reaction takes place by an S_N2 mechanism is often used to relate the configuration of product and starting material in that reaction.

(iii) S_N1 mechanism

$$
\underset{(+)}{\underset{R''}{\overset{R}{R'\!\!-\!\!C\!\!-\!\!Br}}} \longrightarrow \underset{\text{Br}^\ominus}{\underset{R''}{\overset{R'\ R}{C^\oplus}}}
\begin{array}{c} \xrightarrow{\ominus\text{OH}} \underset{(-)}{\underset{R''}{\overset{R}{\text{HO}\!\!-\!\!C\!\!-\!\!R'}}} \\[2em] \xrightarrow{\ominus\text{OH}} \underset{(+)}{\underset{R'}{\overset{R}{R''\!\!-\!\!C\!\!-\!\!OH}}} \end{array}
$$

---

* Hydrolysis of an acetate in which the alcohol–oxygen atom is labelled with ^{18}O fails to result in its replacement, showing that the alkyl–oxygen bond is not broken in the hydrolysis (*cf.* p. 47).

As the carbonium ion formed in the slow, rate-determining stage of the reaction is planar, it is to be expected that subsequent attack by a nucleophilic reagent such as $^{\ominus}$**OH** or the solvent (e.g. H_2O:) will take place with equal readiness from either side of the planar ion leading, in fact, to a 50/50 mixture of species having the same and the opposite configuration as the starting material, i.e. that racemisation will take place yielding an optically inactive ($\pm$) product.

What actually happens, depends on how rapidly the attack by a nucleophile follows on the initial ionisation step. If the second reaction follows closely upon the first, it may be that the receding anion, e.g. **Br**$^{\ominus}$, may still be only a few molecular diameters away and thus attack by an approaching nucleophile is inhibited on the side of the carbonium ion to which the bromine was originally attached. Attack on the 'backside' of the carbonium ion is unaffected, however, and will thus preponderate, leading to more inversion than retention of configuration in the product, i.e. racemisation with some inversion will be observed in the product. Similar results may arise from unsymmetrical solvation of the ion pair produced in the first, rate-limiting stage of the reaction.

What is thus observed in practice, under S_N1 conditions, may range from virtually complete racemisation to almost total inversion of configuration depending on how rapidly attack by a nucleophile follows on the initial ionisation. The most common situation is mainly racemisation attended by some inversion, the relative proportions of the two seen with a particular substrate being profoundly influenced by the conditions under which the reaction is carried out. If the solvent can act as a nucleophile, e.g. H_2O:, attack is likely to be more rapid, because of its very large relative concentration, than if the presence of an added nucleophile, **Y**$^{\ominus}$, is necessary, thus leading to a relatively higher proportion of inversion. A good solvating solvent such as water is particularly effective in this respect because of the rapidity with which the incipient carbonium ion collects a solvating envelope around itself.

(iv) S_Ni mechanism

Despite what has been said above of replacement reactions leading to inversion of configuration or racemisation or, in some cases, a mixture of both, a few cases are known of reactions that proceed with retention of configuration, i.e. in which the starting material and product have the same configuration. One case in which this has been shown to occur is in the replacement of **OH** by **Cl** by the use of thionyl chloride. This reaction has been shown to follow second order kinetics, i.e. rate $\propto$ **[ROH][SOCl$_2$]**, but it clearly cannot proceed according to an unmodified $S_N 2$ mechanism for this would lead to inversion of configuration which is not observed. It has been interpreted mechanistically as follows:

$$R'\!\!-\!\!\overset{\displaystyle R}{\underset{\displaystyle R''}{C}}\!\!-\!\!OH + SOCl_2 \xrightarrow{\ (i)\ } R'\!\!-\!\!\overset{\displaystyle R}{\underset{\displaystyle R''}{C}}\overset{O}{\underset{Cl}{S}}\!\!=\!\!O \xrightarrow{\ (ii)\ } R'\!\!-\!\!\overset{\displaystyle R}{\underset{\displaystyle R''}{C}}\!\!-\!\!Cl + SO_2$$

$$\text{(VIII)}$$

No change in configuration can take place in stage (i) as the **C—O** bond is not broken and in the second stage, where this bond is broken, attack by **Cl** takes place from the same side of the carbon atom because of the orientation of the intermediate (VIII). A close study of the reaction has suggested that the second stage exhibits some $S_N 1$ character in that the breakdown of (VIII) probably proceeds through an ion pair (*cf.* p. 96):

$$R'\!\!-\!\!\overset{\displaystyle R}{\underset{\displaystyle R''}{\overset{\oplus}{C}}} \quad \overset{\ominus O}{\underset{Cl}{S}}\!\!=\!\!O$$

The chlorosulphite anion then breaks down to **SO$_2$** and **Cl$^\ominus$** so rapidly that **Cl$^\ominus$** is available for frontal attack on the carbonium ion before the latter has had time to collapse to the planar state, thus leading to a product having the same configuration as the starting material. Evidence for the involvement of (VIII) as an intermediate is provided by the fact that such alkyl chlorosulphites can actually be isolated, and then shown to undergo extremely ready conversion to alkyl halide and **SO$_2$**. If, however, the reaction is carried out in the presence of base, e.g. pyridine, the hydrogen chloride liberated in forming the chlorosulphite in stage (i) is converted to **Cl$^\ominus$**; this then readily attacks (VIII) 'from the back' with the expulsion of $^\ominus$**OSOCl**. The reaction is now of a normal $S_N 2$ type, proceeding with inversion, and the somewhat specialised $S_N i$ (Substitution Nucleophilic internal) mechanism, with retention of configuration, is no longer observed.

(v) Neighbouring group participation

There are also a number of other cases of replacement reactions in which retention of configuration takes place but these can all be shown to have one feature in common: an atom, close to the carbon undergoing the displacement reaction, which carries a negative charge or has an unshared pair of electrons, i.e. which can act as an 'internal' nucleophilic reagent. Thus in the alkaline hydrolysis of the β-chlorohydrin (IX) the first stage is conversion of the **OH** to the corresponding

alkoxide ion (X). This ion, acting as a nucleophile, then attacks the carbon atom carrying chlorine 'from the back' in an internal S_N2 reaction that results in inversion of configuration at this carbon and the formation of the cyclic intermediate (XI), an epoxide. The three-membered ring is then reopened by the action of $^{\ominus}$OH, attack taking place at the less heavily substituted of the two carbon atoms (i.e. the one that originally carried the chlorine in this case) as this will be the more positive of the two, its electron availability being enhanced by the inductive effect of only one rather than of two alkyl groups. This attack will also be of the S_N2 type, from the side to which the chlorine was originally attached, resulting in a second inversion of configuration at this carbon atom. The reaction is completed by conversion of the alkoxide ion (XII) to the alcohol (XIII). This alcohol will have the same configuration as the original chloride (IX) but it would hardly be true to say that the overall reaction had proceeded with retention of configuration for, in fact, the *apparent* retention of configuration has been brought about by two successive inversions.

This, of course, distinguishes such a reaction type from $S_N i$ which is attended by a *real* retention of configuration.

Evidence for this overall picture of the reaction is provided by the actual isolation of the epoxides postulated as intermediates and the demonstration that they may be hydrolysed under the conditions of the reaction and with results as observed above.

A similar apparent retention of configuration occurs in the hydrolysis of α-halogenated acids in dilute alkaline solution

(XIV)

but here the suggested cyclic intermediate (XIV), an α-lactone, has not actually been isolated. If the reaction is carried out in extremely concentrated base, $[^{\ominus}OH]$ is then large enough to compete effectively with the internal attack by $-CO_2{}^{\ominus}$ and the normal one stage $S_N 2$ reaction then yields a product whose configuration has undergone inversion. At intermediate concentrations of base both reactions take place, i.e. retention of configuration with a varying degree of racemisation is observed.

Even if no stereochemical point is at issue, participation of neighbouring groups can be of interest because of their possible effect on the rate of reaction. Thus $ClCH_2CH_2SEt$ may be hydrolysed approximately 10,000 times as rapidly as $ClCH_2CH_2OEt$ under the same conditions. This is far too large a difference for it to be due to inductive or steric effects in such simple molecules and is thought to arise from the rate-determining formation of a cyclic sulphonium salt (XV) which, being highly strained, undergoes extremely ready and rapid hydrolysis:

$^{\ominus}Cl$ (XV)

The oxygen in the ether, $ClCH_2CH_2OEt$, being more electronegative, does not part with its unshared electrons so readily as sulphur, hence no cyclic salt is formed and the chlorine undergoes hydrolysis by a normal displacement reaction. Suitable nitrogen-containing compounds also show such enhanced ease of hydrolysis proceeding via ethyleneimmonium ions such as (XVI):

$$\text{Me}_2\text{N:} \overset{\curvearrowright}{\underset{\underset{\text{CH}_2}{\diagdown \diagup}}{\text{CH}_2}}\text{Cl} \xrightarrow[\text{determining}]{\text{Rate-}} \text{Me}_2\overset{\oplus}{\underset{\underset{\text{CH}_2}{\diagdown \diagup}}{\text{N}}}\text{CH}_2 \xrightarrow[\text{rapid}]{\ominus\text{OH}} \text{Me}_2\text{N:}\underset{\underset{\text{CH}_2}{\diagdown \diagup}}{\text{CH}_2}\text{OH}$$

$$\ominus\text{Cl} \quad \text{(XVI)}$$

Their hydrolysis normally proceeds less rapidly than that of similar sulphur compounds, however, reflecting the greater stability of the cyclic nitrogen, as compared with the cyclic sulphur, intermediates.

These features are of interest in relation to the classical vesicant agents of chemical warfare such as mustard gas itself, $S(CH_2CH_2Cl)_2$ and the related nitrogen mustards such as $MeN(CH_2CH_2Cl)_2$. The cyclic immonium derivatives obtained as intermediates during the hydrolytic destruction of the latter have the additional hazard of being powerful neurotoxins.

EFFECT OF ENTERING AND LEAVING GROUPS

Changing the nucleophilic reagent employed, i.e. *the entering group*, is not going to alter the rate of an S_N1 displacement reaction, e.g. of a halide, for this reagent does not take part in the rate-determining step of the reaction. With an S_N2 displacement, however, the more strongly nucleophilic the reagent the more the reaction will be promoted. The *nucleophilicity* of a reagent might perhaps be expected to correlate with its basicity in that both involve the availability of electron pairs. This parallel, though having a certain utility, is by no means exact for with an ion such as $Y^\ominus$, nucleophilicity involves electron pair donation usually to carbon, whereas basicity involves donation to hydrogen, the former normally being much more susceptible to steric influences. Further, nucleophilicity involves a kinetic situation, a reaction rate, whereas basicity involves a thermodynamic one, the position of an equilibrium.

The parallel can however be used as a general guide with fair success, particularly if the *attacking atom* of the nucleophiles considered is the same in each case. Thus strong bases such as $EtO^\ominus$ and $HO^\ominus$ are more strongly nucleophilic agents than weak bases such as $MeCO_2{}^\ominus$.

From what has already been said about the effect of change of reagent on the two types of mechanism, it follows that in the displacement of any particular atom or group, the more powerfully nucleophilic the reagent employed the greater is the chance of the reaction proceeding by the S_N2 route. Thus as the series $H_2O:, MeCO_2^\ominus, PhO^\ominus, HO^\ominus$, $EtO^\ominus$ is traversed, it may well be that a displacement reaction of **RHal** which started by being S_N1 with $H_2O:$ or $MeCO_2^\ominus$ has changed over to S_N2 by the time $EtO^\ominus$ is reached.

So far as change of *attacking atom* in a nucleophile is concerned, it is broadly true, within a single group or subgroup of the periodic table, that the larger the atom the greater its nucleophilic reactivity; thus decreasing reactivities $I^\ominus > Br^\ominus > Cl^\ominus > F^\ominus$ and $RS^\ominus > RO^\ominus$ are observed. This is probably due to the fact that as the atom increases in size, the hold the nucleus has on the peripheral electrons decreases, with the result that they become more readily polarisable leading to bonding interaction at greater internuclear distances. Also the larger the ion or group the less its solvation energy, which means the less energy that has to be supplied to it in order to remove, in whole or in part, its envelope of solvent molecules so as to get it into a condition in which it will attack a carbon atom. It is a combination of these two factors which makes the large $I^\ominus$ a better nucleophile than the small $F^\ominus$, despite the fact that the latter is a considerably stronger base than the former.

So far as the *leaving group*, i.e. the one expelled or displaced, in an S_N2 reaction is concerned, the more easily the C-leaving group bond can be distorted the more readily the transition state will be formed, so here again ready polarisability is an advantage. Thus ease of expulsion decreases in the series $I^\ominus > Br^\ominus > Cl^\ominus > F^\ominus$; this, of course, exemplifies the well-known decrease in reactivity seen as we go from alkyl iodide to alkyl fluoride. The fact that $I^\ominus$ can both attack and be displaced so readily means that it is often used as a catalyst in nucleophilic reactions, the desired reaction being facilitated via successive attacks on and displacements from the centre under attack:

$$RCl + H_2O \xrightarrow{\text{Slow}} ROH + H^\oplus + Cl^\ominus$$

$$RCl + I^\ominus \xrightarrow[\text{Fast}]{} Cl^\ominus + RI$$

$$I^\ominus + H^\oplus + ROH$$

The overall effect is thus facilitation of the hydrolysis of **RCl**, which does not readily take place directly, via the easy formation of **RI** ($I^\ominus$ as an effective attacking agent) followed by its ready hydrolysis ($I^\ominus$ as an effective leaving agent).

The best leaving groups are normally the anions of strong acids, e.g. p-**MeC$_6$H$_4$SO$_3$**$^\ominus$ or tosylate: the more basic the leaving group the less easily can it be displaced by an attacking nucleophile; thus strongly basic groups such as **RO**$^\ominus$, **HO**$^\ominus$ and **H$_2$N**$^\ominus$, bound to carbon by small atoms that may not readily be polarised, cannot normally be displaced under ordinary conditions. They can, however, be displaced in acid solution due to initial protonation providing a positively charged species (rather than a neutral molecule) for the nucleophile to attack, and resulting in readier displacement of the much less basic **YH** rather than **Y**$^\ominus$:

$$\overset{..}{R\ddot{O}H} \xrightarrow{\ H\oplus\ } R\overset{\oplus}{\underset{}{\overset{H}{\ddot{O}}}}H \xrightarrow{\ Br\ominus\ } RBr + H_2O$$

Thus even the extremely tightly held fluorine in alkyl fluorides may be displaced by nucleophilic reagents in concentrated sulphuric acid solution. The use of hydrogen iodide to cleave ethers

$$Ph\overset{..}{\ddot{O}}R \xrightarrow{\ H\oplus\ } Ph\overset{\oplus}{\underset{}{\overset{H}{\ddot{O}}}}R \xrightarrow{\ I\ominus\ } PhOH + RI$$

is due to the fact that $I^\ominus$ is the most powerful nucleophile that can be obtained in the strongly acid solution that is necessary to make reaction possible.

The leaving group in an S_N1 reaction determines the reaction rate; the lower the energy of the C-leaving group bond and the greater the tendency of the leaving group to form an anion, the more readily the reaction will proceed via the S_N1 mechanism.

NITROSATION OF AMINES

In the examples considered to date it is carbon that has been undergoing nucleophilic attack but similar attack may also take place on nitrogen as, for example, in the nitrosation of amines where the amine acts as the nucleophile:

$$
\begin{array}{ccccc}
\overset{\displaystyle H}{\underset{\displaystyle H}{RN}}\!:\!\!N\!\!=\!\!O & \longrightarrow & \overset{\displaystyle H}{\underset{\displaystyle H}{RN}}\!\!-\!\!N\!\!=\!\!O & \xrightarrow{-H^{\oplus}} & \overset{\displaystyle H}{RN}\!\!-\!\!N\!\!=\!\!O \\
\end{array}
$$

$$
R^{\oplus} + N\!\!\equiv\!\!N \;\longleftarrow\; R\!\!-\!\!N\!\!\equiv\!\!\overset{\oplus}{N} \;\underset{-H_2O}{\overset{H^{\oplus}}{\longleftarrow}}\; RN\!\!=\!\!N\!\!-\!\!OH
$$

In the familiar reaction of primary amines with nitrites and acid, the species that is acting as the effective nitrosating agent has been shown to depend on the conditions though it is apparently never **HNO₂** itself. Thus at low acidity $N_2O_3(X = ONO)$ obtained by

$$2HNO_2 \rightleftharpoons ONO\text{---}NO + H_2O$$

is thought to be the effective nitrosating agent while as the acidity increases it is first protonated nitrous acid, $H_2\overset{\oplus}{O}\text{---}NO(X = H_2O^{\oplus})$ and finally the nitrosonium ion $^{\oplus}NO$ (*cf.* p. 125), though nitrosyl halides, e.g. **NOCl**, also play a part in the presence of halogen acids. Though the latter are more powerful nitrosating agents than N_2O_3, the reaction with aliphatic amines is nevertheless inhibited by increasing acidity as the nucleophilic **RNH₂** is a relatively strong base which is progressively converted into the unreactive cation, $R\overset{\oplus}{N}H_3$.

With aliphatic primary amines the carbonium ion obtained by breakdown of the highly unstable $RN_2^{\oplus}$ can lead to the formation of a wide range of ultimate products (*cf.* p. 102). The instability of the diazonium cation is due to the very great stability of the N_2 that may be obtained by its breakdown, but with aromatic primary amines some stabilisation of this cation is conferred by delocalisation via the π orbital system of the aromatic nucleus

and diazonium salts may be obtained from such amines provided the conditions are mild. Such diazotisation must normally be carried out under conditions of fairly high acidity (*a*) to provide a powerful nitrosating agent, for primary aromatic amines are not very powerful nucleophiles (due to interaction of the electron pair on nitrogen with the π orbital system of the nucleus, p. 137), and (*b*) to reduce the equilibrium concentration of $\mathbf{ArNH_2}$ by converting it to $\overset{\oplus}{\mathbf{ArNH_3}}$ ($\mathbf{ArNH_2}$ is a very much weaker base than $\mathbf{RNH_2}$, p. 68) so as to avoid as yet undiazotised amine undergoing azo-coupling with the first formed $\overset{\oplus}{\mathbf{ArN_2}}$ (*cf.* p. 133).

Reaction also takes place with the secondary amines but cannot proceed further than the N-nitroso compound, $\mathbf{R_2N-N{=}O}$, while with tertiary aliphatic amines the corresponding nitrosotrialkyl-ammonium cation, $\mathbf{R_3\overset{\oplus}{N}-NO}$, that is formed initially, undergoes ready **C—N** fission to yield relatively complex products. With aromatic tertiary amines such as N-dialkylanilines, however, attack can take place on the activated nucleus (*cf.* p. 125) to yield a **C**-nitroso compound:

OTHER NUCLEOPHILIC DISPLACEMENT REACTIONS

In the discussion of nucleophilic substitution at a saturated carbon atom, the attack on halides by negatively charged ions, e.g. $\mathbf{HO^{\ominus}}$ and $\mathbf{EtO^{\ominus}}$, has been used almost exclusively to illustrate the mechanism of the reactions involved. In fact this type of reaction is extremely widespread in organic chemistry and embraces very many more types than those of which passing mention has been made. Thus other typical examples include:

(i) The formation of a tetralkylammonium salt

$$\mathbf{R_3N{:} + RBr \;\rightarrow\; R_3\overset{\oplus}{N}{:}R + Br^{\ominus}}$$

where the unshared pair of electrons on nitrogen attack carbon with the expulsion of a bromide ion; and also the breakdown of such a salt

$$\mathbf{Br^{\ominus} + R{:}\overset{\oplus}{N}R_3 \;\rightarrow\; BrR + {:}NR_3}$$

in which it is the bromide ion that acts as a nucleophile and $\mathbf{{:}NR_3}$ that is expelled as the leaving group. An exactly similar situation, is of course, met with in the formation of sulphonium salts

$$R_2S: + RBr \rightarrow R_2\overset{\oplus}{S}:R + Br^{\ominus}$$

and in their breakdown.

(ii) The alkylation of reactive methylene groups etc. (*cf.* p. 245):

$$EtO^{\ominus} + CH_2(CO_2Et)_2 \rightarrow EtOH + {}^{\ominus}CH(CO_2Et)_2$$

$$(EtO_2C)_2HC^{\ominus} + RBr \rightarrow (EtO_2C)_2HCR + Br^{\ominus}$$

Here, and in the next two examples, a carbanion or a source of negative carbon is acting as the nucleophile.

(iii) Reactions of acetylene in the presence of strong base, e.g. $NaNH_2$ in liquid NH_3:

$$HC{\equiv}CH + {}^{\ominus}NH_2 \rightarrow HC{\equiv}C^{\ominus} + NH_3$$

$$HC{\equiv}C^{\ominus} + RBr \rightarrow HC{\equiv}CR + Br^{\ominus}$$

(iv) Reaction of Grignard reagents:

$$\overset{\delta + \quad \delta -}{BrMgR + R'Br} \rightarrow Br_2Mg + RR'$$

(v) Decomposition of diazonium salts in water:

$$H_2O: + PhN_2^{\oplus} \rightarrow \overset{\oplus}{HO}:Ph + N_2$$
$$\phantom{H_2O: + PhN_2^{\oplus} \rightarrow} \underset{H}{}$$

$$\overset{\oplus}{HO}:Ph \rightarrow HO:Ph + H^{\oplus}$$
$$\underset{H}{}$$

(vi) The formation of alkyl halides:

$$R\overset{..}{O}H + H^{\oplus} \rightarrow R\overset{\oplus}{O}H$$
$$\phantom{R\overset{..}{O}H + H^{\oplus} \rightarrow} \underset{H}{}$$

$$Br^{\ominus} + R\overset{\oplus}{\underset{H}{O}}H \rightarrow BrR + H_2O$$

(vii) The cleavage of ethers:

$$Ph\overset{..}{O}R + H^{\oplus} \rightarrow Ph\overset{\oplus}{O}R$$
$$\phantom{Ph\overset{..}{O}R + H^{\oplus} \rightarrow} \underset{H}{}$$

$$Ph\overset{\oplus}{\underset{H}{O}}R + I^{\ominus} \rightarrow PhOH + RI$$

(viii) The formation of esters:

$$R'CO_2^{\ominus} + RBr \rightarrow R'CO_2R + Br^{\ominus}$$

(ix) The formation of ethers:

$$RO^{\ominus} + R'Br \rightarrow ROR' + Br^{\ominus}$$

(x) **LiAlH₄** reduction of halides:

$$LiAlH_4 + RBr \rightarrow LiAlH_3Br + HR$$

Here the complex hydride is, essentially, acting as a carrier for hydride ion, $H^{\ominus}$.

(xi) Ring fission in epoxides:

$$Cl^{\ominus} \curvearrowright CH_2\!-\!CH_2 \rightarrow ClCH_2CH_2O^{\ominus}$$

$$ClCH_2CH_2O^{\ominus} + H_2O \rightarrow ClCH_2CH_2OH + {}^{\ominus}OH$$

Here it is the relief of strain achieved on opening the three-membered ring that is responsible for the ready attack by a weak nucleophile.

This is only a very small selection: there are many more displacement reactions of preparative utility and synthetic importance.

It will be noticed from these examples that the attacking nucleophile need not of necessity be an anion with a full-blown negative charge (e.g. $HO^{\ominus}$, $Br^{\ominus}$, $(EtO_2C)_2HC^{\ominus}$) but it must at least have unshared electron pairs available (e.g. $R_3N:$, $R_2S:$) with which to attack a positive carbon or other atom. Equally the species that is attacked may be a cation with a full-blown positive charge (e.g. $R_3\overset{\oplus}{N}:R$), but more commonly it is a neutral molecule (e.g. **RBr**). It must, of course, also be remembered that what is a nucleophilic attack from the point of view of one participant will be an electrophilic attack from the point of view of the other. Our attitude, and hence normal classification of reactions, tends to be formed by somewhat arbitrary preconceptions about what constitutes a reagent as opposed to a substrate (*cf*. p. 31). Overall, the most common nucleophile of preparative significance is probably $HO^\ominus$ or, producing essentially the same result, $H_2O:$, especially when the latter is the solvent and therefore present in extremely high concentration.

Hardly surprisingly, not all displacement reactions proceed so as to yield nothing but the desired product. Side reactions may take place yielding both unexpected and unwanted products, particularly elimination reactions to yield unsaturated compounds: these are discussed subsequently (p. 211).

5 CARBONIUM IONS, ELECTRON-DEFICIENT N AND O ATOMS, AND THEIR REACTIONS

REFERENCE has already been made in the last chapter to the generation of carbonium ions as intermediates in displacement reactions at a saturated carbon atom, e.g. the hydrolysis of an alkyl halide that takes place via the S_N1 mechanism. Carbonium ions are, however, fairly widespread in occurrence and although their existence is normally only transient, they are of considerable importance in a wide variety of chemical reactions.

METHODS OF FORMATION OF CARBONIUM IONS

(i) Direct ionisation

This has already been commented on in the last chapter, e.g.

$$Me_3CCl \rightarrow Me_3C^{\oplus} Cl^{\ominus}$$

$$PhCH_2Cl \rightarrow PhCH_2{}^{\oplus} Cl^{\ominus}$$

$$CH_2{=}CHCH_2Cl \rightarrow CH_2{=}CHCH_2{}^{\oplus} Cl^{\ominus}$$

$$MeOCH_2Cl \rightarrow MeOCH_2{}^{\oplus} Cl^{\ominus}$$

It should be emphasised however that a highly polar, ion-solvating medium is usually necessary and that it is *ionisation* (i.e. the formation of an ion pair) rather than *dissociation* that may then occur.

A particularly striking example is provided by the work of Olah with SbF_5

$$R{-}F + SbF_5 \rightarrow R^{\oplus}SbF_6{}^{\ominus}$$

leading to the formation of simple alkyl cations as crystalline complexes that allow of their ready study by spectroscopic and other means. The use of the same investigator's 'super acids,' such as $SbF_5 \cdot FSO_3H$, allows of the formation of alkyl cations even from alkanes:

$$Me_3C{-}H + SbF_5 \cdot FSO_3H \rightarrow H_2 + Me_3C^{\oplus} SbF_5 \cdot FSO_3{}^{\ominus}$$

The question of how the relative stability, and consequent ease of formation from precursors, of carbonium ions is influenced by their structure will be discussed below (p. 98).

(ii) Protonation

This may, for instance, occur directly by addition to an unsaturated linkage, e.g. in the acid-catalysed hydration of alkenes (p. 162):

$$-CH=CH- \underset{}{\overset{H^\oplus}{\rightleftharpoons}} -CH_2-\overset{\oplus}{C}H- \underset{}{\overset{H_2O}{\rightleftharpoons}} -CH_2-CH_2 \underset{}{\overset{-H^\oplus}{\rightleftharpoons}} -CH_2-CH- \\ \overset{\oplus}{O}H_2 OH$$

This reaction is, of course, reversible and the reverse reaction, the acid-catalysed dehydration of alcohols, is probably more familiar. A proton may also add on to a carbon–oxygen double bond

$$>C=O \underset{}{\overset{H^\oplus}{\rightleftharpoons}} >\overset{\oplus}{C}-OH \underset{}{\overset{Y^\ominus}{\rightleftharpoons}} >C\overset{OH}{\underset{Y}{<}}$$

as in the acid-catalysed addition of some anions, $Y^\ominus$, to an aldehyde or ketone, the addition of proton to the $>C=O$ providing a highly positive carbon atom for attack by the anion. That such protonation does indeed take place is confirmed by the fact that many ketones showed double the theoretical freezing-point depression when dissolved in concentrated sulphuric acid due to:

$$>C=O + H_2SO_4 \rightleftharpoons >\overset{\oplus}{C}-OH + HSO_4{}^\ominus$$

That the ketones undergo no irreversible change in the process may be shown by subsequent dilution of the sulphuric acid solution with water when the ketone may be recovered unchanged.

A similar result may also be obtained by the use of other electron deficient species, i.e. Lewis acids:

$$>C=O + AlCl_3 \rightleftharpoons >\overset{\oplus}{C}-\overset{\ominus}{O}AlCl_3$$

Carbonium ions may also be generated where an atom containing unshared electrons is protonated, the actual carbonium ion being generated subsequently by the removal of this atom:

$$R\text{—}\overset{\cdot\cdot}{O}\text{—}H + H^\oplus \; \rightleftharpoons \; R\text{—}\overset{\oplus}{\underset{H}{O}}\text{—}H \; \rightleftharpoons \; R^\oplus + H_2O$$

This is, of course, one of the steps in the acid-catalysed dehydration of alcohols mentioned above. It may also be encountered in the acid-catalysed decomposition of ethers:

$$R\overset{\cdot\cdot}{O}R + H^\oplus \; \rightleftharpoons \; R\overset{\oplus}{\underset{H}{O}}R \; \rightleftharpoons \; R^\oplus + HOR$$

(iii) Decomposition

The most common example is the decomposition of a diazonium salt, $RN_2^\oplus$:

$$[R\text{—}\overset{\curvearrowleft}{N}{\overset{\oplus}{=}}N \; \leftrightarrow \; R\text{—}\overset{\oplus}{N}{\equiv}N] \; \longrightarrow \; R^\oplus + N_2$$

This may be observed with both aromatic and aliphatic diazonium compounds but, under suitable conditions, these may also undergo decomposition to yield free radicals (p. 283).

The catalysis of a number of nucleophilic displacement reactions of halides by $Ag^\oplus$ is due to 'electrophilic pull' on the halogen atom by the heavy metal cation:

$$Ag^\oplus + Br\text{—}R \; \rightarrow \; AgBr + R^\oplus$$
$$\downarrow$$

The presence of $Ag^\oplus$ may thus have the effect of inducing a shift in mechanistic type from S_N2 to S_N1 but the kinetic picture is often complicated by the fact that the precipitated silver halide may itself act as a heterogeneous catalyst for the displacement reaction.

It should be emphasised that the methods of formation of carbonium ions considered above are not intended to constitute a definitive list. Thus another method of formation arises from the possibility of aromatisation, which is discussed below (p. 100) in connection with tropylium and cyclopropenium cations.

THE STABILITY OF CARBONIUM IONS

The major factor influencing the stability of carbonium ions is that the more the positive charge may be shared among nearby atoms the greater will be the stability of the ion. This is particularly marked where the charge-spreading may take place through the intervention of suitably placed π orbitals, e.g.

$$CH_2{=}CH{-}\overset{\oplus}{C}H_2 \leftrightarrow \overset{\oplus}{C}H_2{-}CH{=}CH_2$$

$$Me{-}\overset{\cdots}{C}{-}\overset{\oplus}{C}H_2 \leftrightarrow Me{-}\overset{\oplus}{O}{=}CH_2$$

leading to carbonium ions characteristically stabilised by delocalisation. It should be emphasised that such stabilisation of the positively charged species by delocalisation can arise not only from internal structural features as above but also through the agency of the oriented dipoles of polar solvent molecules surrounding the carbonium ion pair.

An interesting example of a carbonium ion being more stable than might have been envisaged is the one that can arise from β-phenylethyl derivatives, $PhCH_2CH_2Y$. The carbonium ion, $PhCH_2\overset{\oplus}{C}H_2$, supposedly involved, would not be markedly stabilised by the phenyl group on the β-carbon atom for this is too far away to exert any marked inductive effect and delocalisation via its π orbitals is prevented by the intervening saturated carbon atom. It has, therefore, been suggested on this and other evidence that the carbonium ion involved is actually a bridged structure, a *phenonium* ion

that can stabilise itself by delocalisation through the orbital system of the aromatic nucleus. In the context of, for example, the hydrolysis of $PhCH_2CH_2Br$ this is a further example of neighbouring group participation (p. 86), and by choice of a suitable example it is possible actually to isolate a bridged intermediate, albeit not a carbonium ion:

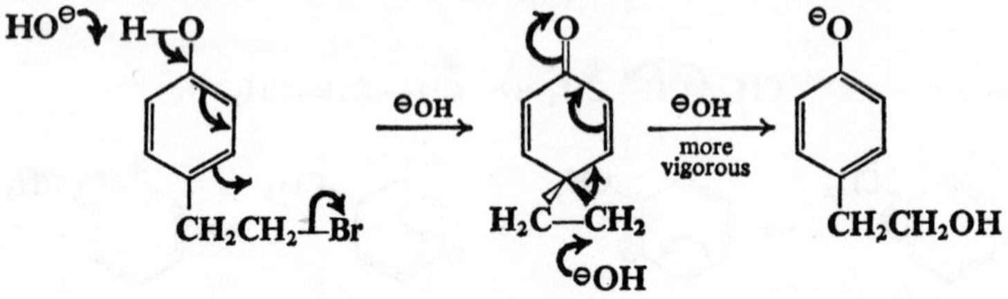

A very highly stabilised carbonium ion may be derived from cycloheptatrienyl (tropylium) bromide which ostensibly has the structure:

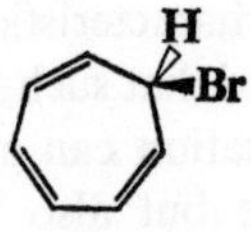

It is found, however, to be highly water-soluble, yielding bromide ions and the evidence is overwhelming that it produces, in solution, a carbonium ion that is so stable that its reaction with water, alcohols, etc., is quite slow. The reason for this quite outstanding stability is that in the tropylium cation, the seven-membered ring possesses six π electrons, which can spread themselves over the seven carbon atoms in delocalised orbitals (*cf.* benzene, p. 13), thereby conferring on the ion quasi-aromatic stability:

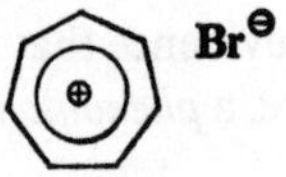

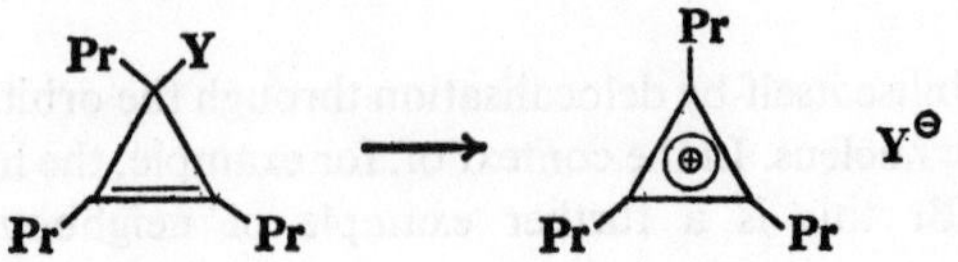

The stabilisation conferred on cyclic systems with $2 + 4n$ π electrons has already been referred to (p. 17) and a particularly interesting case arises when $n = 0$. Thus derivatives of 1,2,3-tripropylcyclopropene are found to yield the corresponding carbonium ion extremely readily and the extraordinary stability of the latter is reflected in the fact that it continues to exist as the carbonium ion to the extent of 50 % even in water at **pH**7! More recently it has even proved possible to isolate a salt of cyclopropene itself

$$\triangle_{\oplus} \quad SbCl_6^{\ominus}$$

as a white crystalline solid.

The simple alkyl carbonium ions have already been seen (p. 78) to follow the stability sequence

$$\overset{\oplus}{Me_3C} > \overset{\oplus}{Me_2CH} > \overset{\oplus}{MeCH_2} > \overset{\oplus}{CH_3}$$

due to the fact that increasing substitution of the carbonium ion carbon atom by methyl groups results in increasing delocalisation of the charge by both inductive and hyperconjugative effects. The particular stability of $Me_3C^{\oplus}$ is borne out by the fact that it may often be formed, under vigorous conditions, by the isomerisation and/or disproportionation of other first-formed carbonium ions, and also by the observation that it remained unchanged after heating at 170° in $SbF_5 \cdot FSO_3H$ for four weeks! Stabilisation here and in other cases requires that the carbonium ion should be planar, for it is only in this state that effective delocalisation can occur. It has indeed been calculated for simple alkyl cations that the planar state (sp^2) is more stable than the pyramidal (sp^3) by ≈ 20 kcal/mole. As planarity is departed from or its attainment inhibited, instability of the ion, with consequent difficulty in its initial formation, increases very rapidly. This has already been seen in the extreme inertness of 1-bromotriptycene, where inability to assume a planar state prevents formation of a carbonium ion with consequent inertness to S_N1 attack (p. 81).

This great preference for the planar state, if at all possible, effectively settles the question of the stereochemistry of simple carbonium ions. A result entirely in accord with expectations in that they take up the same configuration as that already established for the trialkyl borons, R_3B, with which they are isoelectronic.

TYPES OF REACTION UNDERGONE BY CARBONIUM IONS

Essentially, carbonium ions can undergo three main types of reaction:

    (*a*) combination with nucleophile,

    (*b*) elimination of a proton,

(*c*) rearrangement of structure.

It should be noted that (*c*) will result in a further carbonium ion which may then undergo (*a*) or (*b*) before a stable product is obtained. All these possibilities are nicely illustrated in the reaction of sodium nitrite and dilute acid with propylamine (*cf.* p. 90):

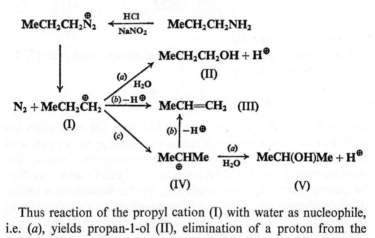

Thus reaction of the propyl cation (I) with water as nucleophile, i.e. (*a*), yields propan-1-ol (II), elimination of a proton from the adjacent carbon atom, (*b*), yields propylene (III), while rearrangement, (*c*), in this case migration of $H^{\ominus}$, yields the isopropyl cation (IV), which can then undergo (*b*) or (*a*) to yield more propylene (III) or propan-2-ol (V), respectively. The products obtained in a typical experiment were propan-1-ol, 7%, propylene, 28% and propan-2-ol, 32%; the greater stability of the iso-, rather than the n-, propyl cation being reflected in the much greater amount of the secondary alcohol produced.

This has not exhausted the possibilities, however, for reaction of either carbonium ion with other nucleophiles present in the system can obviously lead to further products. Thus $NO_2^{\ominus}$ from sodium nitrite may lead to the formation of RNO_2 and $RONO$ (the latter may also arise from direct esterification of first formed ROH), $Cl^{\ominus}$ from the acid may lead to RCl, first formed ROH be converted to ROR and as yet unchanged RNH_2 to $RNHR$. The mixture of products actually obtained is, hardly surprisingly, greatly influenced by the conditions under which the reaction is carried out but it will come as no surprise that this reaction is, in the aliphatic series, seldom a satisfactory preparative method for the conversion of $RNH_2 \rightarrow ROH$!

An analogous situation is observed in the Friedel-Crafts alkylation of benzene with 1-brompropane in the presence of gallium bromide. Here the attacking species, if not an actual carbonium ion, is a highly polarised complex (p. 127) $\overset{\delta+ \ \delta-}{\mathbf{RGaBr_4}}$, and the greater stability of the complex which carries its positive charge on a secondary, rather than a primary, carbon atom, i.e. $\overset{\delta+ \ \ \delta-}{\mathbf{Me_2CHGaBr_4}}$ rather than $\overset{\delta+ \ \ \ \ \delta-}{\mathbf{MeCH_2CH_2GaBr_4}}$, again results in a hydride shift so that the major product of the reaction is actually isopropylbenzene.

That such rearrangements need not always be quite as simple as they look, however, i.e. mere migration of $\mathbf{H}^{\ominus}$, is illustrated by the behaviour with $\mathbf{AlBr_3}$ of propane in which a terminal carbon atom is labelled with $^{13}\mathbf{C}$, when partial transfer of the labelled carbon to the 2-position occurs. This is presumably due to

which *may* happen also in cases such as the above where it is only $\mathbf{H}^{\ominus}$ that has *apparently* moved.

The elimination reactions of carbonium ions will be discussed further below (p. 212) when elimination reactions in general are dealt with, but their rearrangement merits further study.

THE REARRANGEMENT OF CARBONIUM IONS

Despite the apparent confusion introduced above by the isomerisation of propane, the rearrangement reactions of carbonium ions can be divided essentially into those in which a change of actual carbon skeleton does, or does not, take place; the former are the more important but the latter will be briefly mentioned first.

(i) Without change in carbon skeleton

(a) **Allylic rearrangements:** A classical example of this variety may occur where the carbonium ion formed is stabilised by delocalisation, e.g. in the S_N1 solvolysis of 3-chlorobut-1-ene, $MeCHClCH=CH_2$, in ethanol. After formation of the carbonium ion

$$MeCHClCH=CH_2 \rightleftarrows Cl^{\ominus} + \left[\begin{array}{c} Me\overset{\oplus}{C}H\text{---}CH=CH_2 \\ \updownarrow \\ MeCH=CH\text{---}\overset{\oplus}{C}H_2 \end{array} \right]$$

attack by **EtOH** can be at C_1 *or* C_3 and a mixture of the two possible ethers is indeed obtained:

$$\left[\begin{array}{c} Me\overset{\oplus}{C}H\text{---}CH=CH_2 \\ \updownarrow \\ MeCH=CH\text{---}\overset{\oplus}{C}H_2 \end{array} \right] \xrightarrow{\text{EtOH}} \begin{array}{c} MeCH(OEt)CH=CH_2 \\ + \\ MeCH=CHCH_2OEt \end{array} + H^{\oplus}$$

If, however, the reaction is carried out in ethanol with ethoxide ions present as powerful nucleophilic reagents, the reaction proceeds as a straightforward S_N2 displacement reaction, $^{\ominus}OEt$ displacing $Cl^{\ominus}$, and only the one product, $MeCH(OEt)CH=CH_2$, is obtained. Allylic rearrangements have been observed, however, in the course of displacement reactions that are undoubtedly proceeding by a bimolecular process. Such reactions are designated as S_N2' and are believed to proceed:

$$Y^{\ominus} \enspace CH_2=CH\text{---}CH\text{---}Cl \longrightarrow Y\text{---}CH_2\text{---}CH=CH + Cl^{\ominus}$$

S_N2' reactions tend to occur more particularly when there are bulky substituents on the α-carbon atom for these markedly reduce the rate of the competing, direct displacement reaction by the normal S_N2 mode. Allylic rearrangements, by whichever mechanism they may actually be proceeding, are relatively common.

(ii) With change in carbon skeleton

  (a) The neopentyl rearrangement: A good example is the hydrolysis of neopentyl chloride (VI) under conditons favouring the S_N1

mechanism (it may be remembered that the S_N2 hydrolysis of these halides is highly hindered in any case, p. 80); this might be expected to yield neopentyl alcohol (VIII):

$$\underset{(VI)}{\underset{\underset{\displaystyle Me}{|}}{\overset{\overset{\displaystyle Me}{|}}{Me-C-CH_2Cl}}} \xrightarrow{S_N1} \underset{(VII)}{\underset{\underset{\displaystyle Me}{|}}{\overset{\overset{\displaystyle Me}{|}}{Me-C-\overset{\oplus}{C}H_2}}} \xrightarrow{H_2O} \underset{(VIII)}{\underset{\underset{\displaystyle Me}{|}}{\overset{\overset{\displaystyle Me}{|}}{Me-C-CH_2OH}}} + H^{\oplus}$$

In fact no neopentyl alcohol (VIII) is obtained, the only alcoholic product is found to be *t*-amyl alcohol (X); this is due to the initial carbonium ion (VII) rearranging to yield a second one (IX). It will be seen that the latter is a tertiary carbonium ion whereas the former is a primary one, and it is an interesting reflection that the tertiary ion is so much more stable than the primary as to make it energetically worthwhile for a carbon–carbon bond to be broken and for a methyl group to migrate:

$$\underset{(VII)}{\underset{\underset{\displaystyle Me}{|}}{\overset{\overset{\displaystyle Me}{|}}{Me-C-\overset{\oplus}{C}H_2}}} \longrightarrow \underset{(IX)}{\underset{\underset{\displaystyle Me}{|}}{\overset{\overset{\displaystyle Me}{|}}{Me-\overset{\oplus}{C}-CH_2}}} \begin{array}{l} \xrightarrow{H_2O} \underset{(X)}{\underset{\underset{\displaystyle OH}{|}}{\overset{\overset{\displaystyle Me}{|}}{Me-C-CH_2-Me}}} \\ \\ \xrightarrow{-H^{\oplus}} \underset{(XI)}{\overset{Me}{\underset{Me}{}}\!\!>\!\!C=CH-Me} \end{array}$$

Such reactions in which a rearrangement of carbon skeleton takes place are known collectively as Wagner–Meerwein rearrangements. The rearranged carbonium ion (IX) is also able to eliminate $H^{\oplus}$ to yield an alkene, and some 2-methylbut-2-ene (XI) is, in fact, obtained. The rearrangement, with its attendant consequences, can be avoided if the displacement is carried out under conditions to promote an S_N2 reaction path but, as has already been mentioned, the reaction is then very slow.

The possible occurrence of such rearrangements of a compound's carbon skeleton during the course of apparently unequivocal reactions is clearly of the utmost significance in interpreting the results of experiments aimed at structure elucidation. Some rearrangements of this type are highly complex, e.g. in the field of natural products such as the terpenes, and have often rendered the unamibiguous assignment of structure extremely difficult.

It is interesting that if the halide **Me₃CCHClPh** is hydrolysed under S_N1 conditions, no rearrangement like the above takes place for the first formed carbonium ion (XII) can stabilise itself by delocalisation via the π orbitals of the benzene nucleus, and rearrangement such as the above is thus no longer energetically advantageous:

(XII)

(b) Rearrangement of hydrocarbons: Wagner–Meerwein type rearrangements are also encountered in the cracking of petroleum hydrocarbons where catalysts of a Lewis acid type are used. These generate carbonium ions from the straight-chain hydrocarbons (*cf.* the isomerisation of ¹³C labelled propane above), which then tend to rearrange to yield branched-chain products. Fission also takes place, of course, but the branching is of importance as the branched hydrocarbons produced cause less knocking in the cylinders of internal combustion engines than do their straight-chain isomers. It should, however, be mentioned that cracking can also be brought about by catalysts that promote reaction via radical intermediates (p. 262).

Rearrangement of unsaturated hydrocarbons takes place readily in the presence of acids:

$$
\begin{array}{c}
\text{Me} \\
\mid \\
\text{Me—C—CH=CH}_2 \\
\mid \\
\text{Me}
\end{array}
\underset{}{\overset{\text{H}^{\oplus}}{\rightleftharpoons}}
\begin{array}{c}
\text{Me} \\
\mid \\
\text{Me—C—}\overset{\oplus}{\text{C}}\text{H—Me} \\
\mid \\
\text{Me}
\end{array}
$$

$$
\begin{array}{c}
\text{Me} \qquad \text{Me} \\
\diagdown\diagup \\
\text{C=C} \\
\diagup\diagdown \\
\text{Me} \qquad \text{Me}
\end{array}
\underset{}{\overset{-\text{H}^{\oplus}}{\rightleftharpoons}}
\begin{array}{c}
\text{Me} \\
\mid \\
\text{Me—C—CH—Me} \\
\overset{\oplus}{} \mid \\
\text{Me}
\end{array}
$$

This tendency can be a nuisance when acid reagents, e.g. hydrogen halides, are being added preparatively to alkenes: mixed products that are difficult to separate may result or, in unfavourable cases, practically none of the desired product may be obtained.

(c) **The pinacol/pinacolone rearrangement:** Another case of migration of an alkyl group to a carbonium ion carbon atom occurs in the acid-catalysed rearrangement of pinacol (*cf.* p. 190) to pinacolone, $Me_2C(OH)C(OH)Me_2 \rightarrow MeCOCMe_3$:

$$
\begin{array}{c}
\text{Me} \\
\mid \\
\text{Me—C——CMe}_2 \\
\mid \qquad \mid \\
\text{OH} \quad \ddot{\text{O}}\text{H}
\end{array}
\overset{+\text{H}^{\oplus}}{\rightleftharpoons}
\begin{array}{c}
\text{Me} \\
\mid \\
\text{Me—C——CMe}_2 \\
\mid \qquad \overset{\oplus}{\mid} \\
\text{OH} \quad \ddot{\text{O}}\text{H} \\
\qquad\quad \mid \\
\qquad\quad \text{H}
\end{array}
\overset{-\text{H}_2\text{O}}{\rightleftharpoons}
\begin{array}{c}
\text{Me} \\
\mid \\
\text{Me—C—}\overset{\oplus}{\text{C}}\text{Me}_2 \\
\mid \\
\text{OH}
\end{array}
\qquad \text{(XIII)}
$$

$$
\begin{array}{c}
\text{Me—C—CMe}_3 \\
\parallel \\
\text{O}
\end{array}
\overset{-\text{H}^{\oplus}}{\rightleftharpoons}
\begin{array}{c}
\oplus \quad \text{Me} \\
 \mid \\
\text{Me—C—CMe}_2 \\
\mid \\
\overset{}{\text{O}}\text{—H}
\end{array}
$$

The fact that a 1,2-shift takes place at all in (XIII), which is already a tertiary carbonium ion, is probably due to the stabilisation of the rearranged ion that can be effected by delocalisation involving the electron pairs on the oxygen atom, and also to the ready loss of proton from the latter to yield a stable end product.

It might be expected that an analogous reaction would take place with any other compounds that could yield the crucial carbonium ion (XIII) and this is, in fact, found to be the case; thus the corresponding bromohydrin (XIV) and hydroxyamine (XV) yield pinacolone when reacted with $Ag^{\oplus}$ and $NaNO_2/HCl$, respectively:

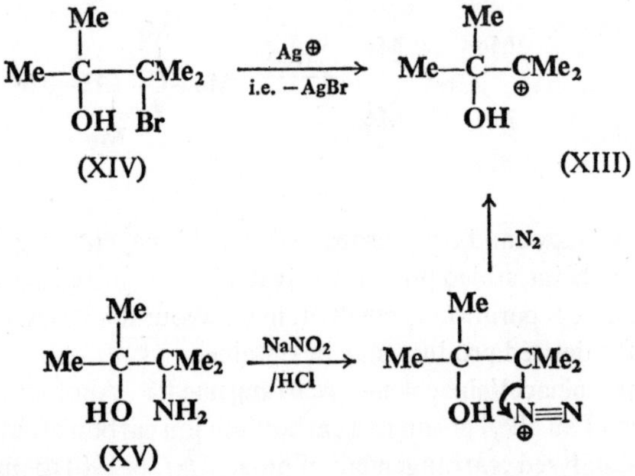

It seems likely that the migration of the alkyl group follows extremely rapidly on the loss of H_2O, $Br^{\ominus}$ or N_2, or probably takes place simultaneously, for in a compound in which the carbonium ion carbon is asymmetric, the ion does not get time to become planar and so yield a racemic product, for the product obtained is found to have undergone inversion of configuration

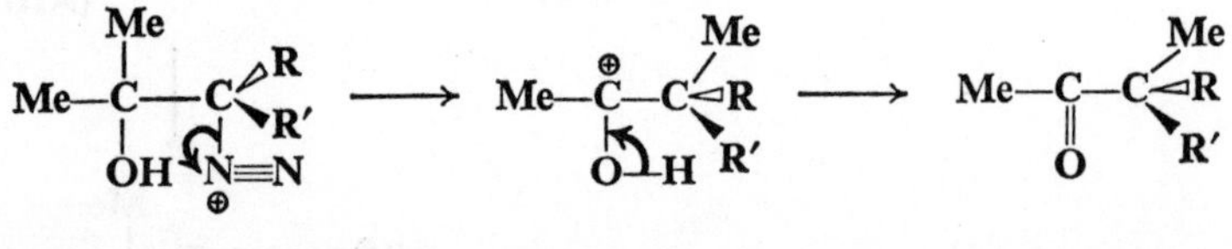

attack taking place 'from the back' in an internal S_N2 type displacement reaction. That the migrating group prefers to move in from the side opposite to that of the leaving group may be demonstrated in cyclic systems where there is restricted rotation about the C_1—C_2 bond; it is then found that compounds in which migrating and leaving groups are *trans* to each other rearrange very much more readily than do those in which the groups are *cis*. It is noteworthy that the migrating alkyl group in this and other cases is migrating with its bonding electrons and so can obviously act as a powerfully nucleophilic reagent. Where the migrating group is asymmetric, it has in certain other cases, though not in this particular one, been shown to retain its configuration as it migrates, indicating that it never actually becomes wholly free from the rest of the molecule; other evidence is also against the migrating group ever becoming free, e.g. no 'crossed product' when two different but very similar pinacols (that undergo rearrangement at approximately the same rate) are rearranged in the same solution: thus the reaction is said to be a typical *intra*molecular, as opposed to *inter*molecular, rearrangement. Indeed, it is probable that the migrating group begins to be attached to the carbonium ion carbon before becoming separated from the carbon atom that it is leaving. A state such as

$$\begin{array}{c} R^{\oplus} \\ \diagdown \\ >\!C\!-\!C\!< \end{array}$$

probably intervenes between the initial and the rearranged carbonium ions (*cf.* bromonium ion structures encountered in the addition of bromine to alkenes, p. 158).

As the migrating group migrates with its electron pair, i.e. as a nucleophile, it might be expected that where the groups on the non-carbonium ion carbon are different, it would be the more nucleophilic of them, i.e. the more powerful electron donor, that would actually migrate. Thus in the example it is the p-MeOC$_6$H$_4$ that migrates in preference to C$_6$H$_5$ owing to the electron-donating effect of the MeO group in the p-position.

Steric factors also play a part, however, and it is found that o-MeOC$_6$H$_4$ migrates more than a thousand times less readily than the corresponding p-substituted group—less readily indeed than phenyl itself—due to its interference in the transition state with the non-migrating groups.

In pinacols of the form

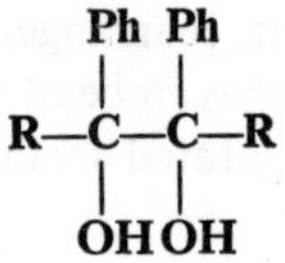

$$\begin{array}{c} \text{Ph} \quad \text{Ph} \\ | \quad\quad | \\ \text{R—C—C—R} \\ | \quad\quad | \\ \text{OH OH} \end{array}$$

Ph will migrate in preference to **R** because of the greater stabilisation it can, by delocalisation, confer on the intervening bridged intermediate (*cf.* p. 99).

(d) **The Wolff rearrangement**: This involves the loss of nitrogen from α-diazoketones (XVI) and their rearrangement to highly reactive ketenes (XVII):

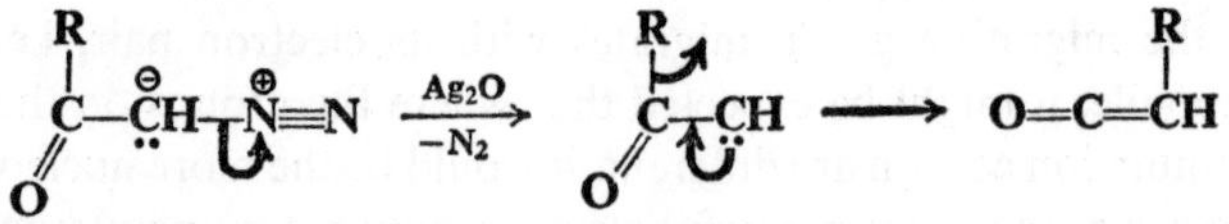

(XVI)   (XVIII)   (XVII)

The intermediate (XVIII) is not a carbonium ion but it is nevertheless an *electron-deficient* species, known as a *carbene*, so the R group migrates with its electron complement complete as in the cases we have already considered. The diazoketone may be obtained by the reaction of diazomethane, CH$_2$N$_2$, on the acid chloride and the Wolff rearrangement is of importance because it constitutes part of the Arndt–Eistert procedure by which an acid may be converted into its homologue:

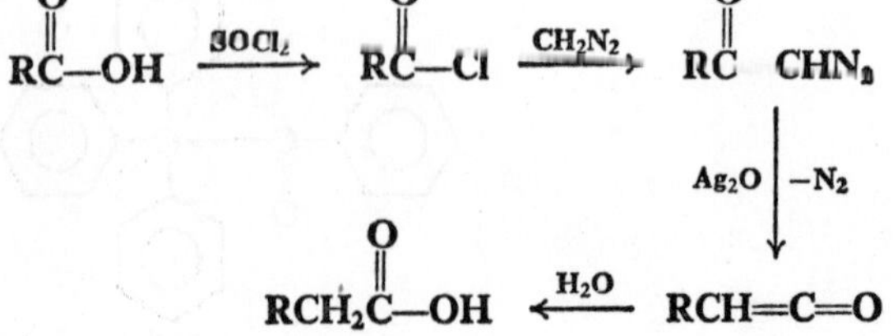

In aqueous solution, the acid is obtained directly by addition of water to the ketene but if the reaction is carried out in ammonia or an alcohol the corresponding amide or ester, respectively, may be obtained directly.

MIGRATION TO ELECTRON-DEFICIENT NITROGEN ATOMS

The reactions involving rearrangement of structure that we have already considered all have one feature in common: the migration of an alkyl or aryl group with its electron pair to a carbon atom which, whether a carbonium ion or not, is electron-deficient. Another atom that can similarly become electron-deficient is nitrogen in, e.g., $R_2\ddot{N}^{\oplus}$ or $R\ddot{N}$ (a *nitrene*, *cf*. carbenes above), and it might be expected that the nitrogen atoms in such species should be able to induce migration to themselves as is observed with $R_3C^{\ominus}$ or $R_2\ddot{C}$. This is indeed found to be the case.

(i) The Hofmann, Curtius and Lossen reactions

A typical example is the conversion of an amide to an amine containing one carbon less by the action of alkaline hypobromite, the Hofmann reaction:

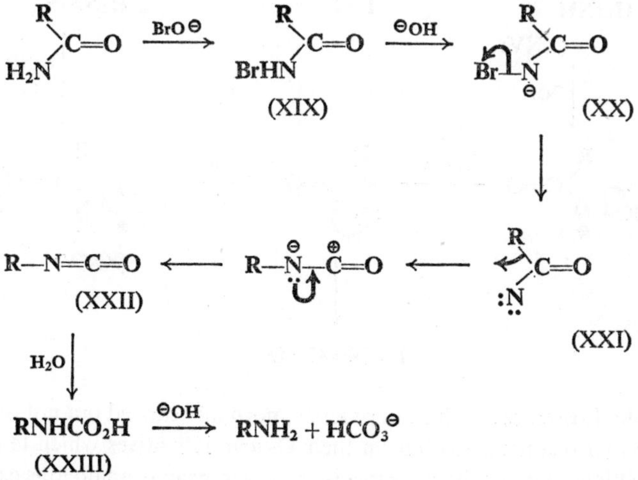

It will be noticed that the species (XXI) has an electron-deficient nitrogen atom corresponding exactly to the electron-deficient carbon atom in the carbene (XVIII) from the Wolff rearrangement, and that the isocyanate (XXII) obtained by the former's rearrangement corresponds closely to the ketene (XVII) obtained from the latter. The reaction is completed by hydration of the isocyanate to yield the carbamic acid (XXIII) which undergoes spontaneous decarboxylation to the amine. The N-bromamide (XIX), its anion (XX) and the isocyanate (XXII) postulated as intermediates can all be isolated under suitable conditions.

The rate-determining step of the reaction is the loss of $Br^{\ominus}$ from the ion (XX) but it is probable that the loss of $Br^{\ominus}$ and the migration of **R** take place simultaneously, i.e. effectively internal S_N2 once again. It might be expected that the more electron-releasing **R** is, the more rapid would be the reaction: this has been confirmed by a study of the rates of decomposition of benzamides substituted in the nucleus by electron-donating substituents.

There are two reactions very closely related to that of Hofmann, namely the Curtius degradation of acid azides (XXIV) and the Lossen decomposition of hydroxamic acids (XXV), both of which also yield amines; all three reactions proceed via the isocyanate as a common intermediate:

$$R\text{—}N{=}C{=}O$$

The Lossen reaction is, in practice, normally carried out not on the free hydroxamic acids but on their **O**-acyl derivatives which tend to give higher yields; the principle is, however, exactly analogous except that now $R'CO_2^{\ominus}$ instead of $HO^{\ominus}$ is expelled from the anion. In the Curtius reaction, the azide is generated as required by the action of sodium nitrite and acid on the hydrazide; if the reaction is carried out in solution in an alcohol instead of in water (nitrous acid being derived from amyl nitrite and hydrogen chloride), the urethane is obtained:

$$\underset{\overset{\|}{R-C-N_3}}{O} \longrightarrow \underset{\overset{\|}{R-C-\overset{\cdot\cdot}{\underset{\cdot\cdot}{N}}}}{O} \longrightarrow R-N=C=O \xrightarrow{R'OH} R-NHCO_2R'$$

In all these cases, the **R** group that migrates conserves its configuration as in the carbon → carbon rearrangements already discussed and, as with them, no mixed products are formed when two different, but very similar, compounds are rearranged in the same solution, showing that the **R** groups never became free in the solution when migrating, i.e. these too are *intra*molecular rearrangements.

(ii) The Beckmann rearrangement

The most famous of the rearrangements in which **R** migrates from carbon to nitrogen is undoubtedly the conversion of ketoximes to N-substituted amides, the Beckmann transformation:

$$\textbf{RR'C=NOH} \rightarrow \textbf{R'CONHR} \quad \text{or} \quad \textbf{RCONHR'}$$

The reaction is catalysed by a wide variety of acidic reagents, e.g. H_2SO_4, P_2O_5, SO_3, $SOCl_2$, BF_3, PCl_5, etc., and takes place not only with the oximes themselves but also with their **O**-esters. Only a very few aldoximes rearrange under these conditions but more can be made to do so by use of polyphosphoric acid as a catalyst. The most interesting feature of the change is, that unlike the reactions we have already considered, it is not the *nature*, e.g. relative electron-releasing ability, but the *stereochemical arrangement* of the **R**, **R'** groups that determines which of them in fact migrates. Thus it is found, in practice, to be always the *anti*-**R** group that rearranges:

(i.e. **R'CONHR**) *only*

Confirmation of this fact requires an initial, unambiguous assignment of configuration to a pair of oximes. This was effected as follows: working with the pair of oximes (XXVI) and (XXVII), it was shown that one of them was cyclised to the isoxazole (XXVIII) on treatment with alkali even in the cold, while the other was but little attacked even under very much more vigorous conditions. The oxime undergoing ready cyclisation was, on this basis, assigned the configuration (XXVI) in which the oxime **OH** group and the nuclear bromine atom are close together and the one resisting cyclisation, the configuration (XXVII), in which these groups are far apart and correspondingly unlikely to interact with each other:

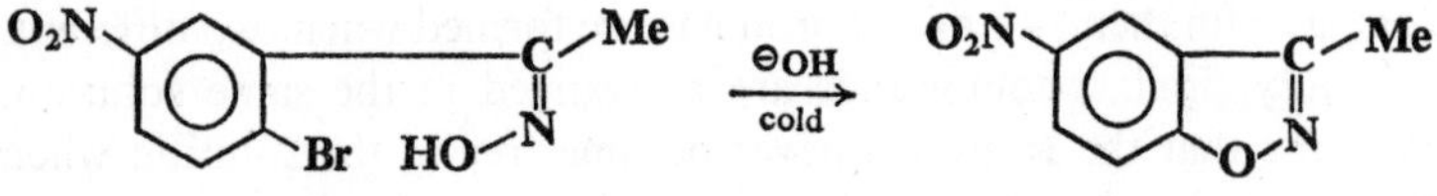

(XXVI)   (XXVIII)

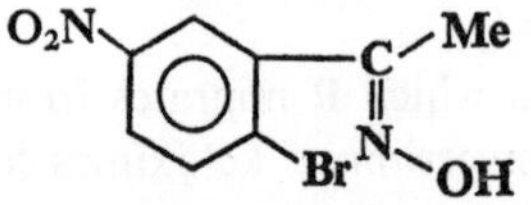

(XXVII)

Subsequently, configuration may be assigned to other pairs of ketoximes by correlation of their physical constants with those of pairs of oximes whose configuration has already been established. Once it had been clearly demonstrated that it was *always* the *anti*-**R** group that migrated in the Beckmann reaction, however, the product obtained by such transformation of a given oxime has normally been used to establish the configuration of that oxime. Thus, as expected, (XXVI) is found to yield only a substituted N-methylbenzamide, while (XXVII) yields only a substituted acetanilide.

That a mere, direct interchange of **R** and **OH** has not taken place has been shown by rearrangement of benzophenone oxime to benzanilide in $H_2{}^{18}O$:

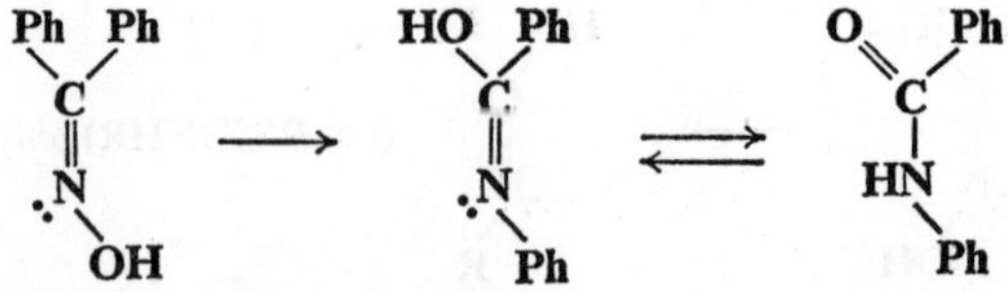

Provided that neither the initial oxime nor the anilide produced will exchange its oxygen for ^{18}O when dissolved in $H_2{}^{18}O$ (as has

114

been confirmed), a mere intramolecular exchange of **Ph** and **OH** cannot result in the incorporation of any ^{18}O in the rearranged product. In fact, however, the benzanilide is found to contain the same proportion of ^{18}O as did the original water so that the rearrangement must involve loss of the **OH** group and the subsequent replacement of oxygen by reaction with water.

The rearrangement is believed to take place as follows:

$$\underset{R'}{\overset{R}{>}}C=N\overset{..}{\cdot}_{OH} \quad \xrightarrow[\text{etc.}]{\overset{RCOCl}{\underset{RSO_2Cl}{}}} \quad \underset{R'}{\overset{R}{>}}C=N\overset{..}{\cdot}_{OX} \qquad (XXX)$$

$$\downarrow H^{\oplus} \qquad\qquad\qquad\qquad \downarrow -OX^{\ominus}$$

$$\underset{R'}{\overset{R}{>}}C=N\overset{..}{\underset{\overset{\oplus}{\underset{H}{OH}}}{}} \quad \xrightarrow{-H_2O} \quad \underset{R'}{\overset{R}{>}}C\overset{\oplus}{=}N\overset{..}{} \quad \longrightarrow \quad \underset{R'}{\overset{\oplus}{C}}=N\overset{..}{\diagup}^{R}$$

$$(XXIX)$$

$$\downarrow H_2O$$

$$\underset{R'-\overset{O}{\overset{||}{C}}-NHR}{} \quad \longleftarrow \quad \underset{R'}{\overset{HO}{>}}C=N\overset{..}{\diagup}^{R} \quad \xleftarrow{-H^{\oplus}} \quad \underset{R'}{\overset{HO\overset{\oplus}{\overset{H}{}}}{>}}C=N\overset{..}{\diagup}^{R}$$

In fairly strong acid, the rearrangement proceeds by protonation of the oxime, followed by loss of water to yield the species (XXIX) having an electron-deficient nitrogen atom; while with acid chlorides, etc., the ester (XXX) is obtained which loses an anion, $^{\ominus}OX$, to yield the same intermediate. Support for the latter interpretation is provided by the fact that such **O**-esters may be prepared separately and shown to undergo the subsequent rearrangement in neutral solvents in the absence of added catalysts. Also, the stronger the acid **XOH**, i.e. the more stable the anion $^{\ominus}OX$, the more readily $^{\ominus}OX$ should be lost to yield (XXIX) and the more rapid the reaction should be. This is borne out by the fact that the rate of reaction increases in the series where $XO^{\ominus}$ is $CH_3CO_2^{\ominus} < ClCH_2CO_2^{\ominus} < PhSO_3^{\ominus}$. That such ionisation is the rate-determining step in the reaction is also suggested by the observed increase in the rate of reaction as the solvent is made more polar.

It is not certain, in either case, whether the fission of the **N—O** bond and migration of **R** are actually simultaneous but, if not, the rearrangement follows extremely rapidly after the fission for it has been demonstrated that the migrating group attacks the *back* of the nitrogen atom, i.e. the side remote from the leaving group, and that if **R** is asymmetric it migrates without undergoing any change of configuration. In addition, no cross-migration of **R** groups has been observed when two different, but similar, oximes are rearranged simultaneously in the same solution, i.e. this is another intramolecular rearrangement in which **R** never becomes wholly detached from the molecule.

After migration of **R**, the rearrangement is completed by attack of water on the positive carbon (it is, of course, at this stage that ^{18}O is introduced in the rearrangement of benzophenone oxime referred to above), followed by loss of proton to yield the enol of the amide which then reverts to the amide proper.

The stereochemical use of the Beckmann rearrangement in assigning configuration to ketoximes has already been referred to and one large-scale application is in the synthesis of the textile polymer, perlon:

Perlon

MIGRATION TO ELECTRON-DEFICIENT OXYGEN ATOMS

It might reasonably be expected that similar reactions could occur in which the migration terminus is an electron-deficient oxygen atom: such rearrangements are indeed known.

(i) The Baeyer-Villiger oxidation of ketones

Treatment of ketones with hydrogen peroxide or organic peracids,

$$RC\text{—}O\text{—}OH,$$
with O double-bonded to C

results in their conversion to esters:

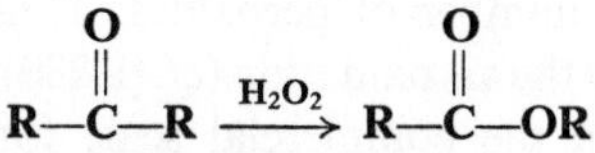

$$\underset{\substack{\| \\ \text{R—C—R}}}{\overset{\text{O}}{}} \xrightarrow{\text{H}_2\text{O}_2} \underset{\substack{\| \\ \text{R—C—OR}}}{\overset{\text{O}}{}}$$

The rate-determining step has been shown to be the acid-catalysed addition of the peracid to the ketone and the reaction is believed to follow the course:

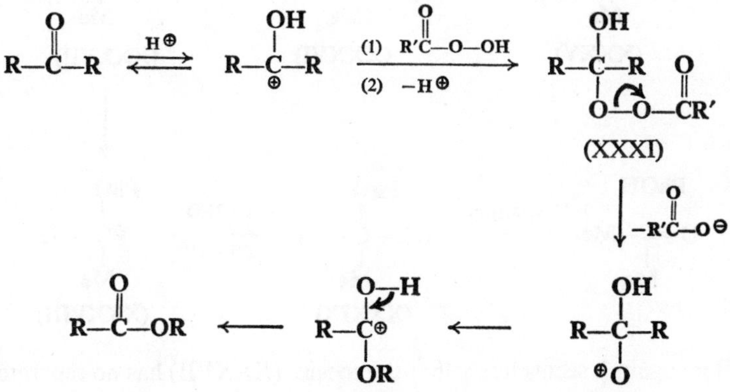

The initial adduct (XXXI) undergoes ready loss of an anion and migration of one of the **R** groups with its electron pair to yield a protonated form (XXXIII) of the final ester (XXXIV). In support of the above mechanism it has been shown by using ^{18}O labelled oxygen that the carbonyl oxygen in the original ketone becomes the carbonyl oxygen in the final ester. The **R** group has been shown to migrate with retention of configuration and bearing in mind that a cation of the **RO**$^{\oplus}$ type, such as (XXXII), is likely to be extremely unstable, it seems probable that the rearrangement occurs simultaneously with loss of the anion as a concerted process reminiscent of the Hofmann reaction (p. 112). When an unsymmetrical ketone is oxidised it is usually the more nucleophilic group that migrates, as in the pinacol/pinacolone rearrangement (p. 109), but as in the latter reaction steric effects may also be involved and can have the effect of markedly changing the expected order of relative migratory aptitude of a series of groups based on their electron-releasing abilities alone.

5

(ii) Rearrangements of peroxides

A rather similar rearrangement is observed during the acid-catalysed decomposition of a number of peroxides. Thus the hydroperoxide (XXXV) obtained by the air oxidation (*cf.* p. 280) of isopropylbenzene (cumene) is used on the commercial scale for the production of phenol and acetone by treatment with acid:

$$
\begin{array}{ccc}
\text{O--ÖH} & \text{O--OH} & \text{O}^{\oplus} \\
| & | & | \\
\text{Ph--C--Me} \underset{\longleftarrow}{\overset{\text{H}^{\oplus}}{\rightleftharpoons}} & \text{Ph--C--Me} \xrightarrow{-\text{H}_2\text{O}} & \text{Ph--C--Me} \\
| & | & | \\
\text{Me} & \text{Me} & \text{Me} \\
\text{(XXXV)} & \text{(XXXVI)} & \text{(XXXVII)}
\end{array}
$$

$$
\begin{array}{ccc}
\text{PhOH} & \text{PhO} & \text{PhO} \\
 & | & | \\
\text{O==C--Me} \xleftarrow{\text{H}^{\oplus}/\text{H}_2\text{O}} & \text{HO--C--Me} \xleftarrow[(2)\ -\text{H}^{\oplus}]{(1)\ \text{H}_2\text{O}} & {}^{\oplus}\text{C--Me} \\
| & | & | \\
\text{Me} & \text{Me} & \text{Me} \\
 & \text{(XXXIX)} & \text{(XXXVIII)}
\end{array}
$$

Here again it seems likely that the species (XXXVII) has no separate existence and that loss of H_2O from (XXXVI) and migration of the phenyl group with its electron pair occur as a concerted process to yield the acetone hemiketal (XXXIX), which then undergoes ready hydrolysis to yield the end-products, phenol and acetone.

In these examples we have been considering the heterolytic fission of peroxide linkages, $-\text{O}:\text{O}- \rightarrow -\text{O}^{\oplus}\ :\text{O}^{\ominus}$, and though this takes place in more polar solvents, the linkage may also undergo homolytic fission to yield free radicals, $-\text{O}:\text{O}- \rightarrow -\text{O}\cdot\ \cdot\text{O}-$, as we shall see below (p. 266).

6 ELECTROPHILIC AND NUCLEOPHILIC SUBSTITUTION IN AROMATIC SYSTEMS

REFERENCE has already been made to the structure of benzene and, in particular, to its delocalised π orbitals (p. 13); the concentration of negative charge above and below the plane of the ring-carbon atoms is thus benzene's most accessible feature:

This concentration of charge might be expected to shield the ring carbon atoms from the attack of nucleophilic reagents and, by contrast, to promote attack by cations, $X^{\oplus}$, or electron-deficient species, i.e. by electrophilic reagents; this is indeed found to be the case.

ELECTROPHILIC ATTACK ON BENZENE

(i) π and σ complexes

It might be expected that the first phase of reaction would be interaction between the approaching electrophile and the delocalised π orbitals and, in fact, so-called π *complexes* such as (I) are formed:

(I)

Thus toluene forms a 1:1 complex with hydrogen chloride at $-78°$, the reaction being readily reversible. That no actual bond is formed between a ring-carbon atom and the proton from **HCl** is confirmed by repeating the reaction with **DCl**; this also yields a π complex, but its formation and decomposition does not lead to the exchange of deuterium with any of the hydrogen atoms of the nucleus, showing that no **C—D** bond has been formed in the complex. Aromatic hydrocarbons have also been shown to form π complexes with species such as the halogens, **Ag$^{\oplus}$**, and, better known, with picric acid, **2,4,6-(O$_2$N)$_3$C$_6$H$_2$OH**, to form stable coloured crystalline adducts whose melting points may be used to characterise the hydrocarbons.

In the presence of a compound having an electron-deficient orbital, e.g. a Lewis acid such as **AlCl$_3$**, a different complex is formed, however. If **DCl** is now employed in place of **HCl**, rapid exchange of deuterium with the hydrogen atoms of the nucleus is found to take place indicating the formation of a σ *complex* (II) in which **H$^{\oplus}$** or **D$^{\oplus}$**, as the case may be, has actually become bonded to a ring-carbon atom. The positive charge is shared over the remaining five carbon atoms of the nucleus via the π orbitals and the deuterium and hydrogen atoms are in a plane at right angles to that of the ring:

(II)

That the π and σ complexes with, e.g. toluene and **HCl**, really are different from each other is confirmed by their differing behaviour. Thus formation of the former leads to no colour change and but little difference in absorption spectrum, indicating that there has been practically no disturbance in the electron distribution in toluene; while if **AlCl$_3$** is present the solution becomes green, will conduct electricity and the absorption spectrum of toluene is modified, indicating the formation of a complex such as (II) as there is no evidence that aluminium chloride forms complexes of the type, **H$^{\oplus}$AlCl$_4$$^{\ominus}$**.

The reaction may be completed by **AlCl$_4$$^{\ominus}$** removing a proton from the σ complex (II) → (IV). This can lead only to exchange of hydrogen atoms when **HCl** is employed but to some substitution of hydrogen by deuterium with **DCl**, i.e. the overall process is electrophilic *substitution*. In theory, (II) could, as an alternative, react by removing **Cl$^{\ominus}$** from **AlCl$_4$$^{\ominus}$** resulting in an overall electrophilic *addition* reaction

(II) → (III) as happens with the π orbital of a simple carbon–carbon double bond (p. 160); but this would result in loss of the stabilisation conferred on the molecule by the presence of delocalised π orbitals involving all six carbon atoms of the nucleus, so that the product, an addition compound, would no longer be aromatic with all that that implies. By expelling $H^\oplus$, i.e. by undergoing substitution rather than addition, the complete delocalised π orbitals are regained in the product (IV) and characteristic aromatic stability recovered:

(III)	(II)	(IV)
Addition		*Substitution*

The gain in stabilisation in going from (II) → (IV) helps to provide the energy required to break the strong **C—H** bond that expulsion of $H^\oplus$ necessitates; in the reaction of, for example, **HCl** with alkenes (p. 160) there is no such factor promoting substitution and addition reactions are therefore the rule.

It might perhaps be expected that conversion of benzene into the σ complex (II), which has forfeited its aromatic stabilisation, would involve the expenditure of a considerable amount of energy, i.e. that the activation energy for the process would be high and the reaction rate correspondingly low: in fact, many aromatic electrophilic substitutions are found to proceed quite rapidly at room temperature. This is because there are two factors operating in (II) that serve to reduce the energy barrier that has to be surmounted in order to effect its formation: first, the energy liberated by the complete formation of the new bond to the attacking electrophile, and, second, the fact that the positively charged σ complex can stabilise itself, i.e. lower its energy level, by delocalisation

(II)

as has indeed been implied by writing its structure as (II).

If we are correct in our assumption that the electrophilic substitution of aromatic species involves such σ complexes as intermediates—and it has proved possible actually to isolate them in the course of some such substitutions (pp. 124, 128)—then what we commonly refer to as aromatic 'substitution' really involves initial *addition* followed by subsequent *elimination*. How this basic theory is borne out in the common electrophilic substitution reactions of benzene will now be considered.

NITRATION

The aromatic substitution reaction that has received by far the closest study is nitration and, as a result, it is the one that probably provides the most detailed mechanistic picture. Preparative nitration is most frequently carried out with a mixture of concentrated nitric and sulphuric acids, the so-called 'nitrating mixture'. The classical explanation for the presence of the sulphuric acid is that it absorbs the water formed in the nitration proper

$$C_6H_6 + HNO_3 \rightarrow C_6H_5NO_2 + H_2O$$

and so prevents the reverse reaction from proceeding. This explanation is unsatisfactory in a number of respects, not least that nitrobenzene, once formed, appears not to be attacked by water under the conditions of the reaction! What is certain is that nitration is slow in the absence of sulphuric acid, yet sulphuric acid by itself has virtually no effect on benzene under the conditions normally employed. It would thus appear that the sulphuric acid is acting on the nitric acid rather than the benzene in the system. This is borne out by the fact that solutions of nitric acid in concentrated sulphuric acid show a four-fold molecular freezing-point depression, which has been interpreted as being due to formation of the four ions:

$$HNO_3 + 2H_2SO_4 \rightleftharpoons {}^{\oplus}NO_2 + H_3O^{\oplus} + 2HSO_4^{\ominus}$$

i.e.

$$HO-NO_2 \xrightarrow{H_2SO_4} HSO_4^{\ominus} + HO\overset{\oplus}{\overset{|}{\underset{H}{-}}}NO_2 \xrightarrow{H_2SO_4} H_3O^{\oplus} + HSO_4^{\ominus} + {}^{\oplus}NO_2$$

The presence of ${}^{\oplus}NO_2$, the *nitronium ion*, in this solution and also in a number of salts has now been confirmed spectroscopically, and some of the salts, e.g. $^{\oplus}NO_2\ ClO_4{}^{\ominus}$, have actually been isolated. Nitric acid itself is converted in concentrated sulphuric acid virtually entirely into $^{\oplus}NO_2$, and there can be little doubt left that this is the effective electrophile in nitration under these conditions. If the purpose of the sulphuric acid is merely to function as a highly acid medium in which $^{\oplus}NO_2$ can be released from $HO{-}NO_2$, it would be expected that other strong acids, e.g. $HClO_4$, would also promote nitration. This is indeed found to be the case, HF and BF_3 also being effective. The poor performance of nitric acid by itself in the nitration of benzene is thus explained for it contains but little $^{\oplus}NO_2$; the small amount that is present is obtained by the two-stage process

$$HO{-}NO_2 + HNO_3 \overset{\text{Fast}}{\rightleftharpoons} NO_3{}^{\ominus} + \overset{\oplus}{\underset{H}{HO}}{-}NO_2$$

$$\overset{\oplus}{\underset{H}{HO}}{-}NO_2 + HNO_3 \overset{\text{Slow}}{\rightleftharpoons} H_3O^{\oplus} + NO_3{}^{\ominus} + {}^{\oplus}NO_2$$

in which nitric acid is first converted rapidly into its conjugate acid and that then more slowly into nitronium ion.

The kinetics of nitration are not easy to follow under normal preparative conditions for the solubility of, for example, benzene in nitrating mixture is sufficiently low for the rate of nitration to be governed by the rate at which the immiscible hydrocarbon dissolves in the mixture. This apart, however, many nitrations can be shown to follow an idealised kinetic law of the form

$$\text{Rate} \propto [Ar{-}H][{}^{\oplus}NO_2]$$

and in the light of what has already been said about energetic considerations it would seem that the rate-determining step is almost certainly the initial attack by $^{\oplus}NO_2$

(V)

rather than the subsequent removal of proton by $HSO_4{}^{\ominus}$ or other anion. That the latter step cannot be rate-determining has been confirmed by studying the nitration of nitrobenzene in which the hydrogen atoms have been replaced by deuterium. Nitrobenzene is chosen rather than benzene itself as the former is more soluble in nitrating mixture so that the overall rate of the reaction is no longer controlled by the rate at which it dissolves. Studies of the relative rates of fission of C—H and C—D in general would lead us to expect an approximately ten-fold drop in nitration rate on going $C_6H_5NO_2 \rightarrow C_6D_5NO_2$. In fact there is no detectable difference in rate, indicating that the fission of the C—H or C—D bond is not involved in the rate-determining stage of nitration.

That species such as (V) are directly involved as transient intermediates in the process of nitration, under some conditions at any rate, has been demonstrated in the nitration of trifluorotoluene with NO_2F/BF_3:

(Va)

By working at $-80°$ (Va) is actually precipitated and may be identified by spectroscopic and other means, raising the temperature to $-15°$ then effects its conversion in essentially quantitative yield to the nitration product that is normally obtained from the reaction; there is little doubt that the high stability of $BF_4^{\ominus}$ is greatly involved in the unusual stabilisation of (Va) that allows of its actual isolation.

It should be emphasised that in discussing rates of aromatic substitution reactions it is the formation of the transition state immediately preceding (Va) that is the controlling factor. We are, however, normally unable to obtain any salient data about such species, and given that the intermediate (Va) and the transition state that immediately precedes it do not differ too greatly from each other in energy level (*cf.* p. 42), they are likely to resemble each other in structure as well: it is thus usual, and reasonable, to take σ complexes such as (Va) as models for the transition states that immediately precede them.

An added significance of the existence of (V) is that the energy that becomes available from the formation of the C—NO_2 linkage may be used to assist in the subsequent fission of the strong C—H bond, which would otherwise be a relatively difficult undertaking.

A further point of preparative significance still requires explanation, however. Highly reactive aromatic compounds, such as phenol, are found to undergo ready nitration even in dilute nitric acid and at a far more rapid rate than can be explained on the basis of the concentration of $^{\oplus}NO_2$ that is present in the mixture. This has been shown to be due to the presence of nitrous acid in the system which nitrosates the reactive nucleus via the *nitrosonium ion*, $^{\oplus}NO$ (or other species capable of effecting nitrosation, *cf.* p. 91):

$$HNO_2 + 2HNO_3 \; \rightleftharpoons \; H_3O^{\oplus} + 2NO_3^{\ominus} + {}^{\oplus}NO$$

(VI)   (VII)   (VIII)

The nitroso-phenol (VII) so obtained is known to be oxidised very rapidly by nitric acid to yield the nitrophenol (VIII) and nitrous acid; thus more nitrous acid is produced and the process is progressively speeded up. No nitrous acid need be present initially in the nitric acid for a little of the latter attacks phenol oxidatively to yield HNO_2. The rate-determining step is again believed to be the formation of the intermediate (VI). Some direct nitration of such reactive aromatic compounds by $^{\oplus}NO_2$ also takes place simultaneously, the relative amount by the two routes depending on the conditions.

HALOGENATION

Halogenation, e.g. bromination, with the halogen itself only takes place in the presence of a catalyst such as $ZnCl_2$, $FeBr_3$, $AlBr_3$, etc. The nature of the catalyst is usually that of a Lewis acid and it acts by inducing some degree of polarisation in the halogen molecule, thereby increasing its electrophilic character, so that its now more positive end attacks the π electrons of the nucleus (see p. 126). After fission of the bromine–bromine bond in forming the σ complex with benzene, the anionic complex, $^{\ominus}Br \cdot FeBr_3$, so obtained then removes the proton to yield bromobenzene.

Halogenation may also be carried out with the aqueous hypohalous acid, $HOHal$, provided that a strong acid is also present. Here the evidence is very strong that in, for example, chlorination, the chlorinating agent is actually $Cl^{\oplus}$, produced as follows:

$$HO{-}Cl \xrightarrow[\text{fast}]{H^{\oplus}} \overset{\oplus}{\underset{H}{HO}}{-}Cl \xrightarrow{\text{Slow}} H_2O + Cl^{\oplus}$$

The further attack on benzene is then exactly analogous to nitration by $^{\oplus}NO_2$. A further similarity between the two is provided by the fact that HOCl alone has, like HNO_3, very little action on benzene; the presence of a further entity, i.e. strong acid, is necessary in either case to release the highly electrophilic species, $Cl^{\oplus}$ or $^{\oplus}NO_2$ by protonation of their 'carrier molecules':

$$\overset{\oplus}{\underset{H}{HO}}{\overset{\frown}{-}}Cl \quad \text{and} \quad \overset{\oplus}{\underset{H}{HO}}{\overset{\frown}{-}}NO_2$$

Further support for the idea that a halonium ion or a positively polarised halogen-containing complex is the effective substituting agent is provided by a study of the reactions of interhalogen compounds with aromatic substances. Thus BrCl leads only to bromination and ICl only to iodination, i.e. it is the *less* electronegative halogen that is introduced, due to:

$$\overset{\delta+}{Br}{\overset{\frown}{\rightarrow}}\overset{\delta-}{Cl}, \text{ etc.}$$

In the absence of a catalyst and in the presence of light, chlorine will add on to benzene; this proceeds by a radical mechanism and will be discussed subsequently (p. 270).

126

SULPHONATION

The intimate details of sulphonation are less well known than those of nitration and there has been a good deal of debate about whether the effective electrophilic agent is the bisulphonium ion, $^{\oplus}SO_3H$, or free SO_3. The weight of evidence for sulphonation under normal conditions, however, resides with the latter produced in the following way:

$$2H_2SO_4 \rightleftarrows SO_3 + H_3O^{\oplus} + HSO_4^{\ominus}$$

The sulphur atom of the trioxide is highly electron-deficient

and it is this atom, therefore, that becomes bonded to a ring carbon atom. Two features of sulphonation that distinguish it from nitration are that it is reversible and that it is slowed down when the hydrogen atoms of an aromatic nucleus are replaced by the heavier, radioactive isotope tritium, ^3H. The latter observation indicates, of course, that the removal of proton from the σ complex of benzene and sulphur trioxide (IX) must, by contrast to nitration, be involved to some extent at least in the rate-determining step of the overall reaction:

(IX)

Practical use is made of the reversibility of the reaction in the replacement of SO_3H by H on treating sulphonic acids with steam. Partly because of low miscibility, the reaction of hot concentrated sulphuric acid with benzene is slow and fuming sulphuric acid in the cold is generally used instead; the more rapid reaction is, of course, due to the concentration of free SO_3 that this acid contains.

FRIEDEL-CRAFTS REACTION

This can be conveniently divided into alkylation and acylation.

(i) Alkylation

The reaction of primary alkyl halides, e.g. **MeCl**, with aromatic compounds in the presence of Lewis acids—such as aluminium halides, BF_3, etc.—closely resembles the mechanism of catalysed halogenation that has already been discussed:

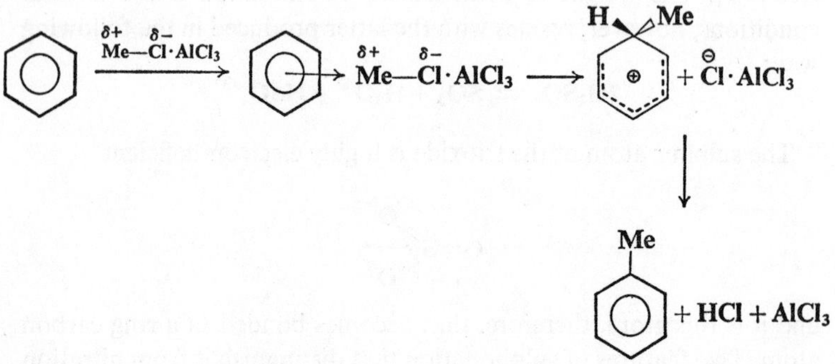

Here too it has proved possible actually to isolate a σ complex in the course of a reaction:

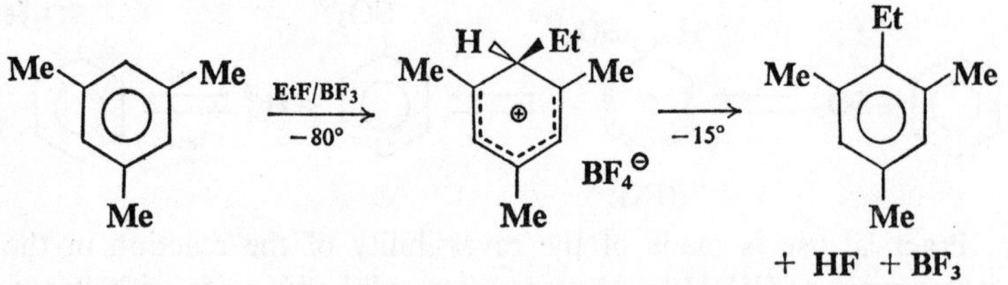

The intermediate is an orange crystalline solid melting with decomposition at $-15°$ to yield the normal alkylated product in essentially quantitative yield.

That polarised complexes between the halide and the aluminium halide are undoubtedly formed is shown by the fact that the halogen of isotopically-labelled aluminium halides is found to exchange with that of the alkyl halide. With secondary and tertiary halides, however, the carbon atom carrying the halogen is increasingly more able to accommodate positive charge (i.e. it will form a more stable carbonium ion, *cf.* p. 101), and there is thus an increasing tendency towards ionisation of the complex to yield $R^\oplus$, in an ion pair (*cf.* p. 96), as the effective electrophilic species. No clear distinction can be made between primary and other halides in the extent to which they form actual carbonium ions, however, as the nature of the catalyst used and of the halogen in the halide also play a part. Thus with the halide Me_3CCH_2Cl in the presence of $AlCl_3$, benzene yields almost wholly $PhCMe_2CH_2Me$, due to isomerisation of the first-formed primary carbonium ion, $Me_3CCH_2^\oplus$, to the tertiary carbonium ion, $Me_2\overset{\oplus}{C}CH_2Me$ (*cf.* p. 105) *before* it reacts with benzene. In the presence of $FeCl_3$ as catalyst, however, the major product is Me_3CCH_2Ph from the unrearranged ion, indicating that it never became fully-formed in the complex. Similarly 1-bromopropane in the presence of gallium bromide, $GaBr_3$, yields isopropylbenzene as the major product (p. 103), whereas 1-chloropropane with aluminium chloride yields very largely propylbenzene.

Alkenes can also be used in place of alkyl halides for alkylating benzene, the presence of an acid being required to generate a carbonium ion; BF_3 is then often used as the Lewis acid catalyst:

$$Me-CH\!=\!CH_2 \xrightarrow{H^\oplus} Me-\overset{\oplus}{C}H-Me \xrightarrow[BF_3]{C_6H_6} PhCHMe_2$$

Several of the usual catalysts, especially $AlCl_3$, also bring about ready dealkylation: i.e. the reaction is reversible. Thus heating of *p*-xylene (X) with hydrogen chloride and $AlCl_3$ results in the conversion of a major part of it to the thermodynamically more stable *m*-xylene (XI) (*cf.* p. 145). The presence of HCl is essential if isomerisation is to take place and it has been suggested that the Me group migrates directly by a Wagner-Meerwein type rearrangement (p. 105) in the initially protonated species:

(X)   (XI)

Alkylbenzenes may, however, also undergo disproportionation, i.e. involving *de*alkylation and *re*alkylation, as witnessed by the behaviour of ethylbenzene with **HF** and excess **BF$_3$**:

The main drawback in the preparative use of the Friedel-Crafts reaction is polyalkylation, however (p. 137).

(ii) Acylation

Acid chlorides or anhydrides in the presence of Lewis acids yield an *acylium ion*, $\overset{\oplus}{\text{RC}}{=}\text{O}$, or in some cases only a highly polarised complex, e.g. $\text{MeCO}\overset{\delta+}{\text{Cl}}\cdot\overset{\delta-}{\text{AlCl}_3}$, which acts as the effective electrophile to form a ketone (XII):

The ketone, once formed, complexes with aluminium chloride

removing it from the sphere of reaction. Thus rather more than one equivalent of the catalyst must be employed, unlike alkylation where only small amounts are necessary. There is however some evidence that such **AlCl$_3$** complexing of the ketone is an essential rather than merely a nuisance feature of the reaction as otherwise the ketone forms a complex with the acylium ion

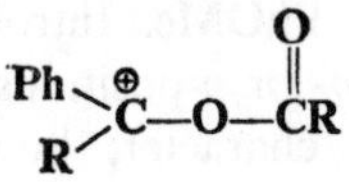

$$Ph\underset{R}{\overset{\oplus}{>}}C-O-\overset{\overset{O}{\|}}{C}R$$

and thus prevents the latter from attacking its proper substrate, in this case C_6H_6.

Rearrangement of **R** does not take place as with alkylation, but if it is highly branched loss of **CO** can occur leading ultimately to alkylation rather than the expected acylation:

$$Me_3C\overset{\curvearrowleft\oplus}{C}=O \longrightarrow CO + Me_3\overset{\oplus}{C} \overset{C_6H_6}{\longrightarrow} PhCMe_3$$

A useful synthetic application of Friedel-Crafts acylation is the use of cyclic anhydrides in a two-stage process to build a second ring on to an aromatic nucleus:

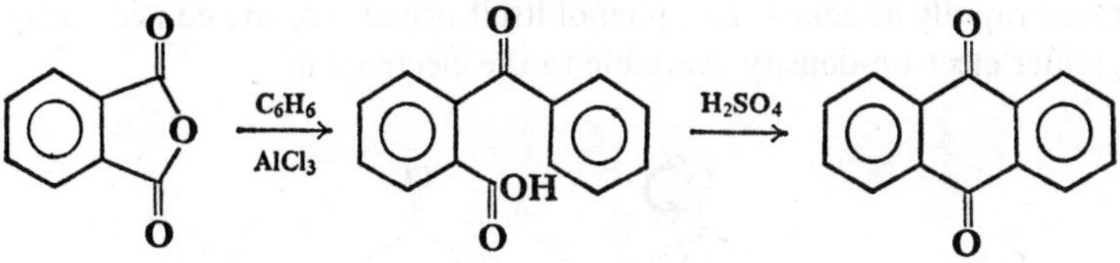

HCOCl is very unstable but formylation may be accomplished by protonating carbon monoxide to yield $H\overset{\oplus}{C}{=}O$, i.e. by use of **CO**, **HCl** and **AlCl₃** (the Gattermann-Koch reaction):

$$C_6H_6 + H\overset{\oplus}{C}{=}O \overset{AlCl_3}{\longrightarrow} PhCHO + H^\oplus$$

DIAZO COUPLING

Another classical electrophilic aromatic substitution is diazo coupling, in which the effective electrophile has been shown to be the diazonium cation:

$$PhN{\equiv}\overset{\oplus}{N} \quad \leftrightarrow \quad Ph\overset{\oplus}{N}{=}\overset{..}{N}$$

This is, however, a weak electrophile compared with species such as $^\oplus NO_2$ and will normally only attack highly reactive aromatic compounds such as phenols and amines; it is thus without effect on the otherwise highly reactive **PhOMe**. Introduction of electron-with-drawing groups into the *o*- or *p*-positions of the diazonium cation enhance its electrophilic character, however, by increasing the positive charge on the diazo group:

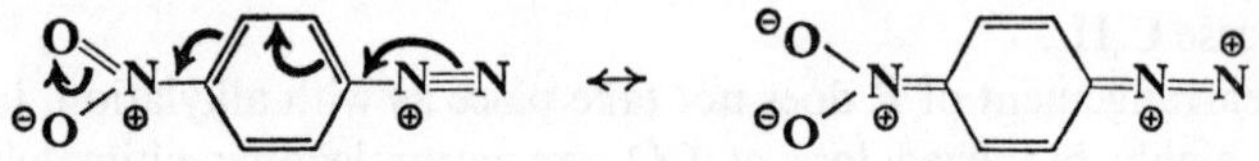

Thus the 2,4-dinitrophenyldiazonium cation will couple with **PhOMe** and the 2,4,6-compound with the hydrocarbon mesitylene. Diazonium cations exist in acid and slightly alkaline solution (in more strongly alkaline solution they are converted into diazohydroxides, **PhN=N—OH** and further into diazotate anions, **PhN=N—O$^\ominus$**) and coupling reactions are therefore carried out under these conditions, the optimum **pH** depending on the species being attacked. With phenols this is at a slightly alkaline **pH** as phenoxide ion is very much more rapidly attacked than phenol itself because of the considerably greater electron-density available to the electrophile:

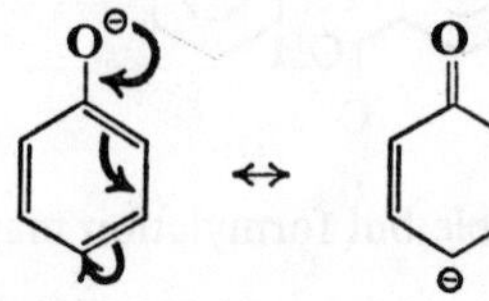

Coupling could take place on either oxygen or carbon and though relative electron-density might be expected to favour the former, the strength of the bond formed is also of significance and as with electro-philic attack on phenols in general it is a **C**-substituted product that normally results:

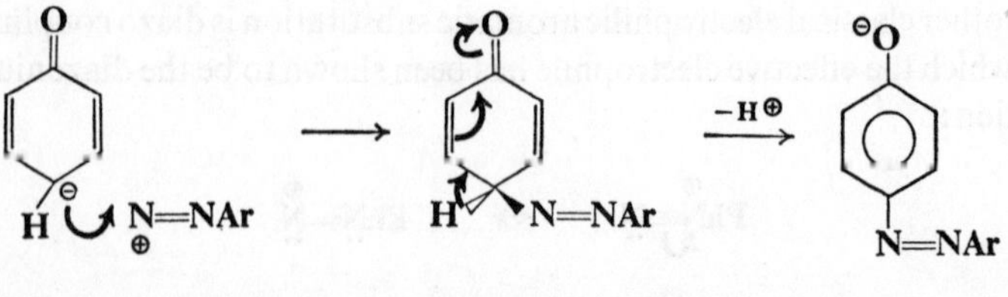

The proton is removed by one or other of the basic species present in solution.

Aromatic amines are in general somewhat less readily attacked than phenols and coupling is often carried out in slightly acid solution, thus ensuring a high [**PhN$_2$**$^\oplus$] without markedly converting the amine, **ArN̈H$_2$**, into the unreactive, protonated cation, **ArN̈H$_3$**—such aromatic amines are very weak bases (*cf.* p. 68). The initial diazotisation of aromatic primary amines is carried out in strongly acid media to ensure that as yet unreacted amine *is* converted to the cation and so prevented from coupling with the diazonium salt as it is formed.

With aromatic amines there is the possibility of attack on either nitrogen or carbon, and, by contrast with phenols, attack takes place largely on nitrogen in primary and secondary amines (i.e. N-alkylanilines) to yield *diazo-amino* compounds:

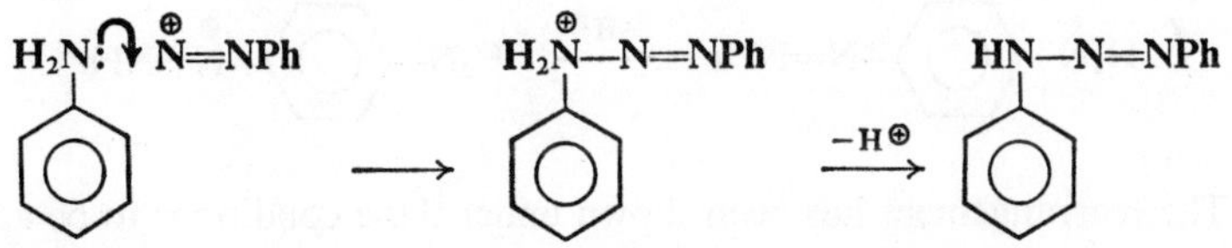

H$_2$N: $\overset{\oplus}{N}$=NPh $\longrightarrow$ H$_2\overset{\oplus}{N}$—N=NPh $\xrightarrow{-H^\oplus}$ HN—N=NPh

With most primary amines this is virtually the sole product, with N-alkylated anilines some coupling may also take place on the benzene nucleus while with tertiary amines (N-dialkaylanilines) only the product coupled on carbon is obtained:

NR$_2$

N=NPh

This difference in position of attack with primary and secondary aromatic amines, compared with phenols, probably reflects the relative electron-density of the various positions in the former compounds exerting the controlling influence for, in contrast to a number of other aromatic electrophilic substitution reactions, diazo coupling is sensitive to relatively small differences in electron density (reflecting the rather low ability as an electrophile of **PhN$_2$**$^\oplus$). Similar differences in electron-density do of course occur in phenols but here control over the position of attack is exerted more by the relative strengths of the bonds formed in the two products: in the two alternative coupled products derivable from amines, this latter difference is much less marked.

The formation of diazoamino compounds by coupling with primary amines does not constitute a preparative bar to obtaining the products coupled on the benzene nucleus for the diazoamino compound may be rearranged to the corresponding *amino-azo* compound by warming in acid:

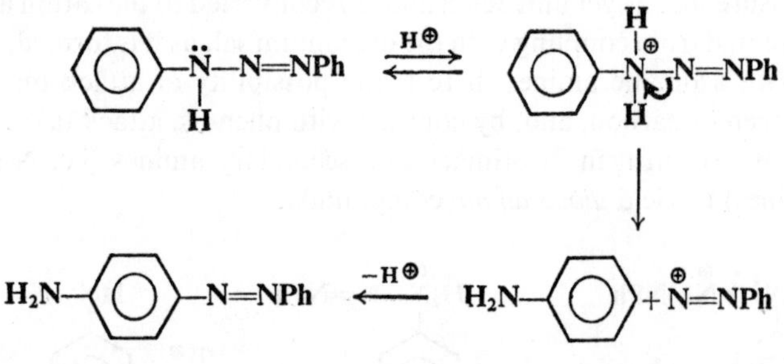

The rearrangement has been shown under these conditions to be an *inter*molecular process, i.e. that the diazonium cation becomes free, for the latter may be transferred to phenols, aromatic amines or other suitable species added to the solution. It is indeed found that the rearrangement proceeds most readily with an acid catalyst plus an excess of the amine that initially underwent coupling to yield the diazoamino compound, it may then be that this amine attacks the protonated diazoamino compound directly with expulsion of PhNH$_2$ and loss of a proton:

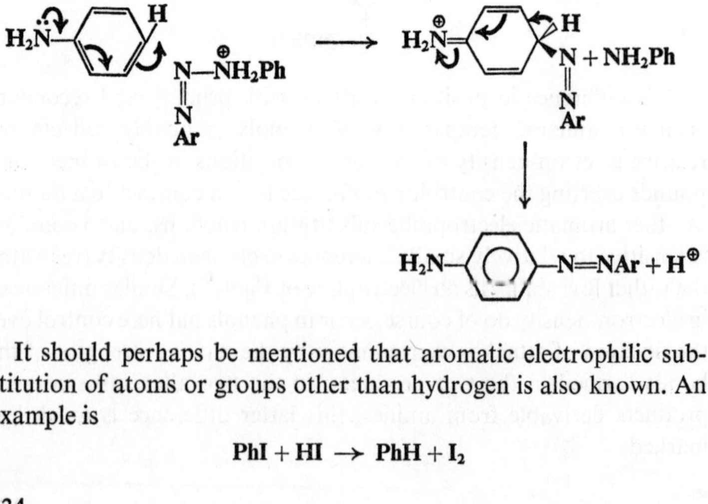

It should perhaps be mentioned that aromatic electrophilic substitution of atoms or groups other than hydrogen is also known. An example is

$$PhI + HI \rightarrow PhH + I_2$$

which shows all the characteristics (in the way of effect of substituents, etc.) of a typical electrophilic substitution reaction, but such displacements are not common and are usually of little preparative importance.

In the face of the wholly polar viewpoint of aromatic substitution that has so far been adopted, it should be emphasised that examples of homolytic aromatic substitution by free radicals are also known (p. 283).

THE EFFECT OF A SUBSTITUENT ALREADY PRESENT

The effect of a substituent already present in a benzene nucleus in governing not only the reactivity of the nucleus towards further electrophilic attack, but also in determining what position the incoming substituent shall enter, is well known. A number of empirical rules have been devised to account for these effects but they can be better explained on the basis of the electron-donating or -attracting powers of the initial substituent.

(i) Inductive effect of substituents

Alkyl groups are electron-donating and so will increase electron-availability over the nucleus. The effect in toluene

(XIII)

probably arises in part from a contribution to the hybrid by forms such as (XIII), i.e. by hyperconjugation (p. 24). The inductive effect of most other substituents, e.g. halogens, OH, OMe, NH_2, SO_3H, NO_2 etc. will be in the opposite direction as the atom next to the nucleus is more electronegative than the carbon to which it is attached, e.g.:

But this is not the only way in which a substituent can affect electron-availability in the nucleus.

(ii) Mesomeric effect of substituents

A number of common substituents have unshared electron pairs on the atom attached to the nucleus and these can interact with its delocalised π orbitals

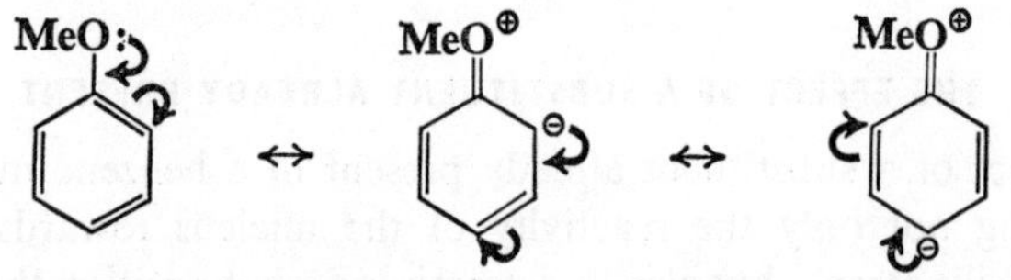

and the same consideration clearly applies to **OH, SH, NH$_2$**, etc.

It will be noticed that electron-availability over the nucleus is thereby increased. An effect in the opposite direction can take place if the substituent atom attached to the nucleus itself carries a more electronegative atom to which it is multiply bonded, i.e. this atom is then conjugated with the nucleus and can interact with its delocalised π orbitals:

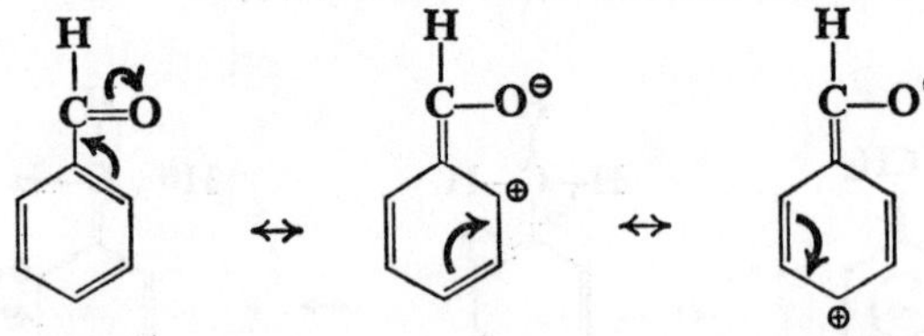

The same consideration clearly applies to **COR, CO$_2$H, SO$_3$H, NO$_2$, CN**, etc. Here it will be seen that electron-availability over the nucleus is thereby decreased.

(iii) The overall effect

A group that, overall, is electron-donating might be expected to lead to more rapid substitution by an electrophilic reagent than in benzene itself, for the electron-density on the ring carbon atoms is now higher; correspondingly, a group that is, overall, electron-withdrawing might be expected to lead to less rapid substitution. This is reflected in the relative ease of attack of oxidising agents, which are, of course, electrophilic reagents (e.g. **KMnO$_4$**), on phenol, benzene and nitrobenzene; phenol is extremely readily attacked with destruction of the aromatic nucleus, while benzene is resistant to attack and nitrobenzene even more so.

It is also reflected in the Friedel-Crafts reaction. Alkylation of benzene leads to an initial product, **PhR**, which is more readily attacked than benzene itself due to the electron-donating substituent **R**. It is thus extremely difficult to stop the reaction at the mono-alkylated stage and polyalkylation is the rule (p. 130). In acylation, however, the initial product, **PhCOR**, is less readily attacked than is benzene itself and the reaction can be stopped at this stage without difficulty. It is indeed often preferable to synthesise a mono-alkyl-benzene by acylation followed by Clemmensen or other reduction, rather than by direct alkylation, because of difficulties introduced during the latter by polyalkylation and possible rearrangement of **R**. The presence of an electron-withdrawing substituent is generally sufficient to inhibit the Friedel-Crafts reaction and, for example, nitrobenzene is often used as a solvent as it readily dissolves $AlCl_3$.

The overall electron-withdrawing effect is clear-cut with, for example, NO_2, for here inductive and mesomeric effects reinforce each other, but with, e.g. NH_2, these effects are in opposite directions

and it is not possible to say, *a priori*, whether the overall effect on the nucleus will be activation or de-activation. Here the direction and magnitude of the dipole moment of **PhY** can be some guide (see p. 138).

The overall electron-donating effect of **OH** and NH_2, as compared with the overall electron-withdrawing effect of **Cl**, reflects the considerably greater ease with which oxygen and nitrogen will release their electron pairs as compared with chlorine; this is more than sufficient to outweigh the inductive effect in the two former cases but not in the latter. It should, however, be remembered that the moments of a number of the composite groups, e.g. **OH**, are not collinear with the axis of the benzene ring and hence the component of the moment

137

Y	μ	Direction in Ph—Y
OH	1·6	
NH$_2$	1·5	
OMe	1·2	←
Me	0·3	
H	0·0	
Cl	1·6	
CHO	2·8	
SO$_3$H	3·8	→
NO$_2$	3·9	

actually affecting the bond to the ring may thus be different from the observed moment of the molecule as a whole:

The relation between electron-availability and ease of electrophilic substitution may, however, be seen by comparing the direction and magnitude of dipole moments (above) with the following relative rates of attack by $^{\oplus}NO_2$:

PhOMe	PhMe	PhH	PhCl	PhNO$_2$
10^3	10	1	3×10^{-2}	$<10^{-4}$

(iv) The position of substitution

In so far as the relative proportion of *o*-, *m*- and *p*-isomers obtained on further electrophilic attack is controlled by the nature of the substituent already present, such control is exerted through the effect of this substituent on the relative energy levels of, and hence relative rates of formation of, the transition states leading to *o*-, *m*- and *p*-attack, respectively: the control is thus kinetic (p. 44). As the transition state usually resembles the related metastable intermediate or σ complex reasonably closely energetically (p. 42), it may be assumed that it also resembles it in structure. Thus structural features that stabilise a particular σ complex might be expected to stabilise the related transition state in a similar way.

Considering the three possibilities for nitration when the initial substituent is a powerfully electron-withdrawing group such as NO_2,

the three alternative σ complexes can each stabilise themselves by delocalisation of their positive charge over the nucleus (*cf.* p. 121). This will, however, clearly be less effective if substitution takes place *o*- or *p*- to the NO_2 already present, for in each case one of the contributing structures (XIV*c* and XVI*b*, respectively) would have to carry a positive charge on a carbon atom which is already bonded to a nitrogen atom carrying a formal positive charge: a far from stable juxtaposition. With the σ complex formed in the course of *m*-substitution (XV) there is no such limitation; this intermediate is thus more stable (lower energy level) than those involved in *o*- or *p*-attack, its rate of formation is thus the greatest of the three and the *m*-product is found to predominate in the final reaction mixture. It should perhaps be mentioned that all three σ complexes are destabilised with respect to the one formed in the nitration of benzene itself and attack

139

on even the *m*-position is thus considerably slower than attack on any position in benzene under parallel conditions.

When the substituent already present is a powerfully electron-donating group such as **OMe** (i.e. in anisole), the three σ complexes

(XVII*a*) (XVII*b*) (XVII*c*) (XVII*d*)

(XVIII*a*) (XVIII*b*) (XVIII*c*)

(XIX*a*) (XIX*b*) (XIX*c*) (XIX*d*)

can each stabilise themselves by delocalisation of their positive charge over the nucleus as before. For attack *o*- or *p*- to the **OMe** already present, however, there is the possibility of further stabilisation by delocalising the charge on to the **OMe** group through the agency of the unshared electron pairs on oxygen (XVII*d* and XIX*c*, respectively): a facility not open to the σ complex resulting from *m*-attack. The *o*- and *p*-isomers are thus found to predominate in the final reaction mixture, the rate of attack being considerably faster than on any position in benzene under parallel conditions.

It should be remembered, however, that whatever the nature of the substituent already present what we are actually considering are the *relative* rates of attack on *o*-, *m*- and *p*-positions and though either *o*/*p*- or *m*-substitution is usually preponderant, neither alternative is of necessity exclusive. Thus nitration of toluene has been found to lead to ≈3% of *m*-nitrotoluene and of *t*-butylbenzene to ≈9% of the *m*-nitro derivative.

The behaviour of chlorobenzene is interesting for although **Cl** is, overall, electron-withdrawing and the nucleus is therefore more difficult to nitrate than is benzene itself (a circumstance normally associated with *m*-directive groups) it does nevertheless substitute *o/p*-. This is due to the fact that though the electron pairs on chlorine in the neutral chlorobenzene molecule itself are considerably more loth than those on oxygen or nitrogen (i.e. in phenol or aniline, respectively) to interact with the π orbital system of the nucleus (p. 138) such interaction is, hardly surprisingly, very markedly enhanced in the positively charged σ complexes leading to *o*- or *p*-, but not in that leading to *m*-, substitution:

o-	*m*-	*p*-

Such an effect may originate as a temporary polarisation, an *electromeric effect* (XX) superimposed on the permanent polarisation of the molecule, prompted by the close approach of the positively charged electrophile, $^{\oplus}NO_2$:

(XX)

The rate of reaction will remain slower than in benzene itself, however, due to the overall deactivation of the nucleus by chlorine's inductive effect in the opposite direction. A very similar situation is encountered in the addition of unsymmetrical adducts to vinyl halides, e.g. $CH_2{=}CHBr$, where the inductive effect controls the *rate*, but mesomeric stabilisation of the carbonium ion intermediate governs the *orientation*, of addition (p. 161).

(v) Conditions of reaction

The conditions under which an electrophilic substitution reaction is carried out can modify or even alter completely the directing effect of a group. Thus phenol is even more powerfully *o/p*-directing in alkaline than in neutral or acid solution, for the species undergoing substitution is then the phenoxide ion (XXI), in which the inductive effect is now reversed compared with phenol itself and, more important, a full blown negative charge is available for interaction with the π orbital system of the nucleus; the electron density over the nucleus is thus notably increased:

(XXI)

Conversely aniline, normally *o/p*-directing, becomes in part at least *m*-directing in strongly acid solution, due to protonation to form the anilinium cation (XXII):

(XXII)

This is due to the fact that there can no longer be any interaction of the unshared electron pair on nitrogen with the delocalised π orbitals of the nucleus, for the former are now involved in bond formation with the proton that has been taken up and the inductive effect, drawing electrons away from the nucleus, is now enormously enhanced by the positive charge on nitrogen.

As soon as more than one saturated atom is interposed between a positive charge and the nucleus, however, its inductive effect falls off very sharply (*cf.* strengths of acids, p. 59) and so does the percentage of the *m*-isomer produced, as seen in the nitration of:

Compound	% *m*-
$\overset{\oplus}{PhNMe_3}$	100
$Ph\overset{\oplus}{CH_2NMe_3}$	88
$PhCH_2\overset{\oplus}{CH_2NMe_3}$	19
$PhCH_2CH_2\overset{\oplus}{CH_2NMe_3}$	5

(vi) *o/p*-ratios

It might, at first sight, be expected that the relative proportions of *o*- and *p*-isomers obtained during substitution of a nucleus containing an *o/p*-directive substituent would be 67 % *o*- and 33 % *p*-, as there are two *o*-positions to be substituted for every one *p*-. Apart from the fact that a little *m*-product is often obtained (the extent to which a position is substituted is merely a matter of *relative* rates of attack, after all), the above ratio is virtually never realised and more often than not more *p*- than *o*-product is obtained. This may be due to the substituent already present hindering attack at the *o*-positions adjacent to it by its very bulk, an interference to which the more distant *p*-position is not susceptible. In support of this it is found that as the initial substituent increases in size from $CH_3 \rightarrow Me_3C$, the proportion of the *o*-isomer obtained on nitration drops markedly (57 % $\rightarrow$ 12 %) while that of the *p*- increases (40 % $\rightarrow$ 80 %). Increase in size of the attacking agent has the same effect; thus substitution of chlorobenzene leads to:

	Group introduced			
	Cl	NO$_2$	Br	SO$_3$H
% *o*-	39	30	11	0
% *p*-	55	70	87	100

That a steric factor is not the only one at work, however, is seen in the nitration of fluoro-, chloro-, bromo- and iodo-benzenes where the percentage of *o*-isomer obtained *increases* as we go along the series, despite the *increase* in size of the substituent. This is due to the fact that the electron-withdrawing inductive effect influences the adjacent *o*-positions much more powerfully than the more distant *p*-position. The inductive effect decreases considerably on going from fluoro- to iodo-benzene (the biggest change being seen in going from fluoro- to chloro-benzene) resulting in easier attack at the *o*-positions *despite* the increasing size of the group already present.

In some cases there is good evidence that the formation of an unexpectedly high proportion of the *o*-isomer results from interaction of the substituent already present with the attacking electrophile, so that the latter is steered into the adjacent *o*-position. Thus in the reaction of phenoxide ion with formaldehyde, initial interaction of CH_2O with the metal ion of the phenoxide/metal ion pair (XXIII) results in the formation of a complex (*cf.* p. 249) within which electron redistribution can take place to yield, specifically, the *o*-isomer:

143

(XXIII)

The o/p-ratio is also influenced considerably by the actual conditions under which the reaction is carried out. This is particularly so when the solvent is changed, either through modifying the nature of the attacking electrophile or as a result of altering the differential stabilisation, by solvent molecules, of the transition states leading to o- and p-products. Changing the temperature may also have an effect, and there are a number of further anomalies that have as yet received no adequate explanation.

KINETIC *v*. THERMODYNAMIC CONTROL

In all that has gone before a tacit assumption has been made: that the proportions of alternative products formed in a reaction, e.g. o-, m- and p-isomers, are determined by their relative rates of formation, i.e. that the control is kinetic (p. 44). This is not, however, always what is observed in practice; thus in the Friedel-Crafts alkylation of toluene with benzyl bromide and $GaBr_3$ (as Lewis acid catalyst) at 25°, the isomer distribution is found to be:

Time (sec)	%o-	%m-	%p-
0·01	40	21	39
10	23	46	31

Even after a very short reaction time (0·01 sec) it is doubtful whether the isomer distribution (in the small amount of product that has as yet been formed) is *purely* kinetically controlled—the proportion of m-isomer is already relatively large—and after 10 sec it clearly is not: m-benzyltoluene, the thermodynamically most stable isomer, predominating and the control now clearly being equilibrium or thermodynamic (p. 44).

This is a situation that must arise where the alternative products are mutually interconvertible under the conditions of the reaction, either by direct isomerisation or by reversal of the reaction to form the starting material which then undergoes new attack to yield a more thermodynamically stable isomer. It is important to emphasise that the relative proportions of alternative products formed will be defined by their relative thermodynamic stabilities *under the conditions of the reaction*, which may possibly differ from those of the isolated molecules. Thus if *m*-xylene is heated at 82° with **HF** and a catalytic amount of **BF₃** the proportions of the three isomeric xylenes in the product resemble very closely those calculated thermodynamically:

	Experiment	Calculated
%*o*-	19	18
%*m*-	60	58
%*p*-	21	24

If, however, an excess of **BF₃** is used the reaction product is found to contain >97% of *m*-xylene; this is because the xylenes can now be converted to the corresponding salts, e.g.

and the equilibrium will therefore be shifted towards the most basic (*m*-) isomer. Cases are also known in which change of temperature can affect the type of control that is operative (p. 146).

ELECTROPHILIC SUBSTITUTION OF OTHER AROMATIC SPECIES

With naphthalene, electrophilic substitution, e.g. nitration, takes place preferentially at the α- rather than the β-position. This can be accounted for by the fact that more effective stabilisation by delocalisation can take place in the metastable intermediate or transition state from α-substitution than that from β-attack (*cf.* benzene with an *o/p*-directive substituent):

More forms can also be written in each case in which the positive charge is now delocalised over the ring, leading to a total of seven forms for the α-intermediate as against six for the β-, but the above, in which the second ring retains intact, fully delocalised π orbitals, are probably the most important and the contrast, between two contributing forms in the one case and one in the other, correspondingly more marked.

The possibility of the charge becoming more widely delocalised in the naphthalene intermediate, as compared with benzene, would lead us to expect more ready electrophilic attack on naphthalene which is indeed observed.

The sulphonation of naphthalene with concentrated H_2SO_4 at 80° is found to lead to almost complete α-substitution, the rate of formation of the alternative β-sulphonic acid being very slow at this temperature, i.e. kinetic control. Sulphonation at 160°, however, leads to the formation of no less than 80% of the β-sulphonic acid, the remainder being the α-isomer. That we are now seeing thermodynamic control is confirmed by the observation that heating pure naphthalene α- *or* β-sulphonic acid in concentrated H_2SO_4 at 160° results in the formation of exactly the same equilibrium mixture as above, containing 80% β-acid and 20% α-acid. The greater thermodynamic stability of the β-acid is due largely to non-bonded interaction in the α-acid between the very bulky SO_3H group and the H atom on the adjacent vertex (8-position) lowering its stability.

The interconversion of α- and β-acids in H_2SO_4 at 160° could result either from a direct intramolecular isomerisation, or by reversal of sulphonation to yield naphthalene which undergoes new attack at the other position. It should be possible to distinguish between these alternatives by carrying out the reaction in $H_2{}^{35}SO_4$, for the former should lead to no incorporation of ^{35}S in the product sulphonic acids, whereas the latter should lead to such incorporation. Experimentally it is found that incorporation of ^{35}S does take place but at a rate slower than that at which the conversion occurs. This could imply either that both routes are operative simultaneously, or that, after reversal of sulphonation, new attack takes place on the resultant naphthalene by the departing H_2SO_4 molecule faster than by surrounding $H_2{}^{35}SO_4$ molecules—the question is still open.

Pyridine (XXIV), like benzene, has six π electrons (one being supplied by nitrogen) in delocalised π orbitals but, unlike benzene, the orbitals will be deformed by being attracted towards the nitrogen atom because of the latter's being more electronegative than carbon. This is reflected in the observed dipole moment of pyridine, and the

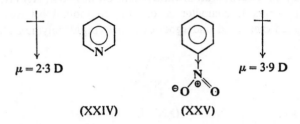

$\mu = 2.3$ D $\qquad\qquad\qquad\qquad\qquad\qquad \mu = 3.9$ D

(XXIV) $\qquad\qquad\qquad$ (XXV)

compound would therefore be expected to have a deactivated nucleus towards electrophilic substitution (*cf.* nitrobenzene (XXV)). The deactivation of the nucleus is considerably increased on electrophilic attack, for the positive charge introduced on nitrogen by protonation or by direct attack on it of the substituting electrophile, withdraws electrons much more strongly:

In fact electrophilic substitution is extremely difficult, sulphonation, for example, requiring twenty-four hours heating with oleum at 230°. Substitution takes place at the β-position (*m*- to the electron-withdrawing centre), the explanation being similar to that already discussed for nitrobenzene (p. 139).

Pyrrole (XXVI) also has delocalised π orbitals but nitrogen has here had to contribute *two* electrons so becoming virtually non-basic (p. 71) and the dipole moment is found to be in the opposite direction to that of pyridine:

$\mu = 1.8$ D

(XXVI)

It is thus referred to as a π *excessive* heterocycle as compared with pyridine which is a π *deficient* one. It behaves like a reactive benzene derivative, e.g. aniline, and electrophilic substitution is very easy.

147

Substitution is complicated, however, by the fact that if protonation is forced on pyrrole in strongly acid solution (this probably takes place on an α-carbon atom rather than on nitrogen, XXVII, *cf.* p. 72), the aromatic character is lost, the compound behaves like a conjugated diene and undergoes extremely rapid polymerisation:

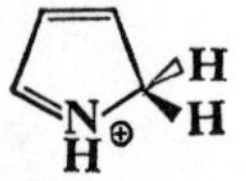

(XXVII)

Electrophilic substitution can, however, be carried out under highly specialised conditions leading to preferential attack at the α-position, reflecting the greater delocalisation, and hence stabilisation, possible in the metastable intermediate leading to α-, as compared with β-, substitution:

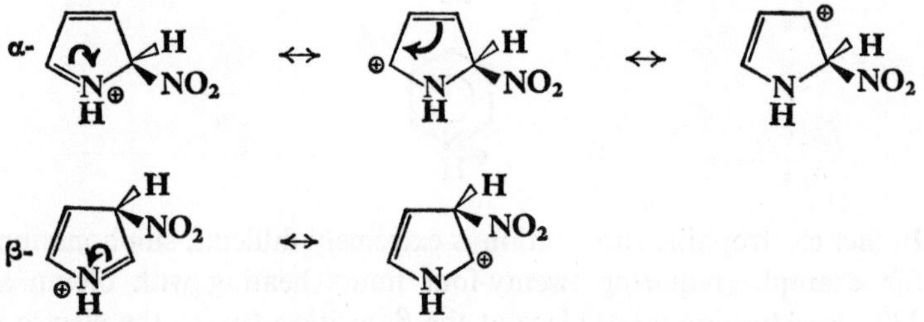

The difference in stability between the two is not very strongly marked, however, reflecting the highly activated state of the nucleus, and ready attack will take place at the β-position if the α- is already substituted. The effect of the N atom in promoting either α- or β-attack by stabilising either intermediate through accommodating its $\oplus$ charge is reminiscent of the similar role of the N atom of the NH_2 group in promoting similar attack on the nucleus in aniline (p. 137).

NUCLEOPHILIC ATTACK ON AROMATIC SPECIES

(i) Substitution of hydrogen

As it is the π electrons that are initially responsible for the normal substitution of benzene being an electrophilic process, the presence of a strongly electron-withdrawing substituent might be expected to render attack by a nucleophile possible provided electron-withdrawal from the nucleus was sufficiently great (*cf.* the addition of nucleophiles to alkenes carrying electron-withdrawing substituents, p. 172). In fact, nitrobenzene can be fused with potash, in the presence of air, to yield *o*-nitrophenol (XXVIII):

(XXIX)

(XXVIII)

The nitro-group is able to stabilise the anionic intermediate (XXIX) (and more importantly the transition state preceding it that it resembles and for which it can therefore serve as a model) by delocalising its charge if $^{\ominus}OH$ enters the *o*- or *p*-positions but not if it goes into the *m*-position. The *o*-attack is likely to be preferred, despite the size of the adjacent NO_2, as the inductive effect of the nitro-group, acting over a shorter distance, will make the *o*-position more electron-deficient than the *p*-. The overall reaction is exactly what we should expect, namely that a substituent promoting attack on the *m*-position by an electrophile would promote *o/p*-attack by a nucleophile.

Once (XXIX) has been formed, it can eliminate $^{\ominus}OH$ (i) and so be reconverted to nitrobenzene as well as being able to eliminate $H^{\ominus}$ (ii) to yield the product (XXVIII). To drive the reaction over to the right an oxidising agent must be present to encourage the elimination of hydride ion and to destroy it as formed. Thus the fusion is either carried out in the air, or an oxidising agent such as potassium nitrate or ferricyanide is added. In the absence of any oxidising agent nitrobenzene itself will act in that capacity; it is thereby converted to azoxybenzene and the yield of nitrophenol correspondingly reduced.

Pyridine behaves in an exactly analogous manner undergoing attack by sodamide (i.e. $^{\ominus}NH_2$, the Tschitschibabin reaction), to yield α-aminopyridine (XXX), a compound of value in the synthesis of sulphapyridine:

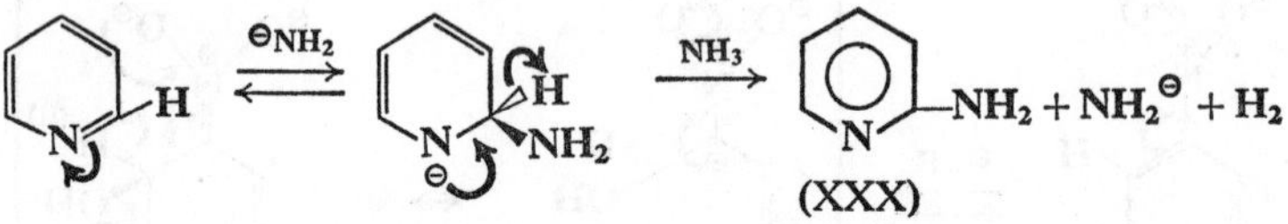

These are analogous to S_N2 reactions but with attack taking place from the side rather than from the back of the carbon atom undergoing nucleophilic attack; they differ also in that this atom never becomes bonded to more than four other atoms at once (*cf.* p. 73). This mechanism is probably sufficiently different from the normal S_N2 for it to be designated specifically as S_N2 (*aromatic*).

(ii) Substitution of atoms other than hydrogen

Aromatic nucleophilic substitution more commonly refers to the replacement of atoms other than hydrogen, i.e. where the atom or atoms displaced constitute a better leaving group than $H^{\ominus}$, e.g. $Cl^{\ominus}$ and N_2, and both S_N1 and S_N2 (*aromatic*) mechanisms are encountered. The only important examples proceeding via the S_N1 mechanism are the replacement reactions of diazonium salts

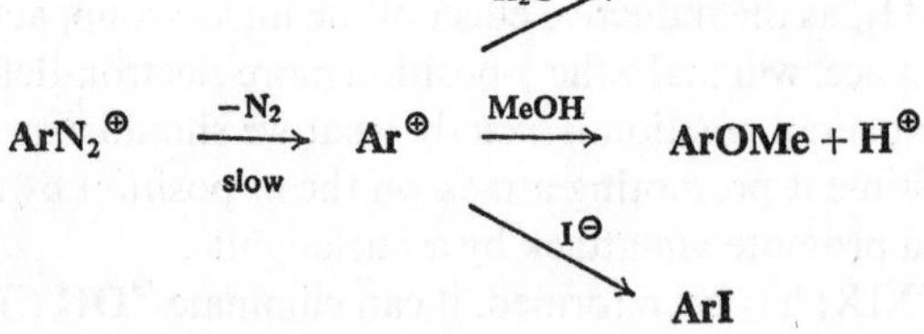

in which the rate-determining step is the elimination of nitrogen from the diazonium cation followed by rapid reaction of the aryl cation with a nucleophile, the rates being the first order in $ArN_2^{\oplus}$ and independent of the concentration of the nucleophile. A number of the reactions of diazonium salts, particularly in less polar solvents, proceed by a radical mechanism, however (p. 283).

The most common example of an S_N2 (*aromatic*) reaction is the

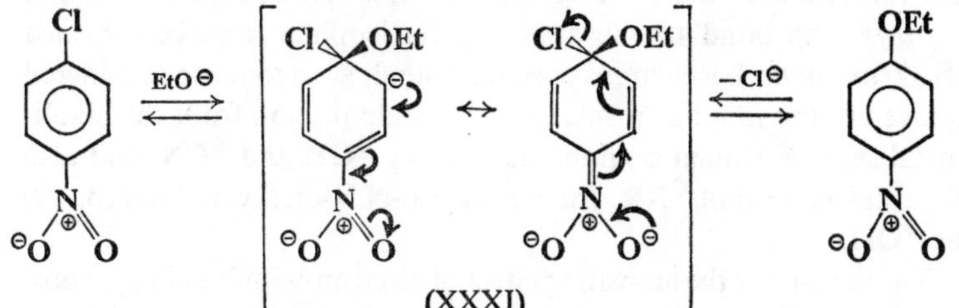

(XXXI)

replacement of an activated halogen atom, kinetic studies in a number of examples supporting the bimolecularity of the reaction. That an actual intermediate such as (XXXI) is formed, unlike aliphatic bimolecular nucleophilic substitution where the bond to the leaving group is being broken as that to the entering group is being formed, is shown by the fact that chlorides and bromides react in a number of cases at essentially the same rate. The breakage of the carbon–halogen bond can thus not be involved in the rate-determining step for a C—Cl bond is more difficult to break than an analogous C—Br one and the chloride would, *a priori*, be expected to react more slowly than the bromide.

Confirmation of the formation of such an intermediate is provided by the actual isolation of the same species (XXXII) from the action of $^{\ominus}$OEt on 2,4,6-trinitroanisole (XXXIII) and $^{\ominus}$OMe on 2,4,6-trinitrophenetole (XXXIV):

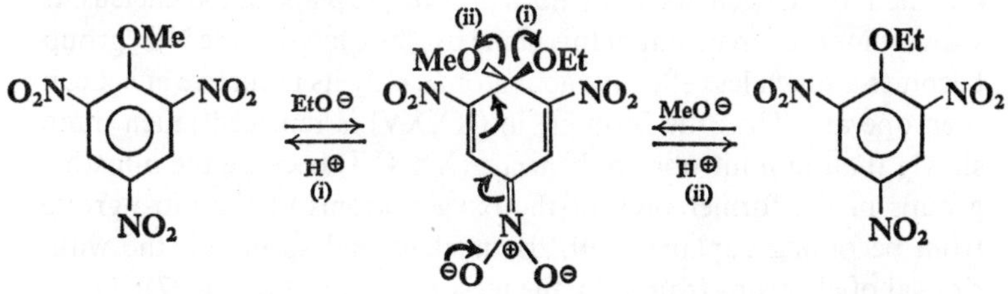

(XXXIII)   (XXXII)   (XXXIV)

It is also found that acidification of the reaction mixture obtained from *either* substrate yields exactly the same proportion of (XXXIII) and (XXXIV).

We have thus now encountered nucleophilic displacement reactions in which the bond to the leaving group is broken (*a*) *before* that to the attacking nucleophile has been formed (S_N1), (*b*) *simultaneously* with

151

the formation of the bond to the attacking nucleophile (S_N2), and (c) *after* the bond to the attacking nucleophile has been formed (S_N2 *(aromatic)*). Reactions essentially analogous to those considered above are the familiar displacement of sulphite ion from the alkali-metal salts of aromatic sulphonic acids by $^\ominus$OH and $^\ominus$CN, and also the displacement of $^\ominus$NR$_2$ from *p*-nitroso-N,N-dialkylanilines (p. 92) by $^\ominus$OH.

The reason for the activating effect of electron-withdrawing groups, especially NO$_2$, on nuclear halogen atoms is their ability to stabilise intermediates such as (XXXI) by delocalisation; it would therefore be expected that nitro-groups would be most effective when *o*- and *p*-to the substituent to be replaced, for in the *m*-position they can only assist in spreading the charge via their inductive effects. The presence of nitro-groups in the 2-, 4- and 6-positions in picryl chloride, $(O_2N)_3C_6H_2Cl$, thus confers almost acid chloride reactivity on the halogen, their effect is so pronounced. 2- and 4-, but not 3-, halogeno-pyridines also undergo ready replacement reactions for exactly the same reasons (the electron-withdrawing group here being the hetero-cyclic nitrogen atom); they do, indeed, resemble the corresponding *o*- and *p*-nitrohalogenobenzenes though the activation of the halogen is slightly less than in the latter.

If nitro-groups are to stabilise, and so assist in the formation of, intermediates such as (XXXI) the *p* orbitals on the nitrogen atom of the NO$_2$ group must be able to become parallel to those on the adjacent nuclear carbon atom. For this to happen the oxygen atoms attached to nitrogen must also lie in or near the plane of the nucleus. If such atoms are forced out of this plane by steric factors, the NO$_2$ group becomes a much less effective activator as only its inductive effect can then operate. Thus the bromine in (XXXV) is replaced much more slowly than in *p*-nitrobromobenzene (XXXVI) because the *o*-methyl groups in the former prevent the oxygen atoms of the nitro-group from becoming coplanar with the nucleus and so inhibit the withdrawal of electrons from it by the mesomeric effect (*cf.* p. 27):

(XXXV)   (XXXVI)

(iii) Replacement of halogen in an unactivated nucleus

The chlorine atom in chlorobenzene only undergoes replacement by nucleophiles such as $^{\ominus}$**OH** under extreme conditions; this is probably due to the fact that the expected S_N2 (*aromatic*) intermediate (XXXVII), not being stabilised like the examples already considered, is correspondingly reluctant to form:

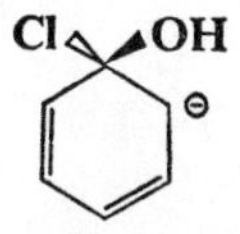

(XXXVII)

It is, however, interesting to find that the reaction of chlorobenzene with $^{\ominus}$**OR** is very much faster in dimethyl sulphoxide than in hydroxylic solvents; this may be due to the fact that alkoxide ions are highly stabilised by hydrogen bonding in hydroxylic solvents and thereby made less effective as attacking nucleophiles.

In the attack of $^{\ominus}$**OH** at 340° on *p*-chlorotoluene (XXXVIII) the expected *p*-cresol anion (XXXIX) is obtained but also the unexpected *m*-compound (XL):

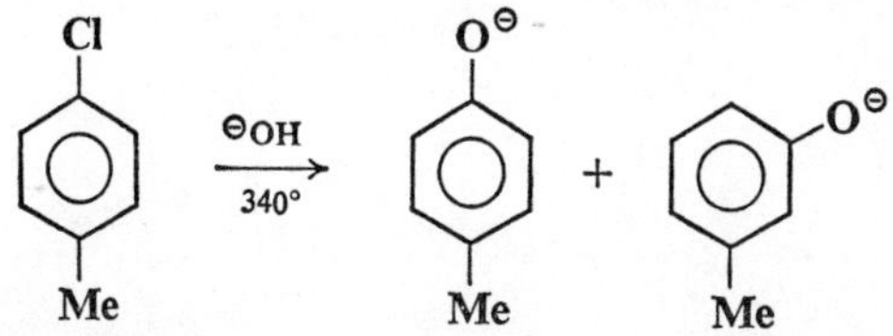

(XXXVIII) (XXXIX) (XL)

It is difficult to see how the latter could arise by a direct displacement, and it can be shown that it does not arise by the subsequent isomerisation of first formed (XXXIX), for the latter is stable under the conditions of the reaction. Exactly the same thing happens in the attack of $^{\ominus}NH_2$ on (XXXVIII) (*cf.* p. 50), but here the unexpected

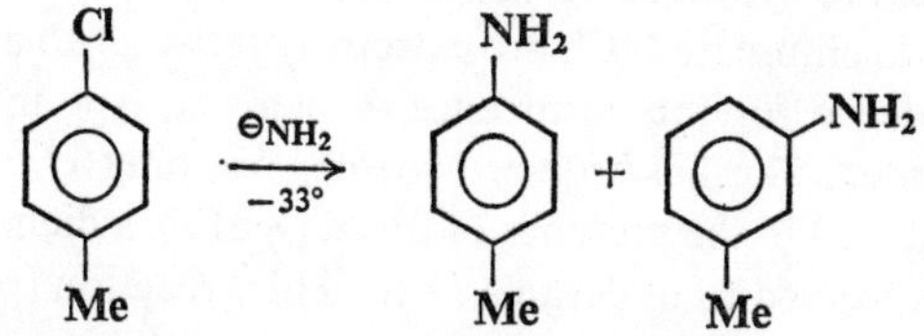

(XXXVIII) (XLI) (XLII)

compound (XLII) is now the major product. Some clue to what may be taking place is provided by the fact that no *o*-product is ever obtained and also that the reaction proceeds very much more readily with $^{\ominus}NH_2$ than with $^{\ominus}OH$, the former being the stronger *base* under the conditions employed. If $^{\ominus}NH_2$ does act as a base, the following sequence could be envisaged

loss of **HCl** leading to the *benzyne* intermediate (XLIII), which then undergoes addition of $^{\ominus}NH_2$ and proton (ex. NH_3), one way round or the other way round, to yield (XLI) and (XLII), respectively. Thus so-called 'substitution' is here *elimination* followed by *addition* (*cf.* p. 122).

Support for such a reaction pathway is provided by the fact that aryl halides lacking hydrogen on the carbon atoms *o*- to the halogen, and so unable to eliminate HCl, are extremely resistant to amination, but direct evidence for the occurrence of benzyne type intermediates is clearly needed. This has been provided by trapping (*cf.* p. 49), benzyne being produced in the presence of furan (XLIV), a diene with which it might be expected to undergo a Diels-Alder reaction (p. 171) to form the stable adduct (XLV):

(XLIV)   (XLV)   (XLVI)

which is indeed formed, but more readily identified by acid hydrolysis to yield the familiar α-naphthol (XLVI). If benzyne itself is produced under conditions where it has nothing suitable to react with, then it dimerises very rapidly to the stable biphenylene (XLVII):

(XLVII)

But one of the most convincing detections of a benzyne is by use of mass spectrometry, a salt of diazotised anthranilic acid (XLVIII) being introduced into the heated inlet of a mass spectrometer. The mass spectrum is found to be a very simple one with m/e peaks at 28, 44, 76 and 152, the 76 peak declining and the 152 peak increasing with time,

(XLVIII)   (76)

the increasing 152 peak showing the progressive dimerisation of benzyne to biphenylene (XLVII).

Further investigation of the conversion of chlorobenzene to phenoxide ion with $^{\ominus}$OH at high temperature (p. 153) shows that part of the conversion takes place via benzyne and part by direct displacement, the relative proportions of the two depending on the conditions employed.

7 ADDITION TO CARBON–CARBON DOUBLE BONDS

As we have already seen (p. 7), a carbon–carbon double bond consists of a strong σ bond plus a weaker π bond, in a different position (I):

$$\underset{\text{(I)}}{\overset{\displaystyle H\diagdown\quad\quad\diagup H}{\underset{\displaystyle H\diagup\quad\quad\diagdown H}{C\!\!-\!\!C}}}$$

The pair of electrons in the π orbital are more diffuse and less firmly held by the carbon nuclei, and so more readily polarisable, than those of the σ bond, leading to the characteristic reactivity of such unsaturated compounds. As the π electrons are the most readily accessible feature of the carbon–carbon double bond, we should expect them to shield the molecule from attack by nucleophilic reagents and this is indeed found to be the case (*cf.* p. 172, however). The characteristic reactions of the system are, hardly surprisingly, found to be initiated by electron-deficient species such as $X^{\oplus}$ and $X\cdot$ (radicals can be considered electron-deficient species as they are seeking a further electron with which to form a bond), cations inducing heterolytic and radicals homolytic fission of the π bond. The former is usually found to predominate in polar solvents, the latter in non-polar solvents especially in the presence of light. Free radical induced additions are discussed subsequently (p. 268) and attention will here be confined to the polar mechanism.

ADDITION OF BROMINE

It might perhaps be supposed that the addition of halogen, e.g. bromine, to a double bond would take place by a simple one-step process

$$\underset{\displaystyle Br\!-\!Br}{\overset{\displaystyle}{>\!\!C\!\!=\!\!C\!\!<}} \longrightarrow \underset{\displaystyle Br\quad Br}{-\overset{|}{\underset{|}{C}}\!-\!\overset{|}{\underset{|}{C}}-}$$

but there are two significant pieces of experimental evidence that serve to refute this: (*a*) such additions are normally found to be stereochemically *trans*—not *cis* as the above formulation would require—as may be seen with cyclopentene (II),

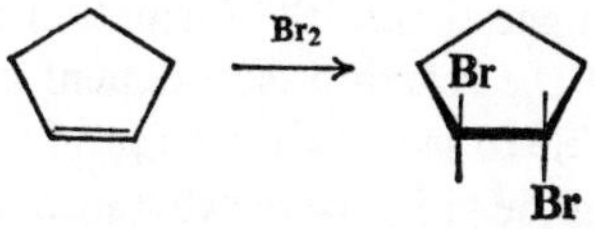

(II)   (III)

and (*b*) carrying out the addition in the presence of nucleophiles, e.g. $Cl^\ominus$, $NO_3^\ominus$, H_2O:, results in the formation, in addition to the expected dibromide, of products in which the added nucleophiles have become bonded to a carbon atom of the original double bond:

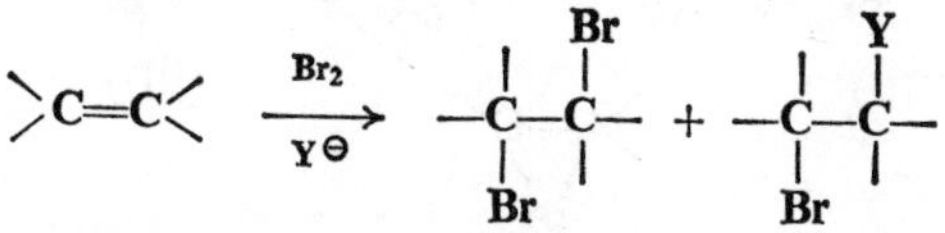

In order to account for these observations addition is envisaged as a stepwise process initiated by the positive end of a bromine molecule that has become polarised either on its close approach to the π electron cloud of the double bond or through other agency in the system, e.g. the glass walls of the container, or an added Lewis acid (Br—Br + $AlBr_3 \rightarrow \overset{\delta+}{Br}$—$\overset{\delta-}{Br} \cdot AlBr_3$, *cf.* p. 125) which is indeed found to speed up the addition considerably:

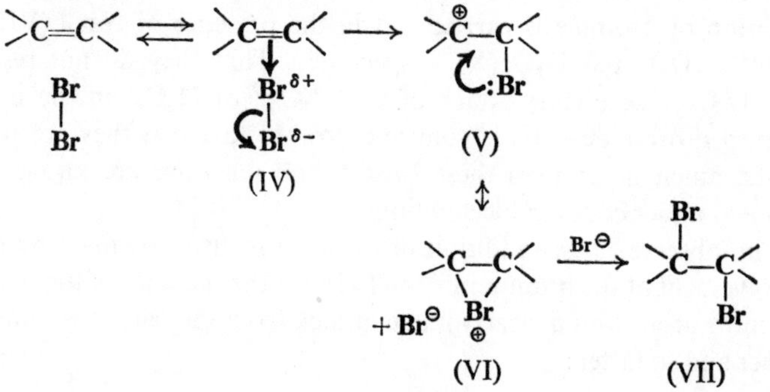

Initial interaction of bromine with the π electrons could form a π complex (IV) leading subsequently to the formation of a σ bond in the carbonium ion (V) or the bromonium ion (VI), the overall addition being completed by attack of $Br^{\ominus}$ to form the product dibromide (VII).

Kinetic evidence suggests that the formation of (V) or (VI) is the rate-limiting stage of the reaction, subsequent attack by $Br^{\ominus}$ being rapid. It is thus possible to see how mixed products result from carrying out the addition in the presence of other nucleophiles, for $Br^{\ominus}$ is in no privileged position, compared with any other nucleophiles present, for effecting the final rapid attack on (V) or (VI) to yield the product:

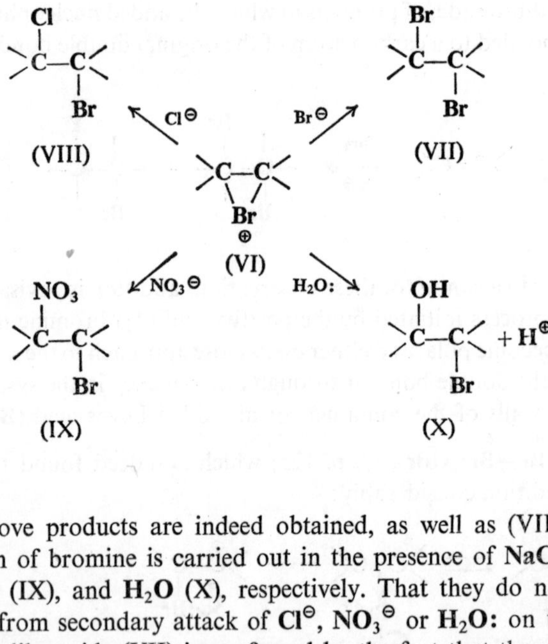

The above products are indeed obtained, as well as (VII), if the addition of bromine is carried out in the presence of NaCl (VIII), NaNO$_3$ (IX), and H$_2$O (X), respectively. That they do not result merely from secondary attack of Cl$^{\ominus}$, NO$_3{}^{\ominus}$ or H$_2$O: on the first-formed dibromide (VII) is confirmed by the fact that they are produced much faster than those substitution reactions are known to proceed under comparable condition.

The observed *trans* addition of bromine is also explained by the involvement of the bromonium ion (VI), for the presence of the bulky bromine atom would clearly make 'attack from the back' very much easier and so faster:

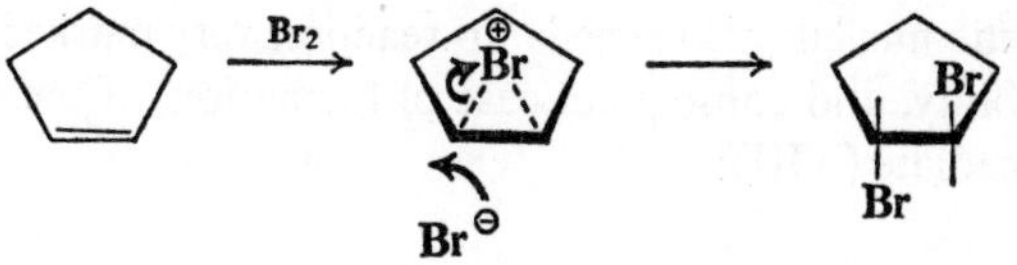

It should be emphasised that (V) and (VI) differ only in electron distribution, i.e. they are canonical states, and that the real structure of the unique intermediate ion probably lies somewhere between them. Acetylenes are also found to undergo preferential, though not exclusive, *trans* addition of halogen; thus acetylene dicarboxylic acid and one mole of bromine yield principally dibromofumaric acid (XI), but some dibromomaleic acid (XII) as well:

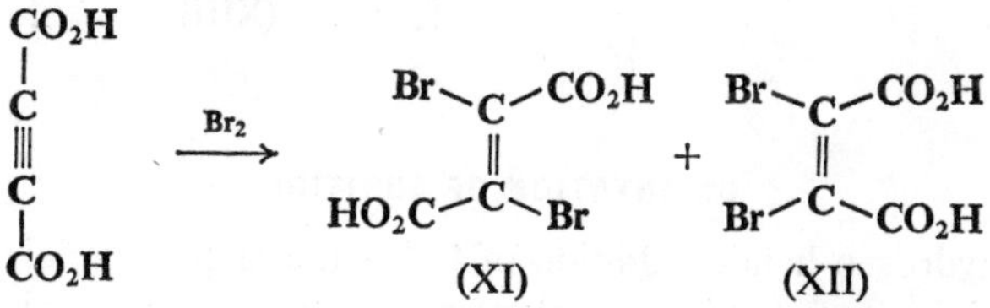

(XI) 70%　　(XII) 30%

EFFECT OF SUBSTITUENTS ON RATE OF ADDITION

If attack by incipient $Br^{\oplus}$ to form a cation, whether (V) or (VI), is the rate-determining step of the reaction, it would be expected that addition would be facilitated by the presence of electron-donating substituents on the double-bond carbon atoms; the following relative rates are observed:

$$\underset{0{\cdot}03}{CH_2{=}CH{-}Br \approx CH_2{=}CH{-}CO_2H} < \underset{1}{CH_2{=}CH_2} < \underset{2}{Me{\rightarrow}CH{=}CH_2}$$

$$< \underset{5}{Me_2C{=}CH_2} < \underset{10}{Me_2C{=}CHMe} < \underset{13}{Me_2C{=}CMe_2}$$

The rate of addition increases with successive introductions of methyl despite access to the double bond becoming progressively more hindered sterically. By contrast, the presence of electron-withdrawing substituents markedly slows down the rate of addition. The presence of a benzene nucleus also speeds up reaction very markedly because of the stability, and consequent ease of formation, of the carbonium ion intermediate (XIII):

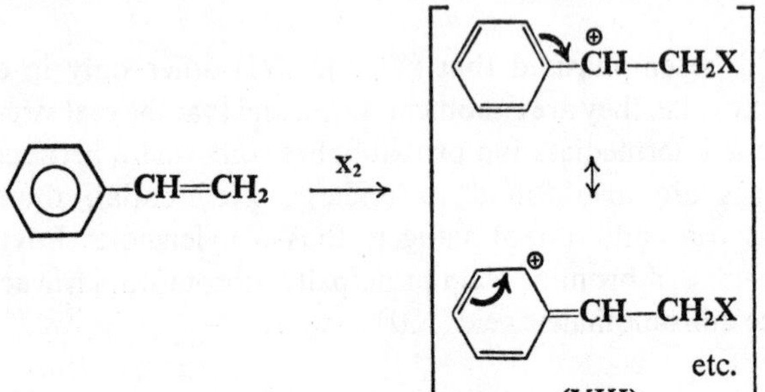

ORIENTATION OF ADDITION

With hydrogen halide, addition of $H^{\oplus}$ is the rate-determining step via an initially formed π complex, the addition being completed by subsequent attack of $Hal^{\ominus}$. In support of this, it is found that ease of addition increases on going $HF \rightarrow HCl \rightarrow HBr \rightarrow HI$, i.e. in order of increasing acid strength. The series also reflects increasing nucleophilicity in the anion, $Hal^{\ominus}$, that completes the attack; but this second stage is not involved in the rate-determining step of the overall addition. When the alkene is unsymmetrical, e.g. propylene, hydrogen bromide can add to form two possible products, **MeCHBrMe** and **MeCH₂CH₂Br**. In practice, however, we should only expect to get the former, exemplifying the greater tendency to form a secondary rather than a primary carbonium ion (*cf.* p. 101):

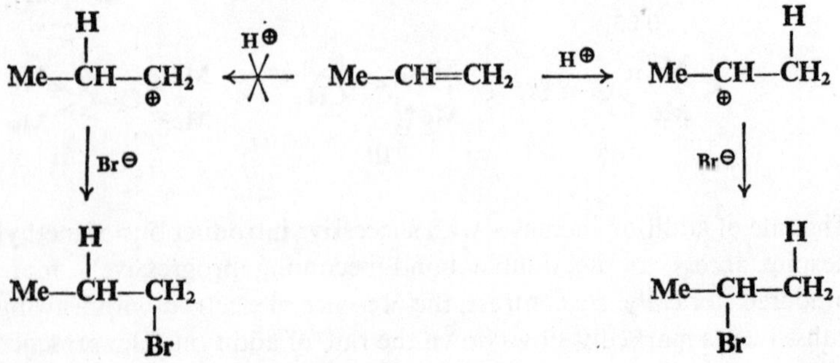

This is found to be the case and is the theoretical justification for the empirical generalisation of Markownikov: 'In the addition of unsymmetrical adducts to unsymmetrical alkenes halogen, or the more negative group, becomes attached to the more highly substituted of the unsaturated carbon atoms.'

The study of the addition of hydrogen halides to alkenes presents a number of experimental difficulties. In solution in water or hydroxylic solvents acid catalysed hydration (p. 162) constitutes a competing reaction, while in less polar solvents radical formation is encouraged and the mechanism changes, resulting with **HBr** in *anti*-Markownikov addition to yield **MeCH$_2$CH$_2$Br** via the preferentially formed intermediate, **MeĊHCH$_2$Br**. The radical mechanism of addition and the reasons for its occurrence are discussed subsequently (p. 271).

Addition of **HBr** to vinyl bromide, **CH$_2$=CHBr**, is also of some interest. Under polar conditions, **CH$_3$CHBr$_2$**, rather than **CH$_2$BrCH$_2$Br**, is obtained reflecting the greater stability of the carbonium ion intermediate (XIVa) rather than (XIVb):

$$
\overset{\displaystyle H}{\underset{}{|}}\quad\quad\quad\overset{\displaystyle H}{\underset{}{|}}\quad\quad\quad\quad\quad\overset{\displaystyle H}{\underset{}{|}}
$$

$$
CH_2\!-\!\overset{\oplus}{C}H\!-\!\ddot{B}r \quad\leftrightarrow\quad CH_2\!-\!CH\!=\!\overset{\oplus}{B}r \quad\quad \overset{\oplus}{C}H_2\!-\!CH\!-\!Br
$$

$$
(XIVa) \quad\quad\quad\quad\quad\quad\quad\quad\quad (XIVb)
$$

Nevertheless the *rate* of addition is, as described above, about thirty times slower than with ethylene, indicating the inductive effect of the bromine atom in reducing overall electron availability at the double bond:

$$CH_2\!=\!CH\!\rightarrow\!Br$$

or more strictly that

$$
\overset{\displaystyle H}{\underset{}{|}}\quad\quad\quad\overset{\displaystyle H}{\underset{}{|}}
$$

$$
CH_2\!-\!\overset{\oplus}{C}HBr \,<\, CH_2\!-\!\overset{\oplus}{C}H_2
$$

because of the inductive effect of the bromine atom.

This very closely resembles what occurs in the electrophilic substitution of chlorobenzene which, as we have seen (p. 141), is predominantly *o*/*p* (preferential stabilisation of the transition states for *o*- and *p*-substitution by interaction of the unshared electron pairs on chlorine with the π orbital system of the nucleus), yet slower than in benzene itself (large inductive effect of chlorine resulting in overall deactivation of the nucleus to electrophilic attack).

The rearrangements of structure that can take place during the addition of acids to alkenes, due to alkyl migrations in the carbonium ion intermediates, have already been referred to (p. 106).

With hypochlorous acid, etc., the adduct polarises in the sense $\overset{\delta-}{HO}\!-\!\overset{\delta+}{Cl}$, thus yielding with propylene

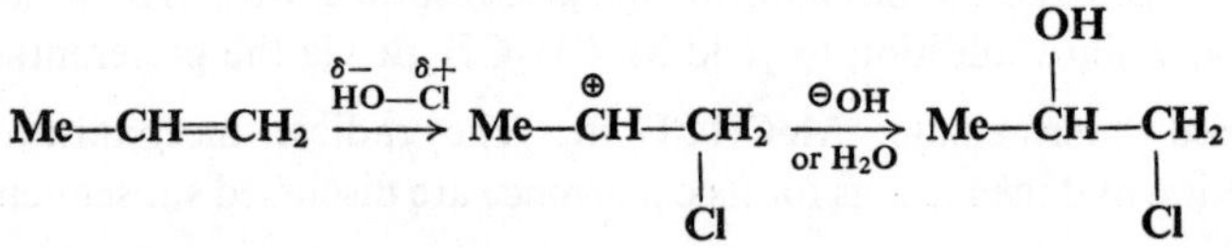

$$Me\!-\!CH\!\!=\!\!CH_2 \xrightarrow{\overset{\delta-}{HO}\!-\!\overset{\delta+}{Cl}} Me\!-\!\overset{\oplus}{CH}\!-\!\underset{\underset{Cl}{|}}{CH_2} \xrightarrow[\text{or } H_2O]{\ominus OH} Me\!-\!\underset{\underset{Cl}{|}}{\overset{\overset{OH}{|}}{CH}}\!-\!CH_2$$

this being in accordance with the Markownikov rule as **OH** is a more negative group than **Cl**. That such additions also proceed *trans* (i.e. initial attack by $X^{\oplus}$ to yield a cationic intermediate) is shown by the conversion of cyclopentene to the *trans* chlorohydrin:

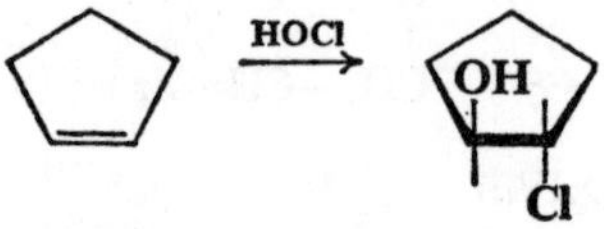

OTHER ADDITION REACTIONS

(i) Hydration

Hydration of a carbon–carbon double bond is, of course, the reversal of the acid-catalysed dehydration of alcohols to alkenes (p. 214):

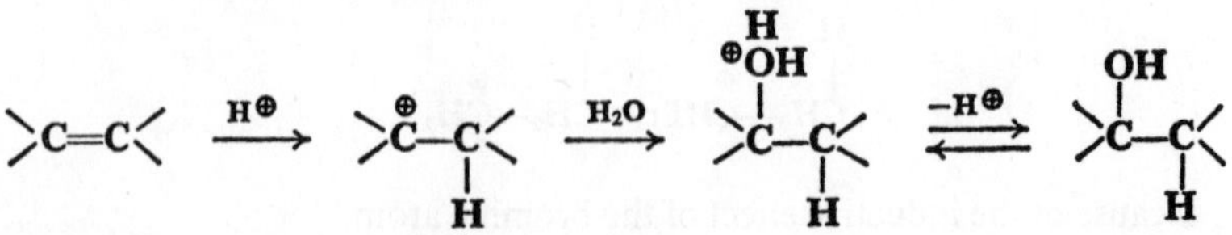

$$\underset{}{>}\!C\!\!=\!\!C\!\!<\ \xrightarrow{H^{\oplus}}\ \underset{\underset{H}{|}}{>}\!\overset{\oplus}{C}\!-\!C\!\!<\ \xrightarrow{H_2O}\ \underset{\underset{H}{|}}{>}\!\overset{\overset{\overset{H}{\oplus OH}}{|}}{C}\!-\!C\!\!<\ \rightleftarrows\ \underset{\underset{H}{|}}{>}\!\overset{\overset{OH}{|}}{C}\!-\!C\!\!<$$

$$(XV)$$

The formation of the carbonium ion (**XV**) is the rate-determining step in the reaction but whether this takes place directly or via the rapid, reversible formation of a π complex (XVI), followed by the slow, rate-determining conversion of the latter to the carbonium ion (XV) is not wholly certain. Hydrogen halides are not normally

$$\text{>C=C<} \underset{\text{fast}}{\overset{\text{H}^{\oplus}}{\rightleftharpoons}} \overset{\text{>C--C<}}{\underset{\overset{|}{\text{H}^{\oplus}}}{}} \overset{\text{Slow}}{\longrightarrow} \overset{\oplus}{\text{>C--C<}}_{\overset{|}{\text{H}}}$$

$$\text{(XVI)} \qquad\qquad\qquad \text{(XV)}$$

used as sources of proton because of their tendency to add on themselves, but the $HSO_4{}^{\ominus}$ ions produced with sulphuric acid are only very weakly nucleophilic and, even if they should add on, the alkyl hydrogen sulphates (XVII) so produced are very readily hydrolysed by water:

$$\overset{\oplus}{\text{>C--C<}}_{\overset{|}{\text{H}}} \overset{\text{HSO}_4{}^{\ominus}}{\longrightarrow} \overset{\overset{\text{OSO}_3\text{H}}{|}}{\text{>C--C<}}_{\overset{|}{\text{H}}} \overset{\text{H}_2\text{O}}{\longrightarrow} \overset{\overset{\text{OH}}{|}}{\text{>C--C<}}_{\overset{|}{\text{H}}}$$

$$\text{(XV)} \qquad\qquad \text{(XVII)}$$

The reaction is of importance for converting petroleum fractions into alcohols and is sometimes brought about by dissolving the alkenes in concentrated sulphuric acid and then diluting the solution with water, or increasingly on the large scale by direct hydration with steam in the presence of acid catalysts. The orientation, being proton-initiated, follows the conventions already discussed and proceeds *trans*; it will yield a primary alcohol only with ethylene (*cf.* p. 160). It is, however, possible to obtain primary alcohols from suitable alkenes by addition of diborane (B_2H_6, generated from $NaBH_4$ and $^{\oplus}EtO:BF_3{}^{\ominus}$), generally referred to as hydroboration, to yield a trialkylboron

$$3MeCH{=}CH_2 + \tfrac{1}{2}B_2H_6 \rightarrow (MeCH_2CH_2)_3B$$

followed by cleavage with alkaline H_2O_2:

$$(MeCH_2CH_2)_3B + 3H_2O_2 \rightarrow 3MeCH_2CH_2OH + B(OH)_3$$

the overall result being anti-Markownikov addition of water. Yields are generally good—hex-1-ene is converted to hexan-1-ol virtually quantitatively—and with cyclic alkenes the overall addition of H_2O is *cis*, thus making the method doubly complementary with the direct, acid-catalysed addition.

(ii) Carbonium ion addition

The carbonium ion intermediate that may result from initial protonation of the double bond in several of the above reactions can itself, of course, act as an electrophile towards a second molecule of alkene. Thus with isobutene (XVIII)

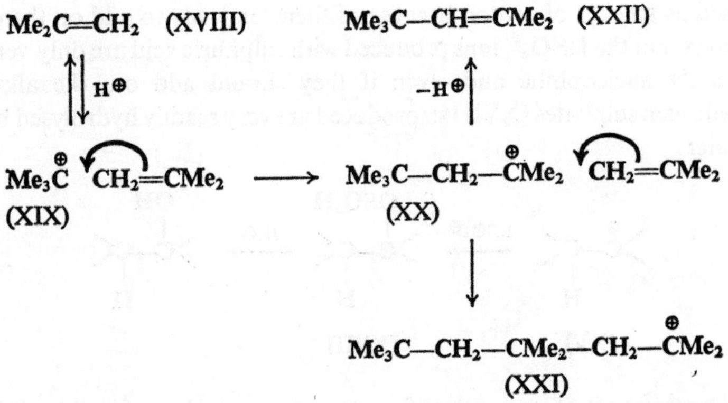

$Me_2C\!=\!CH_2$ (XVIII) $Me_3C\!-\!CH\!=\!CMe_2$ (XXII)

$$Me_3\overset{\oplus}{C} \quad CH_2\!=\!CMe_2 \longrightarrow Me_3C\!-\!CH_2\!-\!\overset{\oplus}{C}Me_2 \quad CH_2\!=\!CMe_2$$
(XIX)   (XX)

$$Me_3C\!-\!CH_2\!-\!CMe_2\!-\!CH_2\!-\!\overset{\oplus}{C}Me_2$$
(XXI)

the first formed carbonium ion (XIX) can add to the double bond of a second molecule to form a second carbonium ion (XX). This in its turn can add on to the double bond of a third molecule to yield (XXI) or, alternatively, lose a proton to yield the alkene (XXII). Such successive additions can lead to unwanted by-products in, for example, the simple addition of hydrogen halides, but they may be specifically promoted to yield polymers by the presence of Lewis acids, e.g. $AlCl_3$, $SnCl_4$, BF_3, as catalysts. Many polymerisations of olefines are radical-induced however (p. 274).

(iii) Hydroxylation

Investigation of the action of osmium tetroxide on alkenes had led to the isolation of cyclic osmic esters (XXIII) which undergo ready hydrolysis to yield the 1,2-diol (see p. 165).

As the hydrolysis results in the splitting of the $Os\!-\!O$ and not the $C\!-\!O$ bonds in (XXIII), no inversion of configuration can take place at the carbon atoms and the glycol produced must, like the cyclic

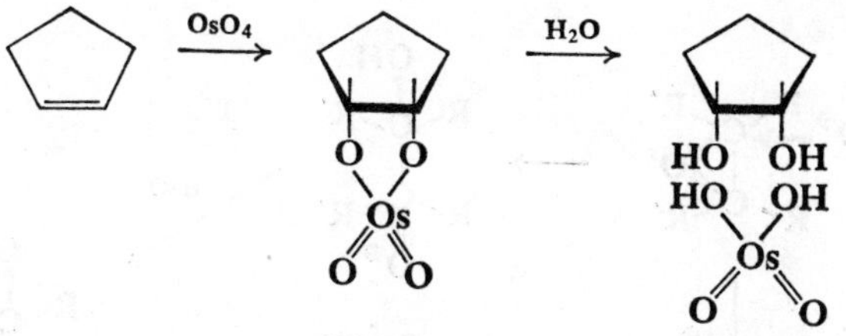

(XXIII)   (XXIV)

osmic ester itself, be *cis*, i.e. this is a stereospecific *cis* addition. The expense and toxicity of osmium tetroxide preclude its large scale use but it can be employed in catalytic amounts in the presence of hydrogen peroxide which reoxidises osmic acid (XXIV) to the tetroxide.

The *cis* glycol is also obtained with alkaline permanganate, the classical reagent for the hydroxylation of double bonds, and though no cyclic permanganic esters have been isolated it is not unreasonable to suppose that the reaction follows a similar course. This is supported by the fact that use of ^{18}O labelled $MnO_4^{\ominus}$ results in *both* oxygen atoms in the resultant glycol becoming labelled, i.e. both are derived from the permanganate and neither from the solvent.

If alkenes are oxidised by peracids, $R-\overset{O}{\overset{\|}{C}}-O-OH$, the result is an alkylene oxide or epoxide (XXV):

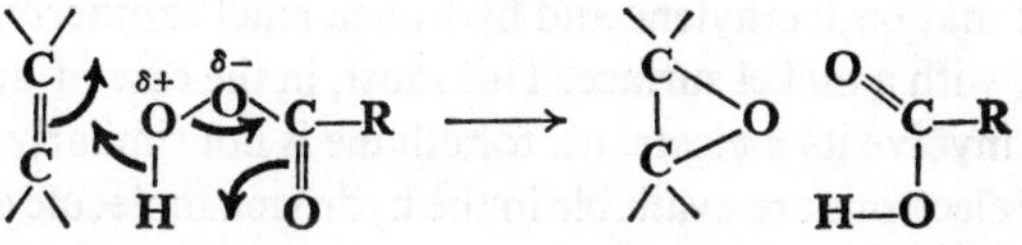

(XXV)

It is possible, however, that in polar solvents the reaction may be initiated by addition of $^{\oplus}OH$ obtained by breakdown of the peracid. The epoxides may be isolated (*cf.* p. 87) and then undergo acid- or base-catalysed hydrolysis (a nucleophilic reaction) to yield the 1,2-diol. As attack must be 'from the back' on the cyclic epoxide, inversion of configuration will take place at the carbon atom attacked so that the *overall* addition reaction to yield the 1,2-diol will be *trans* (see p. 166).

Attack on only one carbon atom has been shown above, but equally easy attack on the other will lead to the mirror image of (XXVI), i.e. the DL-glycol will result from the original *cis* alkene, confirming an overall *trans* hydroxylation.

Thus by suitable choice of reagent, the hydroxylation of alkenes can be stereospecifically controlled to proceed *cis* or *trans* at will.

(iv) Hydrogenation

The addition of hydrogen to alkenes, which takes place only in the presence of suitable metallic catalysts, e.g. **Ni, Pt, Pd**, is usually found to proceed *cis*. The metal atoms at the surface of, for example, a nickel crystal will presumably have unsatisfied valences directed away from the body of the crystal, unlike their fellows in the lump, and it is significant that both ethylene and hydrogen react exothermically, and reversibly, with a nickel surface. This must, in the case of an ethylene molecule, involve its π electrons, for ethane is not similarly adsorbed. No such π electrons are available in the hydrogen molecule either, and its observed exothermic adsorption must clearly involve very considerable weakening of the bond between its hydrogen atoms.

Hydrogenation of ethylene at a nickel surface is thus envisaged as the essentially simultaneous addition of two hydrogen atoms to the same side of the two carbon atoms of the ethylene molecule lined-up alongside: the reason for the observed stereospecific *cis* nature of the addition thus becomes apparent. The resultant molecule of ethane is immediately desorbed and the nickel surface thus freed for a further catalytic cycle. The actual spacings of the metal atoms in the surface must clearly be of importance in determining whether a particular metal will be an effective hydrogenation catalyst or not, and even in a suitable metal, e.g. nickel, one face of a crystal would be expected to be more effective than another depending on how closely their different spacings correspond to the optimium. This is certainly observed in practice where only a relatively small proportion of the total metal surface is found to be catalytically active—the so-called 'active points'.

Stereospecific *cis* hydrogenation has been of the greatest value in confirming molecular structures by synthetic methods and has, hardly surprisingly, been found to occur in the partial hydrogenation of acetylenes as well:

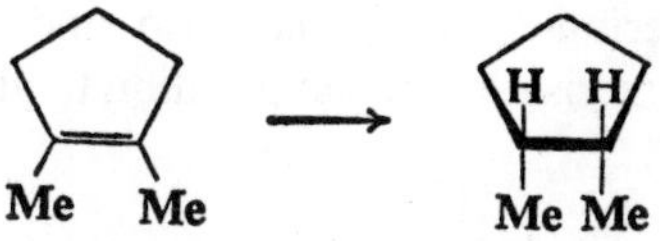

(XXVII)   (XXVIII)

Me—C≡C—Me ⟶

(XXIX)   (XXX)

(v) Ozonolysis

The addition of ozone to alkenes can also be looked upon essentially as an electrophilic addition

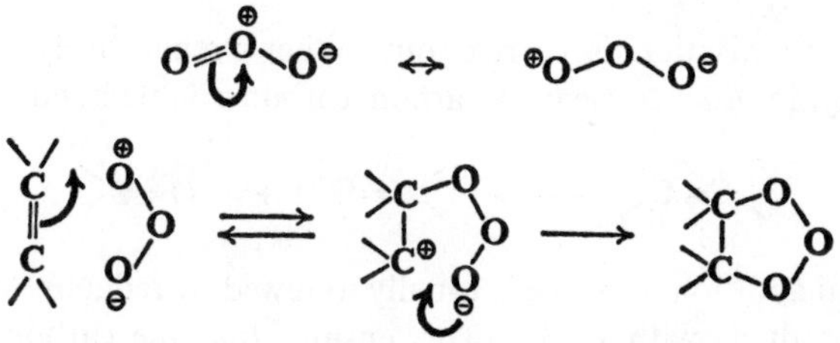

and, in support of this view, it is found that the reaction is catalysed by Lewis acids such as **BF₃**. The primary addition product, the molozonide, has been isolated in a few cases, and by reduction to the corresponding 1,2-diol, shown still to contain an intact **C—C** bond. The molozonide normally enjoys only a transient existence and is found to dissociate readily into two fragments:

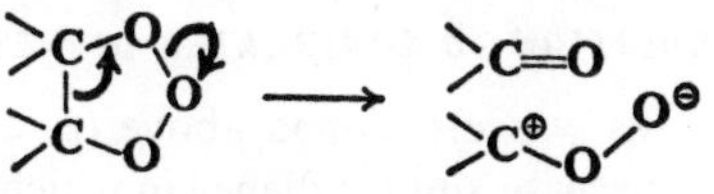

(XXXI)

167

These fragments may recombine to form the usual end-product
(XXXII) of the reaction, generally called the normal ozonide

$$\underset{\substack{\\ \ominus O\!-\!O}}{\overset{\substack{O\ \oplus\\}}{>C\!\!-\!\!C<}} \longrightarrow \underset{\substack{\\ O\!-\!O}}{\overset{\substack{O\\}}{>C\quad C<}}$$

(XXXII)

but the peroxy zwitterion (XXXI) may also undergo alternative
reactions, such as polymerisation or self-addition to yield a dimer:

$$\underset{\substack{\\ O\!-\!O}}{\overset{\substack{O\!-\!O\\}}{>C\qquad C<}}$$

Conversion of molozonide to the normal ozonide is generally quite
rapid—it has been followed spectroscopically in a few cases—and that
it does indeed involve the above decomposition and subsequent
recombination is supported by the fact that if ozonolysis is carried out
in the presence of a foreign aldehyde (one *not* obtainable by ozonolysis
of the alkene), an ozonide incorporating the added aldehyde (resulting
from attack of (XXXI) upon it) is obtained in addition to the expected
one.

When ozonisation is carried out, either preparatively or diag-
nostically, in order to cleave a carbon–carbon double bond

$$>C\!\!=\!\!C< \longrightarrow \ >C\!\!=\!\!O \ + \ O\!\!=\!\!C<$$

the actual addition of ozone is usually followed by reductive cleavage
of the products with **Pd/H$_2$**. This ensures that the carbonyl compounds, especially aldehydes, do not undergo further oxidation as
tends to happen on simple hydrolytic cleavage due to the hydroperoxides (*cf.* p. 280) that are then formed. This is important as one
of the advantages of ozonolysis as a preparative or a diagnostic
method is the ease of isolation and characterisation of the carbonyl
compounds that it yields as end-products.

ADDITION TO CONJUGATED DIENES

Conjugated dienes, as was mentioned above (p. 10), are somewhat
more stable than otherwise similar dienes in which the double bonds
are not conjugated, as is revealed by a study of their respective heats of

hydrogenation (*cf.* p. 15), the delocalisation energy consequent on the extended π orbital system probably being of the order of 6 kcal/mole. Conjugated dienes tend nevertheless to undergo addition reactions somewhat more readily than non-conjugated dienes because the transition state in such reactions, whether the addition is proceeding by a polar or a radical mechanism, is allylic in nature and thus more readily formed (*cf.* pp. 99, 260) than that from an isolated double bond:

$$CH_2=CH-CH=CH_2 \quad \underset{X_2}{\overset{X_2}{\nearrow\searrow}} \quad \begin{array}{l} \overset{\oplus}{CH_2-CH-CH=CH_2} \\ | \\ X \end{array} \quad \begin{array}{l} \cdot \\ CH_2-CH-CH=CH_2 \\ | \\ X \end{array}$$

$$CH_2=CH_2 \quad \underset{X_2}{\overset{X_2}{\nearrow\searrow}} \quad \begin{array}{l} \overset{\oplus}{CH_2-CH_2} \\ | \\ X \end{array} \quad \begin{array}{l} \cdot \\ CH_2-CH_2 \\ | \\ X \end{array}$$

Thus conjugated dienes are reduced to dihydro-derivatives by sodium and alcohol whereas non-conjugated dienes or simple alkenes are unaffected.

It might be expected that in the addition of, for example, chlorine to butadiene, reaction could proceed through a cyclic chloronium ion

$$\begin{array}{c} HC=CH \\ / \quad \backslash \\ H_2C \quad CH_2 \\ \backslash Cl / \\ \oplus \end{array}$$

that would be largely unstrained. That this is *not* formed, however, is shown by the fact that the above addition results in the formation of the *trans* compound

$$\begin{array}{c} ClCH_2 \\ \backslash \\ CH=CH \\ \backslash \\ CH_2Cl \end{array}$$

and not the corresponding *cis* compound that would have been obtained by the attack of $Cl^\ominus$ on the cyclic chloronium ion. Addition thus probably proceeds through a delocalised carbonium ion, *cf.* the addition of hydrogen halide below.

(i) Hydrogen halide

With butadiene itself a proton may initially form a π complex and then a σ complex with hydrogen on a terminal carbon atom (XXXIV). Protonation takes place at C_1 rather than C_2 as the former yields a secondary carbonium ion that is stabilised by delocalisation, whereas the latter would yield a primary carbonium ion (XXXIII) that is not. The resulting allylic cation (XXXIV) can take up $Br^\ominus$ at either C_2 or C_4 leading to 1,2- and 1,4- overall addition, i.e. (XXXV*a*) and (XXXV*b*), respectively:

$$CH_2{=}CH{-}CH{=}CH_2 \xrightarrow{\;H^\oplus\;} \overset{\oplus}{C}H_2{-}CH{-}CH{=}CH_2 \quad (XXXIII)$$

with H substituent below the second carbon.

$$\left[\begin{array}{c} \overset{\oplus}{C}H_2{-}CH{-}CH{=}CH_2 \\ | \\ H \\ \updownarrow \\ CH_2{-}CH{=}CH{-}\overset{\oplus}{C}H_2 \\ | \\ H \end{array} \right] \xrightarrow{\;Br^\ominus\;}$$

(XXXIV)

$$CH_2{-}\underset{|}{CH}{-}CH{=}CH_2 \quad (XXXV a)$$

with Br above the second carbon and H below; *1,2-addition*

$$CH_2{-}CH{=}CH{-}\underset{|}{CH_2} \quad (XXXV b)$$

with Br above the fourth carbon; *1,4-addition*

The presence of conjugation does not make 1,4-addition obligatory: it merely makes it possible, and whether this or 1,2-addition actually takes place is governed by the relative rates of conversion of the cation (XXXIV) to the alternative products, or by the relative stability of these products. By and large, 1,2-addition tends to occur preferentially at lower temperatures in non-polar solvents, i.e. it is the faster of the two reactions and we are here seeing kinetic control of product, while at higher temperatures, with longer reaction times, and in polar solvents, i.e. under conditions that allow of the attainment of equilibrium, it is largely the more thermodynamically stable 1,4-product that is obtained (*cf.* p. 146).

For addition to an unsymmetrical diene the same considerations apply as in the case of mono-alkenes, thus:

$$MeCH{=}CH{-}CH{=}CH_2 \xrightarrow{H^{\oplus}} MeCH{=}CH{-}\overset{\oplus}{C}H{-}CH_2 \longrightarrow products$$
$$\underset{H}{|}$$

$$\underset{\underset{|}{C}H_2{=}\overset{Me}{C}{-}CH{=}CH_2}{} \xrightarrow{H^{\oplus}} CH_2{-}\overset{Me}{\underset{\underset{H}{|}}{C}}{-}CH{=}CH_2 \longrightarrow products$$

(ii) Diels-Alder reaction

The classic example is with butadiene and maleic anhydride

i.e. 1,4-addition, proceeding *cis*, via a cyclic transition state (*cf.* the pyrolysis of esters, p. 230), to yield a cyclic product. It has been used as a diagnostic test for determining whether the double bonds in a diene are conjugated or not (though this is normally more readily determined spectroscopically) and also has considerable synthetic importance. The reaction is promoted by the presence of electron-donating substituents in the diene and of electron-withdrawing substituents in the, so-called, *dienophile*; their absence from the latter normally prevents reaction, i.e. of a simple alkene, though cyclopentadiene will in fact dimerise at room temperature. Other common dienophiles are *p*-benzoquinone, $CH_2{=}CHCHO$ and $EtO_2CC{\equiv}CCO_2Et$. The reaction is also sensitive to steric effects; thus of the three 1,4-diphenyl-butadienes only the *trans/trans* form undergoes reaction with maleic anhydride:

Trans/Trans   *Cis/Trans*   *Cis/Cis*

Similarly the reactivity of the diene is promoted when the double bonds are locked in a *cis* conformation with respect to each other as in cyclopentadiene.

Where there is the possibility of more than one product, depending on which way round the addition takes place, e.g. with maleic anhydride and cyclopentadiene, the major if not the exclusive product is found to have the so-called *endo* structure (XXXVI), rather than the alternative *exo* (XXXVII), despite the latter being the more thermodynamically stable:

(XXXVI)   (XXXVII)

The suggestion has been made that this occurs because of the greater interaction between the π electron systems of the reactants that is possible in the transition state leading to the *endo* form. Though such interaction may well take place, a number of addition products that do not correspond to such maximum overlap have now been discovered, and it is doubtful whether the latter really controls the stereochemistry of addition.

ADDITION OF ANIONS

As has already been seen (p. 148) the introduction of electron-withdrawing groups into an aromatic nucleus tends to inhibit electrophilic substitution and to make nucleophilic substitution possible. The same is true of addition reactions: the introduction of F, NO_2, CN, C=O, CO_2Et, etc., on the carbon atoms of a double bond causes the π electrons to become less available and attack by an anion then becomes possible, though it would not have taken place with the unmodified double bond (see p. 173).

Some of the reactions have important synthetic applications.

(i) Cyanoethylation

The cyano group in acrylonitrile, $CH_2{=}CHCN$, makes the β-carbon atom of the double bond respond readily to the attack of anions or

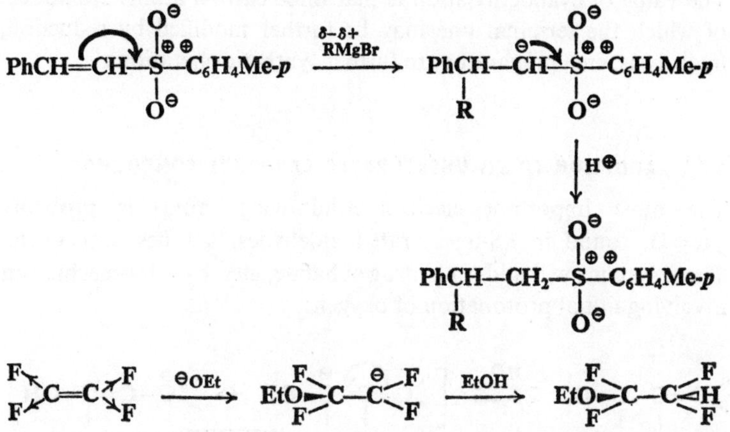

$$PhCH{=}CH{-}\overset{\overset{O^{\ominus}}{\underset{\oplus}{|}}}{\underset{\underset{O^{\ominus}}{|}}{S}}{-}C_6H_4Me\text{-}p \quad\xrightarrow[RMgBr]{\delta^- \delta^+}\quad PhCH{-}\overset{\oplus}{\underset{\underset{R}{|}}{CH}}{-}\overset{\overset{O^{\ominus}}{\underset{\oplus}{|}}}{\underset{\underset{O^{\ominus}}{|}}{S}}{-}C_6H_4Me\text{-}p$$

$$\downarrow H^{\oplus}$$

$$PhCH{-}CH_2{-}\overset{\overset{O^{\ominus}}{\underset{\oplus}{|}}}{\underset{\underset{O^{\ominus}}{|}}{S}}{-}C_6H_4Me\text{-}p$$
$$\underset{R}{|}$$

$$\underset{F}{\overset{F}{>}}C{=}C\underset{F}{\overset{F}{<}} \quad\xrightarrow{\ominus OEt}\quad EtO\underset{F}{\overset{F}{>}}C{-}\overset{\ominus}{C}\underset{F}{\overset{F}{<}} \quad\xrightarrow{EtOH}\quad EtO\underset{F}{\overset{F}{>}}C{-}C\underset{F}{\overset{H}{<}}$$

other powerful nucleophiles, the addition being completed by the abstraction of a proton from the solvent:

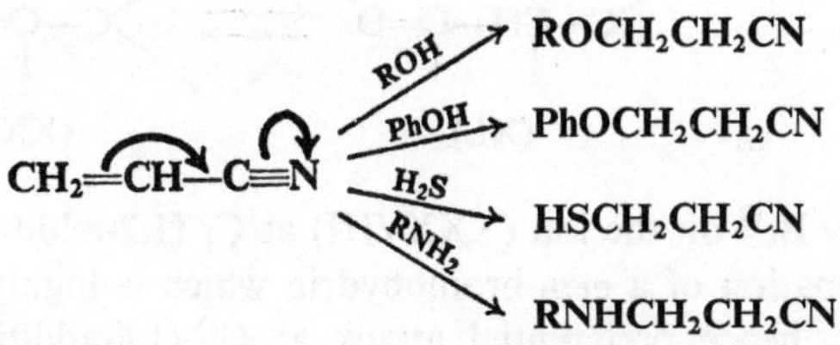

$$CH_2{=}CH{-}C{\equiv}N$$

$$\begin{array}{l} \xrightarrow{ROH}\ ROCH_2CH_2CN \\ \xrightarrow{PhOH}\ PhOCH_2CH_2CN \\ \xrightarrow{H_2S}\ HSCH_2CH_2CN \\ \xrightarrow{RNH_2}\ RNHCH_2CH_2CN \end{array}$$

The reaction is normally carried out in the presence of base in order to obtain an anion from the would-be adduct. Carbon–carbon bonds may also be formed:

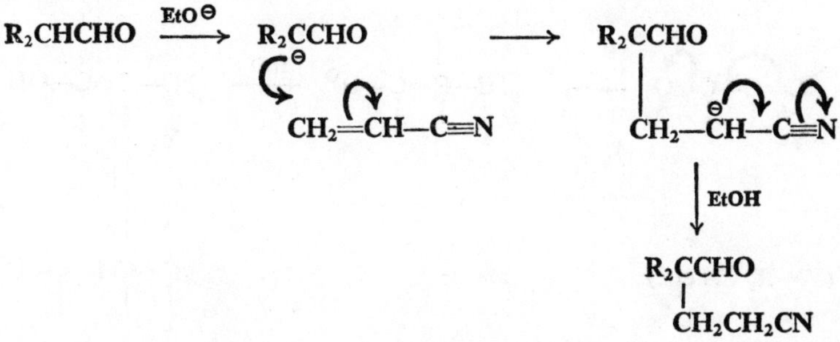

$$R_2CHCHO \xrightarrow{EtO^{\ominus}} R_2\overset{\ominus}{C}CHO \longrightarrow R_2CCHO$$

$$CH_2{=}CH{-}C{\equiv}N \qquad\qquad CH_2{-}\overset{\ominus}{CH}{-}C{\equiv}N$$

$$\downarrow EtOH$$

$$R_2CCHO$$
$$|$$
$$CH_2CH_2CN$$

The value of cyanoethylation is that three carbon atoms are added, of which the terminal one may be further modified by reduction, hydrolysis, etc., preparatory to further synthetic operations.

ADDITION TO αβ-UNSATURATED CARBONYL COMPOUNDS

The most important electron-withdrawing group is probably $\rangle C{=}O$, found in αβ-unsaturated aldehydes, ketones, esters, etc. These systems will add on hydrogen halide, etc., by a 1,4-mechanism involving initial protonation of oxygen:

$$\rangle C{=}C{-}C{=}O \;\underset{}{\overset{H^{\oplus}}{\rightleftharpoons}}\; \left[\rangle C{=}C{-}\overset{\oplus}{C}{-}OH \leftrightarrow \rangle\overset{\oplus}{C}{-}C{=}C{-}OH \right]$$

(XXXVIII)

$$\Big\downarrow Br^{\ominus}$$

$$\underset{\text{(XL)}}{\overset{Br}{\rangle C{-}CH{-}C{=}O}} \;\rightleftharpoons\; \underset{\text{(XXXIX)}}{\overset{Br}{\rangle C{-}C{=}C{-}OH}}$$

Attack by $Br^{\ominus}$ on the ion (XXXVIII) at C_1 (1,2-addition) would lead to formation of a *gem*-bromohydrin which is highly unstable, losing HBr, hence preferential attack at C_3 (1,4-addition) yields (XXXIX) which is, of course, the enol of the β-bromoketone (XL). Addition to αβ-unsaturated acids proceeds somewhat similarly.

With more pronouncedly nucleophilic reagents, e.g. Grignard reagents, $^{\ominus}CN$, etc., overall 1,4-addition will take place without need for initial protonation of the carbonyl oxygen atom:

$$\underset{Y^{\ominus}}{\rangle C{=}C{-}C{=}O} \longrightarrow \underset{Y}{\rangle C{-}C{=}C{-}O^{\ominus}} \overset{H_2O}{\longrightarrow} \underset{Y}{\rangle C{-}C{=}C{-}OH}$$

$$\Big\updownarrow$$

$$\underset{Y}{\rangle C{-}CH{-}C{=}O}$$

$(Y = R, CN \text{ etc.})$

This is what generally happens with αβ-unsaturated ketones—though the reaction may need to be catalysed by copper salts—but with αβ-unsaturated aldehydes direct attack on the carbonyl carbon atom (1,2-addition) is also found to take place because of the more positive character of this atom in aldehydes as compared with ketones (p. 179).

With even more powerful nucleophiles, e.g. $^\ominus$OH, addition at the β-carbon atom takes place, even with αβ-unsaturated aldehydes. This can lead, under suitable conditions, to reversal of the aldol/dehydration reaction (p. 196):

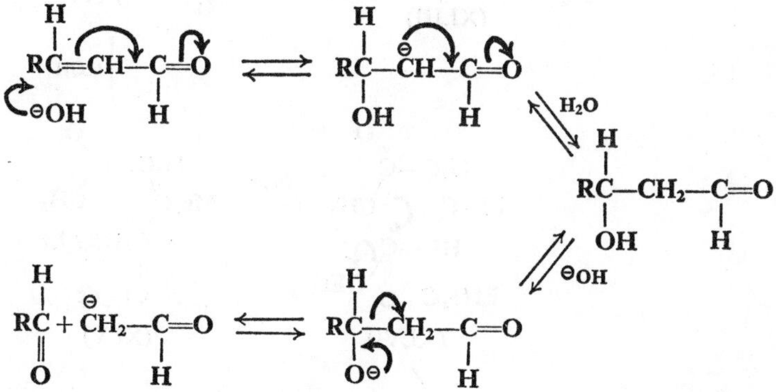

Amines, mercaptans, etc., will also add to the β-carbon atom of αβ-unsaturated aldehydes, ketones and esters. The most important addition reactions of αβ-unsaturated carbonyl compounds, however, are with carbanions in which carbon–carbon bonds are formed.

(i) Michael reaction

The most frequently employed carbanions are probably those derived from diethyl malonate, ethyl acetoacetate, ethyl cyanoacetate and aliphatic nitro-compounds, e.g. CH_3NO_2. Thus in the formation of dimedone (XLI) from diethyl malonate and mesityl oxide (XLII) the carbanion (XLIII) derived from diethyl malonate attacks the β-carbon atom of mesityl oxide to yield the ion (XLIV), which is converted, via its enol, to the ketone (XLV). This constitutes the Michael reaction proper, i.e. the carbanion addition to an αβ-unsaturated carbonyl compound. In this case, however, the reaction proceeds further for (XLV) is converted by $^\ominus$OEt to the carbanion

(XLVI) which cyclises by expelling $^{\ominus}$OEt from the ester group (*cf.* the Dieckmann reaction, p. 199) to yield (XLVII). Hydrolysis and decarboxylation of the β-keto-ester then yields dimedone (XLI):

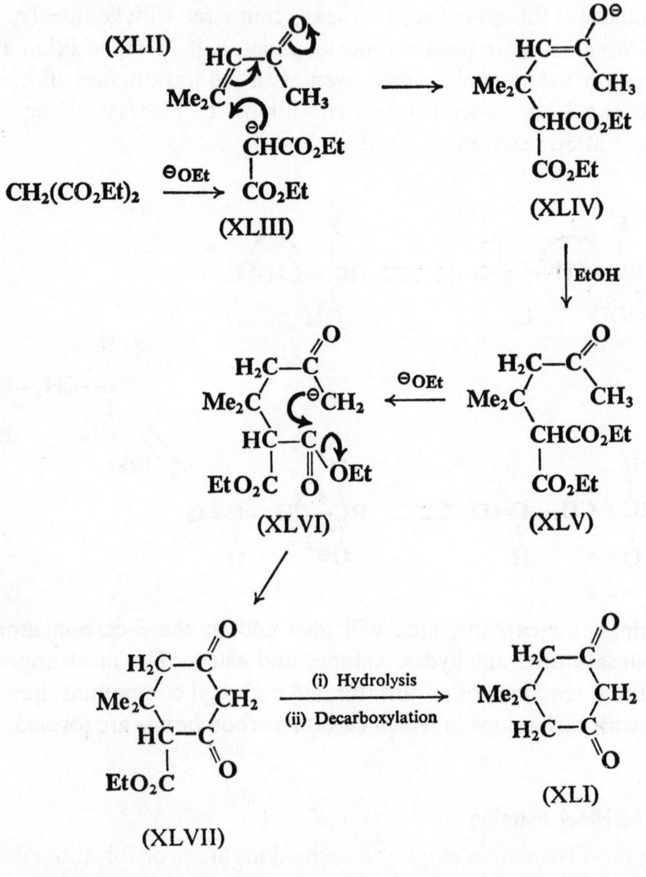

(XLII)

$$CH_2(CO_2Et)_2 \xrightarrow{\ ^{\ominus}OEt\ } (XLIII)$$

(XLIV)

$$\Big\downarrow EtOH$$

(XLVI) $\xleftarrow{\ ^{\ominus}OEt\ }$ (XLV)

(XLVII) $\xrightarrow[\text{(ii) Decarboxylation}]{\text{(i) Hydrolysis}}$ (XLI)

The compound does in fact exist virtually entirely in the enol form:

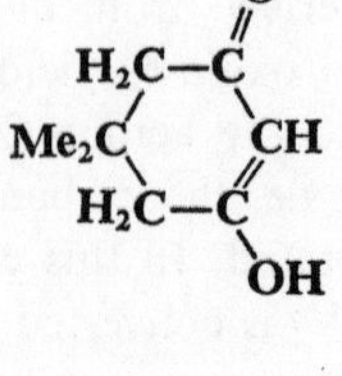

Dimedone is of value as a reagent for the differential characterisation of carbonyl compounds for it readily yields derivatives (XLVIII) with aldehydes but not with ketones, from a mixture of the two:

(XLVIII)

The Michael reaction is promoted by a variety of bases, present in catalytic quantities only, and its synthetic usefulness resides in the larger number of carbanions and $\alpha\beta$-unsaturated carbonyl compounds that may be employed. The Michael reaction is reversible (*cf.* the Claisen ester condensation, p. 198) and the rate-determining step is believed to be the formation of the carbon–carbon bond, i.e. (XLIII) → (XLIV), though this has not been definitely proved.

8 ADDITION TO CARBON–OXYGEN DOUBLE BONDS

BOTH aldehydes and ketones exhibit dipole moments due to the oxygen atom of their carbonyl groups being more electronegative than the carbon; not only is there a $C \rightarrow O$ inductive effect in the σ bond joining the two atoms, the more readily polarisable electrons of the π bond are also affected (*cf.* p. 21) and the real structure of the carbonyl group is probably best represented by something like:

$$\text{>C⇌O} \longleftrightarrow \overset{\oplus}{\text{>C}} \rightarrow \overset{\ominus}{\text{O}} \quad \text{i.e.} \quad \text{C} \rightarrow \text{O}$$

By analogy with $>C=C<$, we should expect $>C=O$ to undergo addition reactions, but whereas polar attack on the former is initiated almost exclusively by electrophiles (p. 156), attack on the latter could clearly be initiated *either* by nucleophilic attack of $Y^\ominus$ or Y: on carbon *or* by electrophilic attack of $X^\oplus$ or X on oxygen. In practice, initial electrophilic attack on oxygen is found to be of little significance except in the case of proton (and Lewis acids), when rapid, reversible protonation is often a prelude to slower, rate-determining attack of the nucleophile on carbon to complete the addition.

Such protonation would clearly enhance the susceptibility of the carbonyl carbon atom to nucleophilic attack

$$\text{>C=O} \underset{\overset{H^\oplus}{\rightleftarrows}}{} \overset{\oplus}{\text{>C}}\text{—OH}$$

and it might therefore be expected that carbonyl addition reactions would be powerfully catalysed by the addition of acid. However, many effective nucleophiles, e.g. $^\ominus CN$, are the anions of weak or very weak acids and if the reaction solution is made too acid their dissociation is suppressed, leading to a very marked drop in the concentration of nucleophile available for attacking the carbonyl carbon, e.g. $^\ominus CN \rightarrow HCN$. Where the nucleophile is not an anion, e.g. $R\overset{..}{N}H_2$, the situation is much the same for any quantity of acid will serve to convert it to the non-nucleophilic species, $\overset{\oplus}{R}NH_3$. In practice it is found that the addition of weak nucleophiles may well require the assistance of a little acid-catalysis in order to take place at all, while powerful nucleophiles are not normally in need of such aid. Apart from actual protonation, the positive character of the carbonyl carbon atom will also be enhanced, albeit to a smaller extent, by formation of a hydrogen-bonded complex between an acid and the carbonyl oxygen atom, and even by hydrogen-bonding of the latter with a hydroxylic solvent:

$$\overset{\delta+}{\underset{}{>}}C\!=\!\!=\!\!\overset{\delta-}{O}\text{---HA} \qquad \overset{\delta+}{\underset{}{>}}C\!=\!\overset{\delta-}{O}\text{···}_{H-O}\!\!\diagup^{R}$$

STRUCTURE AND REACTIVITY

The more positive the carbon atom of a carbonyl group is, the more readily it might be expected to react with a particular nucleophile, its attachment to electron-donating groups would therefore be expected to reduce its reactivity as is observed in the following sequence:

$$\underset{H-\overset{\overset{\displaystyle O}{\|}}{C}-H}{} > R\!\rightarrow\!\overset{\overset{\displaystyle O}{\|}}{C}-H > R\!\rightarrow\!\overset{\overset{\displaystyle O}{\|}}{C}\!\leftarrow\!R \gg$$

$$R\!\rightarrow\!\overset{\overset{\displaystyle O}{\|}}{C}\!\rightarrow\!\ddot{O}R' > R\!\rightarrow\!\overset{\overset{\displaystyle O}{\|}}{C}\!\rightarrow\!\ddot{N}H_2 > R\!\rightarrow\!\overset{\overset{\displaystyle O}{\|}}{C}\!\leftarrow\!\overset{\ominus}{O}$$

The electron-withdrawing inductive effect of oxygen and nitrogen in esters and amides, respectively, is more than outweighed by the tendency of the unshared electron pairs on these atoms to interact with the π orbital of the carbonyl group. Reactivity of the carbonyl carbon atom is also reduced by its attachment to an aromatic nucleus for again interaction takes place, this time with the π orbital system of the nucleus

and the stabilisation arising from this conjugation will be lost on bonding to a nucleophile. Thus benzaldehyde is found to be less

179

reactive than aliphatic aldehydes; this differential is found to be increased by the introduction into the benzene nucleus of electron-donating substituents (e.g. **OH**), and decreased by electron-withdrawing substituents (e.g. **NO₂**). A similar effect is observed with aliphatic aldehydes:

$$O_2N \rightarrow CH_2 \rightarrow \overset{\overset{\displaystyle O}{\|}}{C} - H > Cl \rightarrow CH_2 \rightarrow \overset{\overset{\displaystyle O}{\|}}{C} - H >$$

$$CH_3 \rightarrow \overset{\overset{\displaystyle O}{\|}}{C} - H > CH_3 \rightarrow CH_2 \rightarrow \overset{\overset{\displaystyle O}{\|}}{C} - H$$

Steric effects might also be expected to play a part in the rate of nucleophilic attack on the carbonyl carbon atom for the more crowded the transition state the higher its energy, and the more loath it will be to form. This factor, as well as the electronic ones discussed above, is involved in the lower reactivity towards nucleophiles of aromatic aldehydes compared with aliphatic, and of ketones compared with aldehydes; it is clearly operative also in the series below:

$$Me - \overset{\overset{\displaystyle O}{\|}}{C} - Me > Me - \overset{\overset{\displaystyle O}{\|}}{C} - CMe_3 > Me_3C - \overset{\overset{\displaystyle O}{\|}}{C} - CMe_3$$

In the conversion of a carbonyl compound to its simple addition product, the carbonyl carbon atom changes its hybridisation from sp^2 (planar) to sp^3 (tetrahedral), hence steric effects will be particularly important in influencing the position of the equilibrium

Carbonyl compound + nucleophile ⇌ addition product

where a reaction does not proceed substantially to completion; for any sign of crowding in the starting material is likely to be greatly enhanced in the addition product. This is clearly observed in, for example, cyanohydrin formation (p. 184) where the equilibrium constant for product formation with **MeCHO** is too large to measure, with **PhCHO** is 210, with **MeCOCH₂Me** is 38, and with **PhCOMe** is 0·8.

Variations in the structure of the carbonyl compound, operating through either steric or electronic effects, will generally affect both rate and position of equilibrium in a similar manner; for the transition state for simple addition reactions probably resembles the addition product a good deal more closely than it does the initial carbonyl compound. It is often difficult to separate the operation of steric and of electronic effects, and this is particularly the case with reactions in which the initial nucleophilic addition to carbonyl is followed by an elimination to yield the final product, e.g. oxime formation (p. 186).

Comparison of the relative reactivities of simple aldehydes may be further complicated by the fact that some of them are partially converted to hydrates in aqueous solution (p. 181), or to hemi-acetals in **ROH** (p. 183), the actual concentration of free carbonyl compound varying with the conditions. For a given carbonyl compound, the position of equilibrium with respect to the addition product is controlled, under comparable conditions, by the nature of the nucleophile: the stronger the bond that can be formed to the carbonyl carbon atom, the more the equilibrium is driven over towards products. Thus we observe increasing equilibrium concentrations of product in the series:

$$\textbf{ROH} < \textbf{RNH}_2 < {}^{\ominus}\textbf{CN}$$

A group of characteristic addition reactions will now be studied in more detail.

ADDITION REACTIONS

(i) Hydration

Many carbonyl compounds form hydrates in solution:

$$\underset{\substack{\|\\ \text{R—C—H}}}{\text{O}} + \text{H}_2\text{O} \ \rightleftarrows \ \underset{\substack{|\\ \text{OH}}}{\overset{\substack{\text{OH}\\|}}{\text{R—C—H}}}$$

Thus it has been shown that the percentage hydration at 20° of formaldehyde, acetaldehyde and acetone is 99·99, 58 and $\approx 0\%$, respectively. The latter is confirmed by the fact that if acetone is dissolved in $\text{H}_2{}^{18}\text{O}$, when the following equilibrium could, theoretically be set up,

$$\text{Me}_2\text{C}{=}\text{O} + \text{H}_2\overset{18}{\text{O}} \ \underset{\longleftarrow}{\overset{\longrightarrow}{\quad}} \ \text{Me}_2\text{C} \underset{{}_{18}\text{OH}}{\overset{\text{OH}}{\big\langle}} \ \underset{\longleftarrow}{\overset{\longrightarrow}{\quad}} \ \text{Me}_2\overset{18}{\text{C}}{=}\text{O} + \text{H}_2\text{O}$$

7

no ^{18}O is incorporated into the acetone. In the presence of a trace of acid or base, however, while no equilibrium concentration of the hydrate can be detected, the incorporation of ^{18}O occurs too rapidly to measure, indicating that a hydrate must now, transiently be formed. The acid or base catalysis is presumably proceeding:

$$Me_2\overset{\delta+}{C}{=}\overset{\delta-}{O}\cdots HA \quad \underset{}{\overset{H_2O}{\rightleftharpoons}} \quad Me_2C\overset{OH}{\underset{\underset{H}{OH}}{\overset{\oplus}{<}}}$$

$$HA \Big\updownarrow \qquad\qquad\qquad\qquad \Big\updownarrow -H^{\oplus}$$

$$Me_2C{=}O \quad \underset{\ominus OH}{\rightleftharpoons} \quad Me_2C\overset{O^{\ominus}}{\underset{OH}{<}} \quad \underset{H_2O}{\rightleftharpoons} \quad Me_2C\overset{OH}{\underset{OH}{<}}$$

The acid catalysis exhibited in this case is *general acid catalysis*, that is to say the hydration is catalysed by any acid species present in the aqueous solution and not solely by $H_3O^{\oplus}$ as is so often the case.

The fact that such catalysis is necessary with acetone, but not with the aldehydes, reflects the less positive nature of the carbonyl carbon atom of the ketone, which necessitates initial attack of $^{\ominus}OH$ (or by $H^{\oplus}$ on oxygen), whereas with the aldehydes H_2O: will attack the more positive carbon atom directly. Thus **MeCHO** hydrates at **pH 7** but, hardly surprisingly in the light of the above, it hydrates a great deal more rapidly at either **pH 4 or 11**.

The presence of electron-withdrawing substituents in the alkyl groups makes hydration easier and stabilises the hydrate once formed; thus glyoxal (I), chloral (II) and triketohydrindene (III) all form isolable, crystalline hydrates:

$$\underset{(I)}{\overset{O\quad O}{\underset{}{H{-}C{\leftarrow}C{-}H}}} \quad \overset{H_2O}{\longrightarrow} \quad \underset{OH}{\overset{O\quad OH}{H{-}C{-}C{-}H}}$$

$$\underset{\underset{Cl}{(II)}}{\overset{Cl\quad O}{Cl{\leftarrow}C{\ll}C{-}H}} \quad \longrightarrow \quad \underset{Cl\quad OH}{\overset{Cl\quad OH}{Cl{-}C{-}C{-}H}}$$

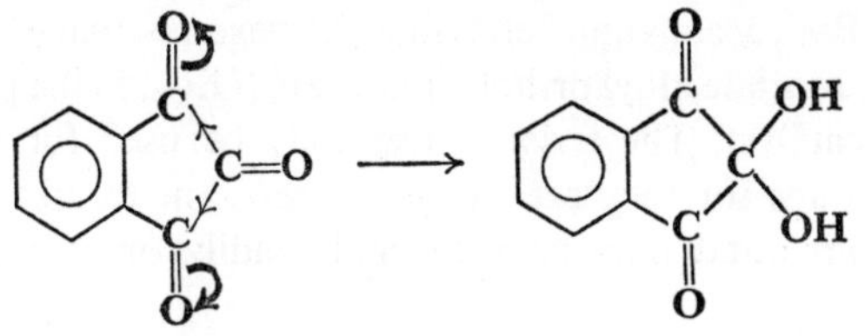

(III)   (IV)

(IV) is ninhydrin, the well-known colour reagent for the detection and estimation of α-amino acids. The hydrates are probably further stabilised by hydrogen bonding between the hydroxyl groups and the electronegative oxygen or chlorine atoms attached to the adjacent α-carbon:

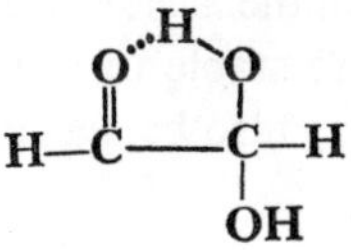

(ii) ROH

Aldehydes with alcohols, in the presence of dry hydrogen chloride, yield acetals:

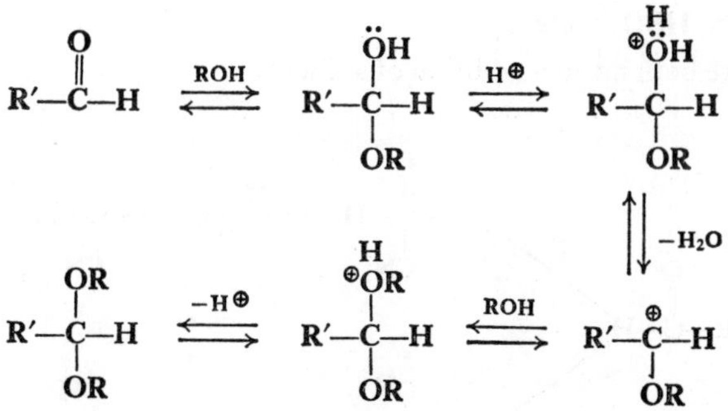

An equilibrium with the hemi-acetal is often set up on dissolving aliphatic aldehydes in alcohols (*cf.* hydration in water, p. 181), and in exceptional cases, e.g. with Br_3CCHO and **EtOH**, the hemi-acetal may actually be isolated; conversion to the acetal proper does not take place in the absence of acid catalysts, however.

With ketones the carbonyl carbon atom is not sufficiently positive to undergo initial attack by **ROH** and ketals cannot readily be made

183

in this way. Both acetals and ketals may, however, be made by reaction with the appropriate alkyl orthoformate, $HC(OR)_3$, in the presence of NH_4Cl as catalyst. These derivatives may be used for protecting carbonyl groups for they are extremely resistant to alkali, but the carbonyl compound may be recovered readily on treatment with dilute acid.

(iii) RSH

Mercaptans will react with aldehydes *and* ketones to yield thioacetals, $R'CH(SR)_2$, and thioketals $R'_2C(SR)_2$, respectively. The successful attack on the carbonyl carbon atom of ketones indicates the greater tendency of **RSH** than **ROH** to form an effective nucleophile, $RS^\ominus$, i.e. the greater acidity of thiols than the corresponding alcohols. These derivatives offer, with the acetals, differential protection of the carbonyl group for they are stable to acid but readily decomposed by $HgCl_2/CdCO_3$. They may also be decomposed by Raney nickel

$$R'_2C{=}O \;\rightarrow\; R'_2C(SR)_2 \xrightarrow{\ Ni/H_2\ } R'_2CH_2$$

the overall reaction offering a preparative method of value for the reduction of $-CHO \rightarrow -CH_3$ and ${>}CO \rightarrow {>}CH_2$.

(iv) $^\ominus CN$, $HSO_3{}^\ominus$, etc.

These are both normal addition of anions:

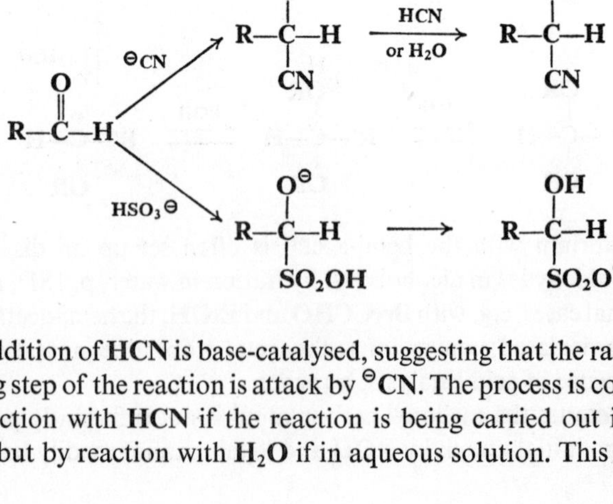

The addition of **HCN** is base-catalysed, suggesting that the rate-determining step of the reaction is attack by $^\ominus CN$. The process is completed by reaction with **HCN** if the reaction is being carried out in liquid **HCN** but by reaction with H_2O if in aqueous solution. This reaction has provided a great deal of the kinetic data on the addition of anions to carbonyl compounds, while the addition of bisulphite has afforded much evidence on the relative steric effect, on such addition, of groups attached to the carbonyl carbon atom. There is evidence that the effective attacking agent in the formation of bisulphite derivatives is actually the more powerfully nucleophilic $SO_3^{2\ominus}$ even under conditions in which its concentration relative to $HSO_3^{\ominus}$ is very small. As is expected, the relative ease of addition, and stability of the derivative once formed, is considerably less with ketones than with aldehydes. The structure of the bisulphite addition compounds was long a matter of dispute but their sulphonic acid character was ultimately confirmed by investigation of their Raman spectra.

Halide ion will add to a $\rangle C{=}O$ group in the presence of acid, but the equilibrium is so readily reversible that the resultant 1,1-halohydrin cannot be isolated. If the reaction is carried out in alcohol, however, the α-halogenoether so produced may be isolated provided the solution is first neutralised:

α-Chloromethyl ether

(v) Amine derivatives

Reaction with NH_3, RNH_2 or, in actual practice, $HONH_2$, $NH_2CONHNH_2$, $PhNHNH_2$ or $2,4\text{-}(NO_2)_2C_6H_3NHNH_2$ is the normal method by which liquid aldehydes and ketones are characterised. If a number of these reactions are followed spectroscopically it is often found that the $\rangle C{=}O$ absorption disappears very rapidly, and may even have gone completely before any $\rangle C{=}N\langle$ absorption, characteristic of the product, is observable: clearly an intermediate is being formed. On the basis of this and other evidence, the formation of oximes, semicarbazones and probably phenylhydrazones is believed to involve initial rapid attack of $R\ddot{N}H_2$ on the carbonyl compound to

185

form the adduct (V), followed by rate-determining, acid-catalysed
dehydration of the latter to yield the product (VI):

$$R-\overset{\displaystyle H}{\underset{\displaystyle H}{N}}\!:\!\overset{\displaystyle O}{\underset{}{C}}- \longrightarrow R-\overset{\displaystyle H}{\underset{\displaystyle H}{\overset{\oplus}{N}}}-\overset{\ominus O}{\underset{}{C}}- \xrightarrow[\text{Fast}]{} R-\overset{\displaystyle HO}{\underset{\displaystyle H}{N}}-\overset{}{\underset{}{C}}-$$

(V)

$$R-N{=}C\!\!<\, + H_2O$$

(VI)

If the acidity of the solution is increased, however, the rate of dehydration is naturally accelerated and the initial formation of (V) is
slowed owing to increasing conversion of the reactive nucleophile
$R\ddot{N}H_2$ into its unreactive conjugate acid, $R\overset{\oplus}{N}H_3$; initial attack of the
nucleophile on the carbonyl compound may then become the rate-
determining step of the overall reaction. The fact that oxime formation may also be catalysed by bases at higher **pH** is due to the
dehydration step being subject to base—as well as acid—catalysis:

$$>\!C{=}O \xrightarrow{:NH_2OH} >\!\overset{\displaystyle OH}{\underset{\displaystyle NOH}{\underset{\displaystyle H}{C}}} \xrightarrow{\ominus OH} >\!\overset{\displaystyle \overset{OH}{C}}{\underset{\displaystyle \underset{\ominus}{NOH}}{C}} \longrightarrow >\!C{=}NOH$$

With ammonia some few aldehydes (e.g. chloral) yield the simple
aldehyde ammonia, $R'CH(OH)NH_2$, but these derivatives more often
react further to yield polymeric products. With primary amines, the
derivatives obtained from both aldehydes and ketones eliminate
water spontaneously, as above, to yield the Schiff base, e.g.
$R'CH{=}NR$ (VI).

(vi) Hydride ions

(a) **LiAlH$_4$ reductions:** Here the complex hydride ion $AlH_4^{\ominus}$, in its
capacity as a nucleophile, is acting as a donor of **H**, with its electron
pair, to the carbonyl carbon atom:

$$R_2C\overset{\curvearrowleft}{=}O \quad \longrightarrow \quad R_2C\overset{\ominus}{-}OAlH_3 \quad \xrightarrow{3R_2C=O}$$
$$\overset{\curvearrowleft}{H}\overset{\ominus}{AlH_3} \qquad\qquad\quad H$$

$$\left(\begin{matrix} R_2C-O \\ | \\ H \end{matrix}\right)_4 Al \quad \xrightarrow{H_2O} \quad \begin{matrix} R_2C-OH \\ | \\ H \end{matrix}$$

With esters, the initial reaction is a nucleophilic 'displacement' (actually addition/elimination), followed by reduction as above:

$$\underset{H\overset{\ominus}{AlH_3}}{R-\overset{\overset{O}{\|}}{C}-OR'} \longrightarrow R-\overset{\overset{\ominus O}{|}}{\underset{H}{C}}-OR' \longrightarrow \underset{\underset{+\,{}^{\ominus}OR'}{H}}{R-\overset{\overset{O}{\|}}{C}} \xrightarrow[\text{(2) }H_2O]{\text{(1) }AlH_4{}^{\ominus}} R-\overset{\overset{OH}{|}}{\underset{H}{C}}-H$$

A similar reduction takes place with amides ($RCONH^{\ominus}$ being obtained by a preliminary removal of proton by $AlH_4{}^{\ominus}$, i.e. an 'active hydrogen' reaction) via an addition/elimination stage,

$$R-\overset{\overset{O}{\|}}{C}-\overset{\ominus}{N}H \xrightarrow{AlH_4{}^{\ominus}} R-\overset{\curvearrowleft}{\underset{H}{C}}\overset{\ominus O}{\underset{}{}}\overset{\curvearrowright}{N}H \longrightarrow O^{2\ominus}+R-CH=NH$$
$$\qquad\qquad\qquad\qquad (VII)$$
$$R-CH_2-NH_2 \xleftarrow{H_2O} R-CH_2-\overset{\ominus}{N}H \quad\Big\downarrow AlH_4{}^{\ominus}$$

the Schiff base being obtained as it is easier to eliminate $O^{2\ominus}$ than $HN^{2\ominus}$ from (VII). LiAlH$_4$ may obviously not be employed in hydroxylic solvents ('active hydrogen' reaction) or in those that are readily reduced and ether or tetrahydrofuran $(CH_2)_4O$ is, therefore, commonly used. NaBH$_4$ may be used in water or alcohol but is, not surprisingly, a less reactive reagent and will not reduce amides.

(b) **Meerwein-Ponndorf reduction:** This is essentially the reduction of ketones to secondary alcohols with aluminium isopropoxide in isopropanol solution (see p. 188).

Hydride ion, $H^{\ominus}$, is transferred from aluminium isopropoxide to the ketone (VIII) via a cyclic transition state and an equilibrium thereby set up between this pair on the one hand and the mixed alkoxide (IX) plus acetone on the other. That there is indeed such a specific transfer of hydrogen may be demonstrated by using $(Me_2CDO)_3Al$ when deuterium becomes incorporated in the α-position of the resultant carbinol, $RCD(OH)R'$.

187

$$(Me_2CHO)_2Al \overset{O}{\underset{O}{\overset{CMe_2}{\underset{CR_2}{\rlap{\big\rfloor}\,\rlap{\big\lfloor}\,H}}}} \rightleftharpoons (Me_2CHO)_2Al \overset{O=CMe_2 \quad H}{\underset{O}{CR_2}}$$

$$\text{(VIII)} \qquad\qquad \text{(IX)}$$

$$\Big\updownarrow \; \overset{\text{xs.}}{Me_2CHOH}$$

$$(Me_2CHO)_2Al \overset{O=CMe_2 \quad H}{\underset{O}{CMe_2}} + R_2CHOH$$

$$\text{(X)}$$

Acetone is the lowest boiling species in the system, so by distilling the mixture the equilibrium is displaced to the right, the secondary alcohol (X) being freed from the alkoxide (IX) by the excess isopropanol present. Because the establishment of this equilibrium is the crucial stage, the reaction is, naturally, very highly specific in its action and $>C=C<$, $-C\equiv C-$, NO_2, etc., undergo no reduction. The reaction may be reversed, $RCH(OH)R' \rightarrow RCOR'$, by use of aluminium *t*-butoxide and a large excess of acetone to displace the equilibrium to the left.

(c) **Cannizzaro reaction:** The disproportionation of aldehydes lacking any α-hydrogen atoms (i.e. **PhCHO, CH₂O** and **R₃CCHO**) to acid anion and primary alcohol in the presence of concentrated alkali, is also a hydride transfer reaction. In its simplest form the reaction rate $\propto [PhCHO]^2[^{\ominus}OH]$ and the reaction is believed to follow the course:

$$\overset{\ominus}{O}H \atop Ph-\overset{O}{\underset{}{C}}-H \;\rightleftharpoons\; Ph-\overset{OH}{\underset{O^{\ominus}}{C}}-H \quad \overset{O}{\underset{H}{C}}-Ph \;\longrightarrow\; Ph-\overset{OH}{\underset{O}{C}} + H-\overset{O^{\ominus}}{\underset{H}{C}}-Ph$$

$$\text{(XI)}$$

$$\downarrow$$

$$Ph-\overset{O^{\ominus}}{\underset{O}{C}} + H-\overset{HO}{\underset{H}{C}}-Ph$$

$$\text{(XII)}$$

Rapid, reversible addition of $^{\ominus}$OH to one molecule of aldehyde results in transfer of hydride ion to a second; this is almost certainly the rate-determining step of the reaction. The acid and alkoxide ion (XI) so obtained then become involved in a proton exchange to yield the more stable pair, alcohol and acid anion (XII), the latter, unlike the alkoxide ion, being able to stabilise itself by delocalisation of its charge. That the migrating hydride ion is transferred directly from one molecule of aldehyde to another and does not actually become free in the solution is shown by carrying out the reaction in **D$_2$O**, when no deuterium becomes attached to carbon in the alcohol as it would have done if the migrating hydride ion had become free and so able to equilibrate with the solvent. In some cases, e.g. with formaldehyde in very high concentrations of alkali, a fourth-order reaction takes place: rate $\propto$ **[CH$_2$O]^2[$^{\ominus}$OH]^2**. This is believed to involve formation of a doubly charged anion and transfer of hydride ion by this to a second molecule of aldehyde to yield carboxylate and alkoxide ions

Intramolecular Cannizzaro reactions are also known, e.g. gly-oxal → hydroxyacetate (glycollate) anion:

As expected

$$\text{Rate} \propto \text{[OHCCHO][}^{\ominus}\text{OH]}$$

and no deuterium attached to carbon is incorporated in the glycollate produced on carrying out the reaction in **D$_2$O**.

(vii) Electrons from dissolving metals

(a) **Magnesium or sodium and ketones:** Magnesium, usually in the form of an amalgam to increase its reactivity, will donate one electron each to two molecules of a ketone to yield an adduct (XIII). This contains two unpaired electrons which can then unite to form a carbon–carbon bond yielding the magnesium salt (XIV) of a pinacol; subsequent acidification yields the free pinacol (XV):

$$
\begin{array}{l}
R_2C{=}O \\
\qquad :Mg \longrightarrow \\
R_2C{=}O
\end{array}
\qquad
\begin{array}{l}
R_2\overset{\bullet}{C}{-}O \\
\qquad\qquad\diagdown Mg \longrightarrow \\
R_2\overset{\bullet}{C}{-}O
\end{array}
\qquad
\begin{array}{l}
R_2C{-}O \\
\quad|\quad\diagdown Mg \xrightarrow{\ H^{\oplus}\ } \\
R_2C{-}O
\end{array}
\qquad
\begin{array}{l}
R_2C{-}OH \\
\quad| \\
R_2C{-}OH
\end{array}
$$

$$\qquad\qquad\qquad\quad (XIII)\qquad\qquad\qquad (XIV)\qquad\qquad (XV)$$

This reaction is unusual in involving initial attack on oxygen rather than carbon. Pinacol itself is $Me_2C(OH)C(OH)Me_2$, but the name has come to be used generally for such tertiary-1,2-diols. The reaction is most readily seen when sodium is dissolved, in the absence of air, in ethereal solutions of aromatic ketones, the blue, paramagnetic radical anion of the sodium ketyl, $Ar_2\overset{\bullet}{C}{-}\overset{\ominus}{O} \leftrightarrow Ar_2\overset{\ominus}{C}{-}\overset{\bullet}{O}$, then being in equilibrium with the dianion of the corresponding pinacol:

$$
\begin{array}{l}
Ar_2C{-}O^{\ominus} \\
\quad| \\
Ar_2C{-}O^{\ominus}
\end{array}
$$

In the presence of proton donors such as **EtOH**, $H^{\oplus}$ becomes attached to the negative carbon atom of the ketyl, a further electron is donated to its oxygen atom by the sodium, and the resultant alkoxide ion finally picks up a proton from ethanol to yield Ar_2CHOH, i.e. overall reduction to the secondary alcohol takes place.

(b) **Sodium and esters:** Sodium will donate an electron to an ester to yield the radical anion (XVI), two molecules of which unite (XVII) (*cf.* pinacol formation above) and expel $EtO^{\ominus}$ to yield the α-diketone (XVIII). Further electron donation by sodium yields the diradical anion (XIX), which again forms a carbon–carbon bond (XX). Acidification yields the α-hydroxyketone or *acyloin* (XXI):

$$2R-\overset{O}{\underset{\|}{C}}-OEt \xrightarrow{2Na\cdot} \begin{array}{c} R-\overset{O^{\ominus}}{\underset{|}{C}}-OEt \\ R-\overset{|}{\underset{|}{C}}-OEt \\ O^{\ominus} \end{array} \longrightarrow \begin{array}{c} R-\overset{O^{\ominus}}{\underset{|}{C}}-OEt \\ R-\overset{|}{\underset{|}{C}}-OEt \\ O^{\ominus} \end{array} \longrightarrow \begin{array}{c} R-\overset{O}{\underset{\|}{C}} \\ R-\overset{}{\underset{\|}{C}} \\ O \end{array}$$

(XVI)   (XVII)   (XVIII)

$$\Big\downarrow 2Na\cdot$$

$$\begin{array}{c} R-\overset{O}{\underset{\|}{C}} \\ R-CH \\ OH \end{array} \rightleftharpoons \begin{array}{c} R-\overset{OH}{\underset{|}{C}} \\ R-\overset{|}{\underset{|}{C}} \\ OH \end{array} \xleftarrow{H^{\oplus}} \begin{array}{c} R-\overset{O^{\ominus}}{\underset{\|}{C}} \\ R-\overset{}{\underset{\|}{C}} \\ O^{\ominus} \end{array} \longleftarrow \begin{array}{c} R-\overset{O^{\ominus}}{\underset{|}{C}}\cdot \\ R-\overset{|}{\underset{|}{C}}\cdot \\ O^{\ominus} \end{array}$$

(XXI)   (XX)   (XIX)

This acyloin condensation is much used for the ring-closure of long chain dicarboxylic esters, $EtO_2C(CH_2)_nCO_2Et$, in the synthesis, in high yield, of large cyclic hydroxyketones.

A larger proportion of sodium in the presence of some **EtOH** as proton donor results in the reaction following a different course. It has been suggested that the larger proportion of sodium donates *two* electrons to the ester, so forming the dianion (XXII) which abstracts a proton from **EtOH** to yield (XXIII); the latter then expels $EtO^{\ominus}$ to form the aldehyde (XXIV). Repetition of the sequence yields the alkoxide (XXV), which is converted on acidification into the primary alcohol (XXVI):

$$R-\overset{O}{\underset{\|}{C}}-OEt \xrightarrow{2Na\cdot} R-\overset{O^{\ominus}}{\underset{\ominus}{C}}-OEt \xrightarrow{EtOH} R-\overset{O^{\ominus}}{\underset{H}{\overset{|}{C}}}-OEt \longrightarrow R-\overset{O}{\underset{H}{\overset{\|}{C}}}$$

(XXII)   (XXIII)   (XXIV)

$$\Big\downarrow 2Na\cdot$$

$$R-\overset{OH}{\underset{H}{\overset{|}{C}}}-H \xleftarrow{H^{\oplus}} R-\overset{O^{\ominus}}{\underset{H}{\overset{|}{C}}}-H \xleftarrow{EtOH} R-\overset{O^{\ominus}}{\underset{H}{\overset{|}{C}}}^{\ominus}$$

(XXVI)   (XXV)

This is the Bouveault-Blanc reduction of esters, but has now been largely displaced by $LiAlH_4$ (p. 187).

When, in addition, the Claisen ester condensation is considered below (p. 198), something of the complexity of the products that may result from the reaction of sodium on esters will be realised!

(viii) Carbanions and negative carbon

The importance of these reactions resides in the fact that carbon–carbon bonds are formed; many of them are thus of great synthetic importance.

(a) **Grignard reagents:** The actual structure of Grignard reagents themselves is still a matter of some dispute. Phenyl magnesium bromide has, however, been isolated in crystalline form as the compound $C_6H_5MgBr \cdot 2Et_2O$, in which C_6H_5, **Br** and the two molecules of ether are arranged tetrahedrally about the magnesium atom. In solution the actual structure of the Grignard reagent appears to vary with the solvent and there is evidence of the occurrence, under differing conditions, of all the species:

$$2R^{\ominus} MgHal^{\oplus}$$

$$2RMgHal \rightleftharpoons 2R \cdot \cdot MgHal$$

$$R_2Mg \cdot MgHal_2 \rightleftharpoons R_2Mg + MgHal_2$$

Thus free radical reactions with them are known, but in most of their useful synthetic applications they tend to behave as though polarised in the sense $\overset{\delta- \ \delta+}{RMgHal}$, i.e. as sources of negative carbon if not necessarily of carbanions as such.

In reactions with carbonyl groups it appears that *two* molecules of Grignard reagent are involved in the actual addition. One molecule acts as a Lewis acid with the oxygen atom of the carbonyl group, thus enhancing the positive nature of the carbonyl carbon atom, and so promotes attack on it by the **R** group of a second molecule of Grignard reagent via a cyclic transition state:

If such a course is followed it might be expected that Grignard reagents of suitable structure, i.e. those having hydrogen atoms on a β-carbon, might undergo conversion to alkenes as a side-reaction, transfer of hydride ion to the positive carbon atom of the carbonyl group taking place:

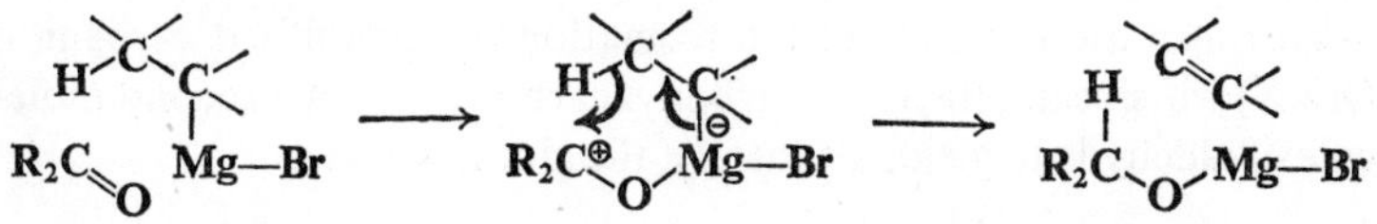

This is, indeed, observed in practice; the ketone is in part reduced to the secondary alcohol in the process, via a cyclic transition state closely resembling that already encountered in the Meerwein-Ponndorf reaction (p. 187).

If Grignard additions do proceed via initial attack of one molecule of the reagent on the carbonyl oxygen, it might be expected that the reaction would be promoted if a more effective Lewis acid were introduced into the solution, for this would co-ordinate preferentially to yield an even more positive carbonyl carbon atom. Thus introduction of $MgBr_2$ has been observed to double the yield of tertiary alcohol in the reactions of some ketones with Grignard reagents.

Grignard reagents are however being increasingly superseded by other organo-metallic compounds for preparative addition of this kind, particularly by the more reactive lithium derivatives.

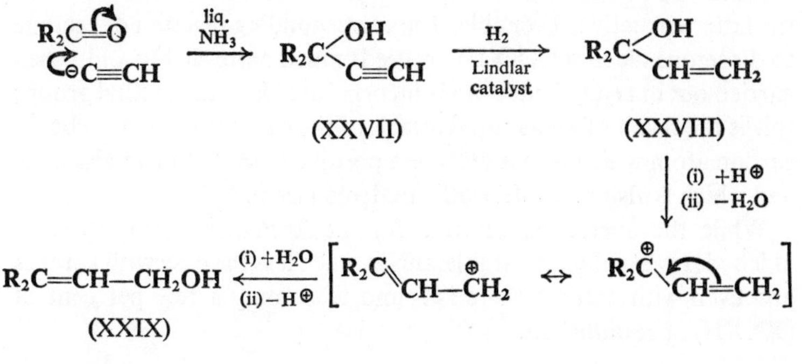

(b) **Acetylide ion:** A very useful reaction is the addition of acetylide ion to carbonyl compounds, the reaction often being carried out in liquid ammonia in the presence of sodamide to convert acetylene into its carbanion (p. 233). Hydrogenation of the resultant acetylenic carbinol (XXVII) in the presence of Lindlar catalyst (partially poisoned palladium) yields the alkene (XXVIII); the latter undergoes an acid-catalysed allylic rearrangement (p. 34) to the primary alcohol (XXIX), which, as the corresponding halide, can be made to undergo further synthetic reactions. The series constitutes a useful preparative sequence (see p. 193).

(c) **Aldol condensations:** The action of bases on an aldehyde having α-hydrogen atom results in the formation of a stabilised carbanion (XXX) which can attack the carbonyl carbon atom of a second molecule of aldehyde to yield, ultimately, the aldol (XXXI):

$$CH_3-\underset{\underset{H}{|}}{C}=O$$

$$(i) \quad \ominus OH$$

$$Me \rightarrow \underset{\underset{H}{|}}{\overset{\overset{O}{||}}{C}} \quad \left[\underset{\underset{H}{|}}{CH_2-C}=O \right] \quad \overset{(ii)}{\rightleftharpoons} \quad Me-\underset{\underset{H}{|}}{\overset{\overset{\ominus}{O}}{C}}-CH_2-\underset{\underset{H}{|}}{C}=O$$

$$\left[CH_2=\underset{\underset{H}{|}}{C}-O^{\ominus} \right] \qquad \qquad H_2O$$

$$(XXX) \qquad \qquad MeCH(OH)CH_2CHO$$

$$(XXXI)$$

The forward reaction (ii) and the reversal of (i) are essentially in competition with each other but, as carrying out the reaction in D_2O fails to result in the incorporation of any deuterium in the methyl group, (ii) must be so much more rapid than the reversal of (i) as to make the latter virtually irreversible. The corresponding reaction of acetone to diacetone alcohol (XXXII) proceeds much more slowly and, when carried out in D_2O, deuterium is incorporated into the methyl group; this is the result of a less rapid attack of the carbanion on a carbonyl carbon atom which is markedly less positive than that in an aldehyde and which is also more sterically hindered (see p. 195).

While the overall equilibrium for the formation of aldols from aldehydes is usually favourable, this is much less the case with ketones and even with acetone there is found to be only a few per cent of (XXXII) at equilibrium.

$$CH_3-\overset{\overset{\displaystyle}{|}}{\underset{Me}{C}}=O$$

(i) $\ominus OH$

$$Me\rightarrow\overset{\overset{\displaystyle O}{||}}{C}\left[\begin{array}{c}\overset{\ominus}{C}H_2-\overset{\overset{\displaystyle}{|}}{\underset{Me}{C}}=O\\[2mm]\updownarrow\\[2mm]CH_2=\overset{\overset{\displaystyle}{|}}{\underset{Me}{C}}-O^{\ominus}\end{array}\right]\overset{(ii)}{\rightleftharpoons}Me-\overset{\overset{\displaystyle \ominus O}{|}}{\underset{Me}{C}}-CH_2-\overset{\overset{\displaystyle}{|}}{\underset{Me}{C}}=O$$

$\underset{Me}{\overset{\uparrow}{|}}$

$\Big\updownarrow H_2O$

$Me_2C(OH)CH_2COMe$

(XXXII)

Aldehydes having no α-hydrogen atoms cannot form carbanions and they, therefore, undergo the Cannizzaro reaction (p. 188) with concentrated alkali; but as this reaction is slow, such aldehydes are often able to function as carbanion acceptors. Thus formaldehyde, in excess, reacts with acetaldehyde:

$$\begin{array}{c}CH_2=O\\\overset{\ominus}{C}H_2CHO\end{array}\rightleftharpoons\begin{array}{c}CH_2O^{\ominus}\\CH_2CHO\end{array}\overset{H_2O}{\rightleftharpoons}\begin{array}{c}CH_2OH\\CH_2CHO\end{array}$$

$\Big\updownarrow \ominus OH$

$$HOCH_2-\overset{\overset{\displaystyle CH_2OH}{|}}{\underset{CH_2OH}{C}}CHO \overset{\text{(i) }\ominus OH}{\underset{\substack{\text{(ii) } CH_2O\\\text{(iii) } H_2O}}{\rightleftharpoons}}\begin{array}{c}CH_2OH\\CHCHO\\CH_2OH\end{array}\overset{\text{(i) } CH_2O}{\underset{\text{(ii) } H_2O}{\rightleftharpoons}}\begin{array}{c}CH_2OH\\CHCHO\\\ominus\end{array}$$

$\ominus OH \Big| /CH_2O$

$$HCO_2^{\ominus}+HOCH_2-\overset{\overset{\displaystyle CH_2OH}{|}}{\underset{CH_2OH}{C}}-CH_2OH$$

(XXXIII)

In the last stage **(HOCH$_2$)$_3$CCHO**, which can no longer form a carbanion, undergoes a cross Cannizzaro reaction with formaldehyde to yield pentaerythritol **(XXXIII)** and formate anion. The reaction proceeds this way rather than to yield $(HOCH_2)_3CCO_2^{\ominus}$ and CH_3OH, as the carbonyl carbon atom of formaldehyde is the more positive of the two aldehydes so that it is attacked preferentially by $^{\ominus}OH$, with resultant transfer of $H^{\ominus}$ to $(HOCH_2)_3CCHO$ rather than the other way round.

A further useful synthetic reaction is the base-catalysed addition of aliphatic nitro-compounds to carbonyl groups:

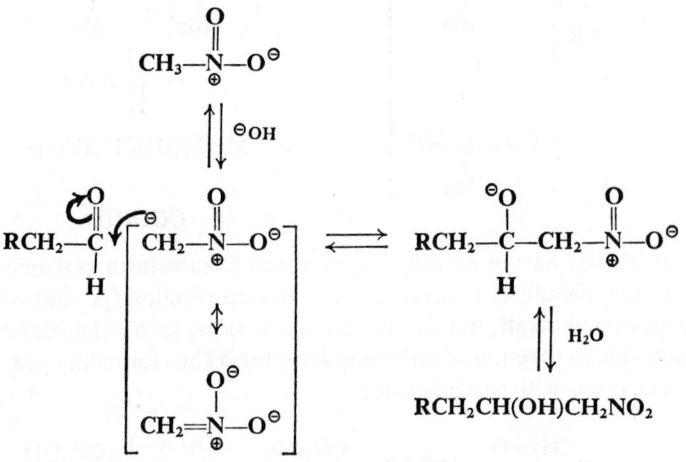

Here the aldehyde itself can also form a carbanion and aldol formation could be a competing reaction, but the carbanion from the nitro-compound tends to be the more stable (due to the more effective delocalisation of its charge) and is thus formed more readily, resulting in the preponderance of the above reaction.

The elimination of water from a hydroxy compound usually requires acid-catalysis (p. 214) but the possibility of carbanion formation in the first formed aldol, coupled with the presence of a group in the adjacent β-position that can be readily expelled as an anion, results in an easy 'attack from the back' by an electron pair:

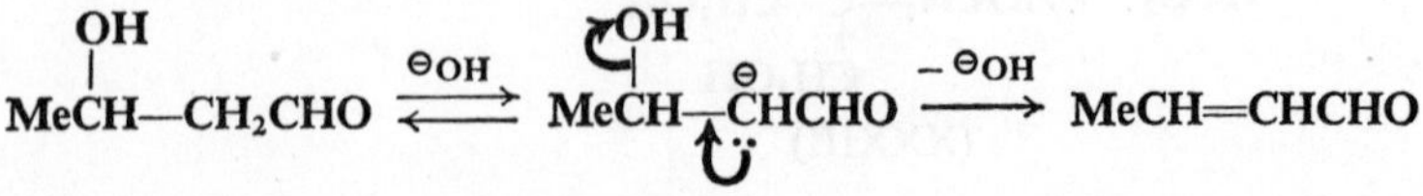

Thus the dehydration of aldols is subject to base-catalysis and carbanion additions are often followed by elimination of water resulting in an overall condensation reaction. The successive additions of carbanion, followed by elimination of water induced by strong base, result in the formation of low molecular weight polymers from simple aliphatic aldehydes; if the process is to be halted at the simple aldol, a weak base such as K_2CO_3 is used. A preparative use of carbanion addition followed by elimination is seen in the Claisen-Schmidt condensation of aromatic aldehydes with aliphatic aldehydes or ketones in the presence of 10% aqueous **KOH**:

$$\underset{\text{Ph--C--H}}{\overset{\text{O}}{\|}} \quad \begin{array}{l} + \quad \overset{\ominus}{\text{CH}_2\text{COMe}} \quad \longrightarrow \quad \text{PhCH=CHCOMe} \\[2mm] + \quad \overset{\ominus}{\text{CH}_2\text{CHO}} \quad \longrightarrow \quad \text{PhCH=CHCHO} \end{array}$$

With aliphatic aldehydes, self-condensation can, of course, constitute an important side reaction. The presence of electron-donating groups in the aromatic nucleus will reduce the positive nature of the carbonyl carbon atom, and $p\text{-MeOC}_6\text{H}_4\text{CHO}$ is found to react at only about one-seventh the rate of benzaldehyde.

(d) Perkin reaction: Closely related to the above is the Perkin reaction for the synthesis of $\alpha\beta$-unsaturated acids from aromatic aldehydes and aliphatic acid anhydrides in the presence of an alkali metal salt of the corresponding acid, e.g. for cinnamic acid, $\text{PhCH=CHCO}_2\text{H}$:

$$\underset{\substack{\text{H} \\ \text{(XXXIV)}}}{\overset{\text{O}}{\underset{}{\text{Ph--C}}}} \overset{\ominus}{\underset{\underset{\text{MeCO}}{\diagdown\text{O}}}{\text{CH}_2\text{CO}}} \xrightarrow[\text{(ii) Protonation}]{\text{(i) Addition}} \underset{\substack{\text{H} \\ \text{(XXXV)}}}{\overset{\text{OH}}{\underset{}{\text{Ph--C--CH}_2\text{CO}}}} \underset{\text{MeCO}}{\diagdown\text{O}}$$

$$\Big\downarrow -\text{H}_2\text{O}$$

$$\begin{array}{c} \text{Ph--CH=CHCO}_2\text{H} \\ + \\ \text{MeCO}_2\text{H} \end{array} \xleftarrow{\text{H}_2\text{O}} \underset{\substack{\text{MeCO} \\ \text{(XXXVI)}}}{\text{Ph--CH=CHCO}} \diagdown\text{O}$$

In acetic anhydride solution, acetate ion is a sufficiently strong base to remove a proton from the activated α-position of the anhydride to yield the carbanion (XXXIV), which adds to the carbonyl group of the aldehyde, the product, after protonation, being the aldol-like species (XXXV). Under the conditions of the reaction (*ca.* 140°), (XXXV) undergoes dehydration in the presence of acetic anhydride and the resultant mixed anhydride (XXXVI), on being poured into water at the end of the reaction, is hydrolysed to cinnamic and acetic acids. The reaction is a general one, depending only on the presence of a CH_2 group in the α-position of the anhydride. That the reaction follows the above course is confirmed by the fact that aromatic aldehydes will, in the presence of suitable basic catalysts, react with anhydrides but not with the corresponding acid anions, and aldol-like intermediates such as (XXXV) have in some cases been isolated.

(e) **Claisen ester condensation:** This too is effectively an aldol type reaction, e.g. with ethyl acetate:

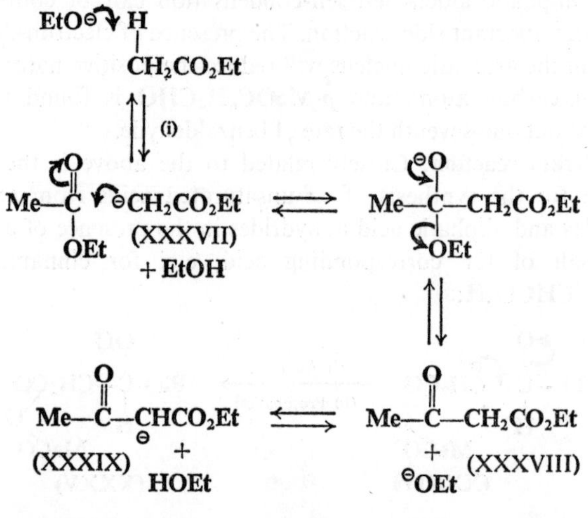

Normally a mole of sodium is employed as a source of the sodium ethoxide catalyst required, but only a little ethanol need be added initially as more is liberated as soon as the reaction starts, with the formation of the carbanion (XXXVII). This adds to the carbonyl group of a second molecule of ester, followed by the expulsion of $EtO^\ominus$ to yield the β-keto ester (XXXVIII), which is finally converted into the carbanion (XXXIX), i.e. the 'sodio-derivative' (hence the need for the employment of a mole of sodium). The formation of (XXXIX) is an essential feature of the reaction for it helps to drive the equilibrium (i) over to the right; this is made necessary by the fact that the carbanion (XXXVII) is not as highly stabilised as, for example

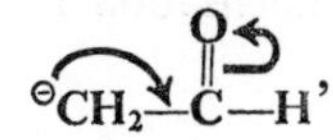

$\overset{\ominus}{CH_2}\!-\!\overset{\overset{\displaystyle O}{\|}}{C}\!-\!H$, and is consequently more reluctant to form. This is

reflected in the fact that R_2CHCO_2Et does not undergo the reaction in the presence of $EtO^{\ominus}$ despite the fact that it has an α-hydrogen atom and so could form a carbanion, because the product $R_2CHCOCR_2CO_2Et$ cannot be converted to a carbanion such as (XXXIX) and so fails to drive the equilibrium over to the right.

Such esters can however be made to condense satisfactorily in the presence of very strong bases such as $Ph_3C^{\ominus}Na^{\oplus}$, for here the initial carbanion formation is essentially irreversible:

$$R_2CHCO_2Et + Ph_3C^{\ominus} \rightarrow R_2\overset{\ominus}{C}CO_2Et + Ph_3CH$$

Crossed condensation of two different esters is not always practicable because of the formation of mixed products, but it can be of synthetic value, particularly where one of the esters used is incapable of forming a carbanion, e.g. $(CO_2Et)_2$, HCO_2Et and $PhCO_2Et$. The complexity of the alternative reactions that can take place from the action of sodium on esters has already been referred to (p. 192).

Where the two ester groups are part of the same molecule and cyclisation can therefore result, the condensation is known as a Dieckmann reaction:

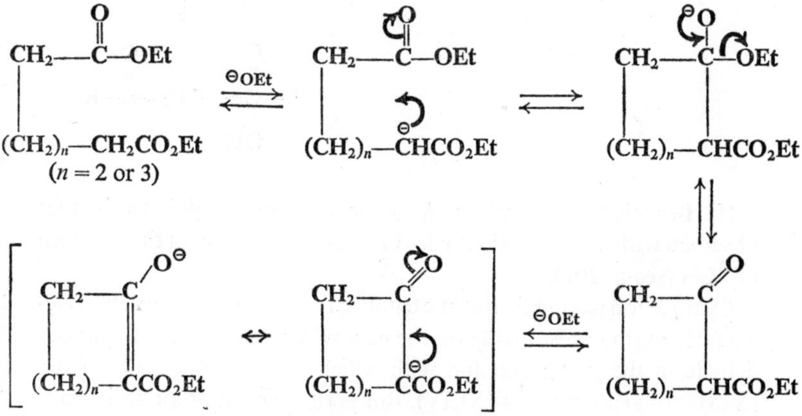

Here as in the simple Claisen ester condensation it is necessary, if ethoxide is used as the catalyst, to be able to form the anion of the final β-keto ester in order to drive the overall reaction in the desired direction.

Like all aldol-type reactions, the Claisen ester condensation is reversible

$$\text{Me}-\overset{\overset{\text{O}}{\|}}{\text{C}}-\text{CH}_2\text{CO}_2\text{Et} + {}^{\ominus}\text{OEt} \;\rightleftharpoons\; \text{Me}-\overset{\overset{\text{O}\ominus}{|}}{\underset{\underset{\text{OEt}}{|}}{\text{C}}}{-}\text{CH}_2\text{CO}_2\text{Et}$$

$$\text{Me}-\overset{\overset{\text{O}}{\|}}{\underset{\underset{\text{OEt}}{|}}{\text{C}}} + {}^{\ominus}\text{CH}_2\text{CO}_2\text{Et}$$

i.e. the so-called 'acid decomposition' of β-keto esters. An exactly analogous fission of β-diketones also takes place:

$$\text{R}-\overset{\overset{\text{O}}{\|}}{\text{C}}-\text{CH}_2-\overset{\overset{\text{O}}{\|}}{\text{C}}-\text{R} + {}^{\ominus}\text{OEt} \;\rightleftharpoons\; \text{R}-\overset{\overset{\text{O}\ominus}{|}}{\underset{\underset{\text{OEt}}{|}}{\text{C}}}{-}\text{CH}_2-\overset{\overset{\text{O}}{\|}}{\text{C}}-\text{R}$$

$$\text{R}-\overset{\overset{\text{O}}{\|}}{\underset{\underset{\text{OEt}}{|}}{\text{C}}} + {}^{\ominus}\text{CH}_2-\overset{\overset{\text{O}}{\|}}{\text{C}}-\text{R}$$

(f) Benzoin condensation: Another carbanion addition is that observed with aromatic aldehydes in alcoholic solution in the presence of ${}^{\ominus}$CN (see p. 201).

Cyanide ion is a highly specific catalyst for this reaction, its effectiveness depending presumably on the ease with which it adds to benzaldehyde in the first place and with which it is finally expelled from (XLIII) to yield benzoin (XLIV). But perhaps most of all, it depends on its electron-withdrawing power which promotes the ready release, as proton, of the hydrogen atom attached to carbon in (XL) to yield the carbanion (XLI), rather than the transfer of this hydrogen with its electron pair to a second molecule of benzaldehyde as in the Cannizzaro reaction (p. 188). The observed kinetics of the reaction

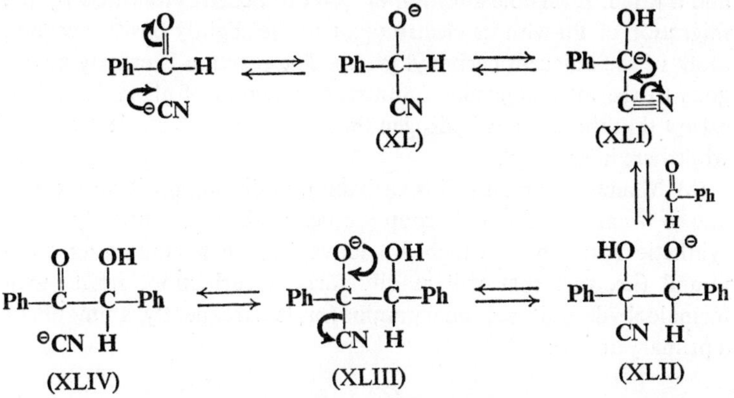

$$\text{Rate} \propto [\text{PhCHO}]^2[^{\ominus}\text{CN}]$$

support the above formulation and the rate-determining step is believed to be the reaction of the carbanion (XLI) with a second molecule of aldehyde to yield (XLII).

(g) **Benzilic acid change:** An interesting intramolecular reaction, which can be looked upon as essentially a carbanion addition, is the base-catalysed conversion of benzil to benzilic acid anion, $\text{PhCOCOPh} \rightarrow \text{Ph}_2\text{C(OH)CO}_2^{\ominus}$:

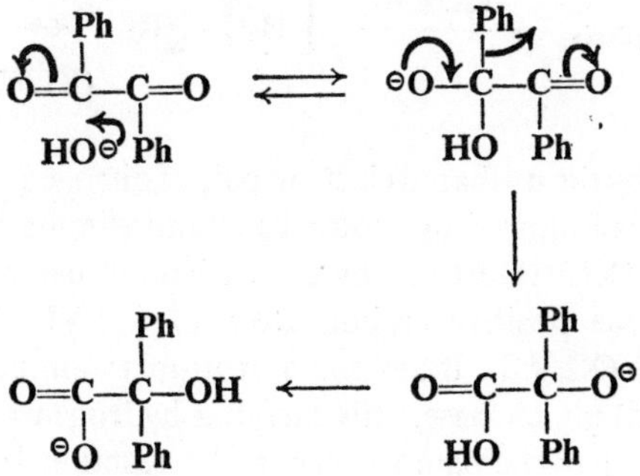

It is found that

$$\text{Rate} \propto [\text{PhCOCOPh}][^{\ominus}\text{OH}]$$

201

and a rapid, reversible addition of $^\ominus OH$ to benzil is followed by the migration of **Ph** with its electron pair to the slightly positive carbon atom of the adjacent carbonyl group. The reaction is exactly analogous to the intramolecular Cannizzaro reaction of glyoxal (p. 189) except that there it was hydrogen that migrated with its electron pair while here it is phenyl.

(h) **Mannich reaction:** This carbanion addition, albeit an indirect one so far as the carbonyl group is concerned, is an extremely useful synthetic reaction in which an active hydrogen containing compound (i.e. one that will readily form a carbanion) reacts with formaldehyde and a secondary amine (or, less frequently, ammonia or a primary amine:

$$\text{>C—H} + CH_2O + HNR_2 \longrightarrow \text{>C—CH}_2\text{—NR}_2 + H_2O$$

$$\text{Rate} \propto [\text{>C—H}]\,[CH_2O]\,[R_2NH]$$

The reactions to form the so-called Mannich bases (XLVII) are believed to proceed as follows, e.g. with **PhCOMe**:

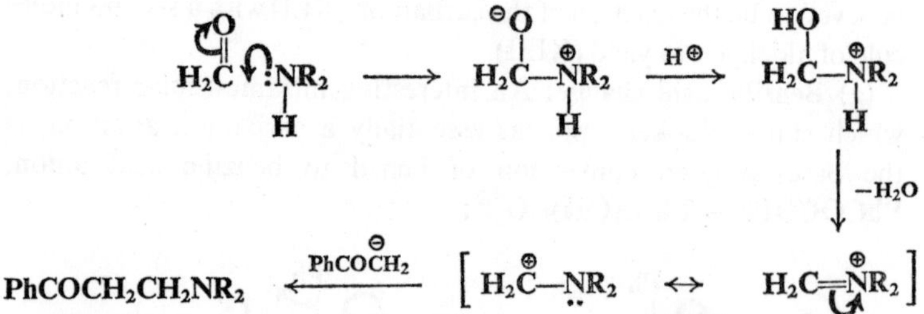

Initial attack by the unshared electron pair of nitrogen on the carbonyl carbon atom is followed by protonation and elimination of water to yield the ion (XLVI). Attack by the carbanion derived from acetophenone on the positive carbon atom of (XLVI) then yields the Mannich base (XLVII). If ammonia or primary amines are used, the first formed Mannich base, still carrying hydrogen on the nitrogen atom, can itself participate further in the reaction leading to more complex products, hence the preference for secondary amines.

The Mannich base formed will readily eliminate R_2NH, hence the synthetic usefulness of the reaction:

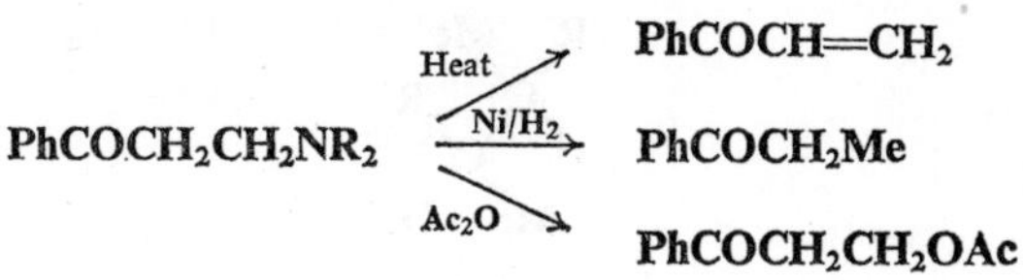
Heat
Ni/H2
Ac2O
PhCOCH2CH2NR2
PhCOCH=CH2
PhCOCH2Me
PhCOCH2CH2OAc

STEREOCHEMISTRY OF ADDITION TO CARBONYL COMPOUNDS

Whether the stereochemical mode of addition to a carbon–oxygen double bond is *cis* or *trans* clearly has no meaning for, unlike a carbon–carbon double bond (p. 157), different products will not be obtained by the two mechanisms because of free rotation about the **C—O** single bond that results:

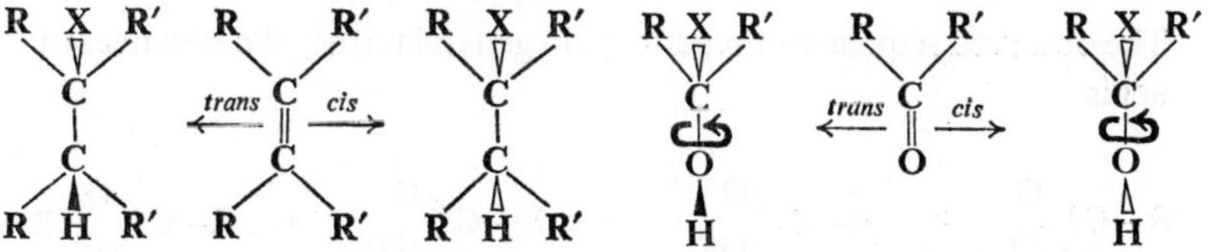
trans
cis
trans
cis

A new asymmetric centre has been introduced but, as always a racemate will be produced. If, however, an asymmetric centre is already present, e.g. **RCHMeCOMe**, and if the addition is carried out on one of the pure optical isomers, the addition is taking place in an asymmetric environment and different quantities of the two possible products are often formed. This is due to the preferential formation of that isomer whose production involves a transition state in which steric interaction is at a minimum. Thus where **R** is a large group and the reaction is, for example, addition of a Grignard reagent, the initial attack of the reagent, as a Lewis acid on oxygen, will yield a complex (XLVIII) in which the now complexed oxygen atom will be as far away from the bulky **R** group as possible. As the nucleophile now attacks the carbon atom of the carbonyl complex it will tend to move in preferentially from the side on which its approach is hindered only by hydrogen rather than by the bulkier methyl group:

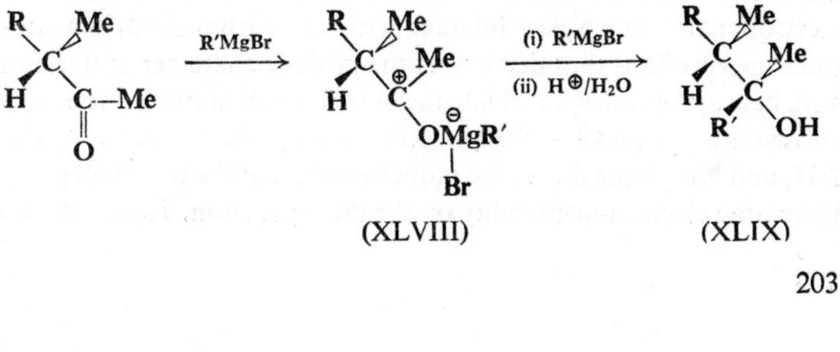
R'MgBr
(i) R'MgBr
(ii) H⊕/H2O

(XLVIII) (XLIX)

$$
\begin{array}{c}
R \quad Me \\
\diagdown C \diagup R' \\
H \diagdown C \diagup \\
Me \quad OH \\
(L)
\end{array}
$$

The overall result being the formation of (XLIX) rather than (L) as the major product. The above argument is essentially the working rule enunciated by Cram which has been found to forecast accurately the major product from a large number of such addition reactions.

NUCLEOPHILIC ATTACK ON CARBOXYLIC ACID DERIVATIVES

The observed sequence of reactivity, in general terms, of derivatives of acids

$$
R-C\diagup_{Cl}^{O} \;>\; R-C\diagup_{OEt}^{O} \;>\; R-C\diagup_{NH_2}^{O} \;>\; R-C\diagup_{N\diagdown R}^{O}\diagdown R
$$

is in accord with the view that their characteristic reactions, e.g. alkaline hydrolysis, can be looked upon as nucleophilic addition followed by elimination:

$$
R-\underset{OH}{\overset{O}{C}}-Y \;\longrightarrow\; R-\underset{OH}{\overset{O}{C}}-Y \;\longrightarrow\; R-\underset{OH}{\overset{O}{C}}+{}^{\ominus}Y
$$

Though it should be said that the difference between an addition/ elimination and a direct displacement reaction may be hardly apparent if the elimination follows sufficiently rapidly on the initial addition.

The observed reactivity sequence is due to the fact that although chlorine, oxygen and nitrogen exert an electron-withdrawing inductive effect on the carbonyl carbon atom, they all have unshared electron pairs which can interact with the carbonyl carbon atom (mesomeric effect) thus decreasing the positive character of this atom and, hence, the ease with which nucleophiles will attack it. This effect increases as we go $Cl \to OEt \to NH_2 \to NR_2$, the difference between NH_2 and NR_2 being due to the inductive effect of the two alkyl groups increasing electron-availability on the nitrogen atom. There may also

204

be a slight fall in the reactivity of any one derivative as the **R** group of the acid is changed from methyl to an alkyl-substituted methyl group as its slightly greater inductive effect also reduces the positive nature of the carbonyl carbon atom.

The reactivity sequence is well illustrated by the fact that acid chlorides react readily with alcohols and amines to yield esters and amides, respectively, while esters react with amines to give amides, but the simple reversal of any of these reactions on an amide though possible is usually very difficult. These reactions, involving as they do the conversion of the sp^2 hybridised carbonyl carbon atom in the starting material into the sp^3 state in the intermediate, are markedly susceptible to steric hindrance and derivatives of tertiary acids, e.g. R_3CCO_2Et, are found to be highly resistant to nucleophilic attack.

(i) Base induced reactions

The example that has been the subject of most investigation is almost certainly the alkaline hydrolysis of esters. This has been shown to be a second-order reaction and, by the use of ^{18}O labelling, it is seen to involve *acyl-oxygen* cleavage (*cf.* p. 47) in most cases

$$\underset{RC-OR'}{\overset{O}{\|}} \quad \overset{\ominus OH}{\longrightarrow} \quad \underset{RC-O^\ominus}{\overset{O}{\|}} + H-\overset{18}{O}R'$$

the labelled oxygen appearing in the alcohol but not in the acid anion from the hydrolysis. The reaction is believed to proceed:

$$\underset{\underset{\ominus OH}{|}}{\overset{\curvearrowleft O}{R-C-OEt}} \rightleftharpoons \underset{\underset{OH}{|}}{\overset{\ominus O}{R-C-OEt}} \rightleftharpoons \underset{\underset{OH}{|}}{\overset{O}{\|}}{R-C} + {}^\ominus OEt \longrightarrow \underset{\underset{O^\ominus}{|}}{\overset{O}{\|}}{R-C} + HOEt$$

The rate-determining step is almost certainly the initial attack of $^\ominus OH$ on the ester and the overall reaction is irreversible due to the insusceptibility of $RCO_2{}^\ominus$ to attack by EtOH or $EtO^\ominus$. The alkaline hydrolysis of amides, **RCONHR'**, follows a very similar course in which it is $R\overset{\ominus}{N}H$ that undergoes expulsion. The action of $^\ominus OR$ in place of $^\ominus OH$ on an ester results in *transesterification* to yield RCO_2R' and the action of amines on esters to form amides also follows an essentially similar course:

205

$$R-\overset{\overset{\displaystyle O}{\parallel}}{\underset{\underset{\displaystyle :NH_2R}{}}{C}}-OEt \rightleftharpoons R-\overset{\overset{\displaystyle \ominus O}{|}}{\underset{\underset{\displaystyle \oplus NH_2R}{}}{C}}-OEt \overset{-H\oplus}{\rightleftharpoons} R-\overset{\overset{\displaystyle \ominus O}{|}}{\underset{\underset{\displaystyle NHR}{}}{C}}-OEt \rightleftharpoons R-\overset{\overset{\displaystyle O}{\parallel}}{\underset{\underset{\displaystyle NHR}{}}{C}}+\ominus OEt$$

It has been shown that the conjugate base of the amine, $R\overset{\ominus}{N}H$ does not play any significant part in the amide formation. Both trans-esterification and amide formation from the ester are reversible, unlike alkaline ester hydrolysis, as the carboxylate anion is not involved.

The reactions of acid chlorides show a number of resemblances to the nucleophilic displacement reactions of alkyl halides, proceeding by uni- and bi-molecular mechanisms, the actual path followed being markedly affected by the polarity and ion-solvating ability of the medium (*cf*. p. 75) as well as by the structure of the substrate. The reactions of acid anhydrides are in many ways intermediate between those of acyl halides, in which the group that is ultimately expelled shows a considerable readiness to be lost as an anion, and esters in which the leaving group normally requires assistance for its ultimate displacement.

(ii) Acid catalysed reactions

Esters also undergo acidic hydrolysis, initial protonation being followed by nucleophilic attack by H_2O:; acyl-oxygen cleavage is again observed:

$$R-\overset{\overset{\displaystyle O}{\parallel}}{C}-OEt \overset{H\oplus}{\rightleftharpoons} R-\overset{\overset{\displaystyle OH}{|}}{\underset{\underset{\displaystyle H_2O:}{}}{C^\oplus}}-OEt \rightleftharpoons R-\overset{\overset{\displaystyle OH}{|}}{\underset{\underset{\displaystyle H_2\overset{\oplus}{O}}{}}{C}}-OEt$$

$$R-\overset{\overset{\displaystyle O}{\parallel}}{\underset{\underset{\displaystyle OH}{}}{C}} \overset{-H\oplus}{\rightleftharpoons} R-\overset{\overset{\displaystyle O-H}{|}}{\underset{\underset{\displaystyle HO}{}}{C^\oplus}} \overset{-EtOH}{\longleftarrow} R-\overset{\overset{\displaystyle OH}{|}}{\underset{\underset{\displaystyle HO}{}}{\underset{}{C}}}\overset{\oplus}{\underset{H}{O}}Et$$

Unlike alkaline hydrolysis, the overall reaction is experimentally reversible and esters are commonly made by protonation of the carboxylic acid followed by nucleophilic attack of **ROH**, an excess of the latter normally being employed so as to displace the equilibrium in the desired direction. Esters, **RCO₂R**, also undergo ester exchange with **R'OH** under these conditions. Esters of tertiary alcohols, however, have been shown by ^{18}O labelling experiments to undergo *alkyl-oxygen* cleavage on hydrolysis

$$R_3CO-\overset{\overset{\displaystyle O}{\|}}{C}-R' \underset{}{\overset{H^\oplus}{\rightleftarrows}} R_3C\overset{\overset{\displaystyle HO}{|}}{\underset{\oplus}{C}-R'} \rightleftarrows R_3C^\oplus + O{=}\overset{\overset{\displaystyle HO}{|}}{C}-R'$$

$$\Big\updownarrow H_2O$$

$$R_3COH \overset{-H^\oplus}{\rightleftarrows} R_3C-\overset{\oplus}{\underset{H}{O}}H$$

reflecting the tendency of the tertiary alkyl group to form a relatively stable carbonium ion. Similar alkyl-oxygen cleavage also tends to occur with esters of secondary alcohols that yield the most stable carbonium ions, e.g. **Ph₂CHOH**. Attempts at ester-exchange with esters of such alcohols lead not surprisingly to acid + ether rather than to the expected new ester:

$$R_3CO-\overset{\overset{\displaystyle O}{\|}}{C}-R \underset{}{\overset{H^\oplus}{\rightleftarrows}} R_3C^\oplus + O{=}\overset{\overset{\displaystyle HO}{|}}{C}-R$$

$$\Big\updownarrow R'OH$$

$$R_3COR' \overset{-H^\oplus}{\rightleftarrows} R_3\overset{\oplus}{\underset{H}{C}OR'}$$

Acid catalysed esterification or hydrolysis is found to be highly susceptible to steric hindrance, thus 2,4,6-trimethylbenzoic acid does not undergo esterification under the normal conditions (*cf.* p. 28). This is due to the fact that such relatively bulky *ortho* substituents force the initially protonated carboxyl group (LI) out of the plane of the benzene nucleus (see p. 208).

Attack by the nucleophile, **ROH**, apparently needs to occur from a direction more or less at right angles to the plane in which the protonated carboxyl group lies and such line of approach is now blocked, from either side, by a bulky methyl group: no esterification thus takes place. It is found, however, that if the acid is dissolved in concentrated

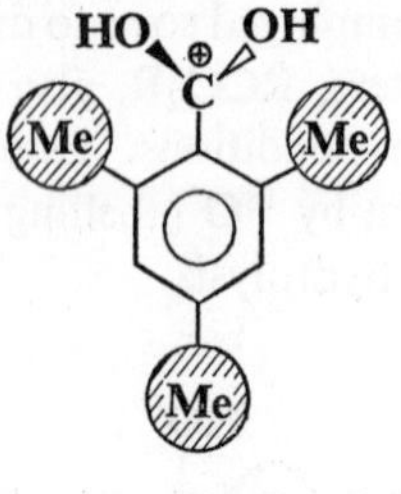

$$HO \underset{\underset{C}{\oplus}}{\overset{OH}{\diagdown\diagup}}$$

Me Me

Me

(LI)

H_2SO_4 and the resultant solution poured into cold methanol, ready esterification takes place. Similarly the methyl ester, which is highly resistant to acid hydrolysis under normal conditions, may be reconverted to the acid merely by dissolving it in concentrated H_2SO_4 and then pouring this solution into cold water.

The clue to what is taking place is provided by the fact that dissolving 2,4,6-trimethylbenzoic acid in H_2SO_4 results in a four-fold depression of the latter's freezing point due to the ionisation:

$$RCO_2H + 2H_2SO_4 \rightleftarrows R\overset{\oplus}{C}{=}O + H_3O^{\oplus} + 2HSO_4^{\ominus}$$

(LII)

The resultant acylium ion (LII) would be expected to undergo extremely ready nucleophilic attack, e.g., by **MeOH**, and as $-\overset{\oplus}{C}{=}O$ has a linear structure

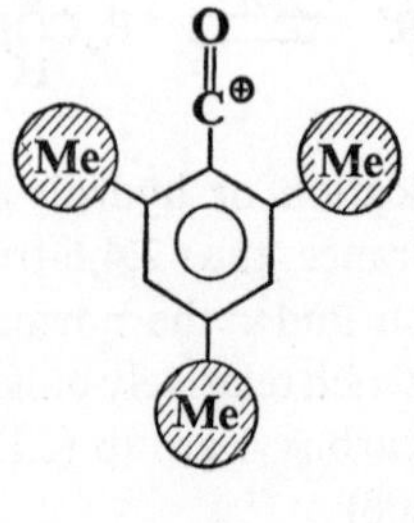

$$\overset{O}{\underset{\overset{\oplus}{C}}{\|}}$$

Me Me

Me

(LII)

attack by **MeOH** can take place at right angles to the plane of the benzene ring and is thus not impeded by the two flanking methyl groups. The same acylium ion is obtained on dissolving the methyl ester in H_2SO_4 and this undergoes equally ready attack by H_2O to yield the acid. Benzoic acid itself and its esters do not form acylium ions under these conditions, however. This is probably due to the fact that whereas protonated benzoic acid can stabilise itself by delocalisation (LIII),

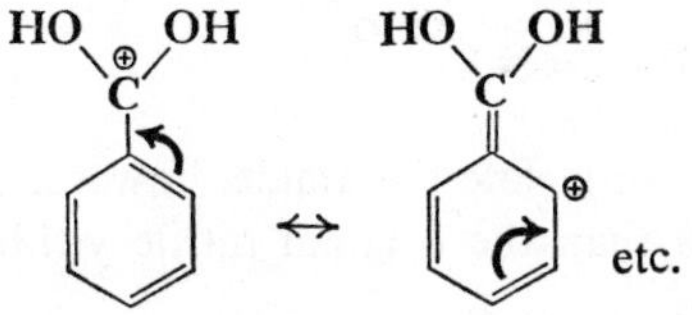

protonated 2,4,6-trimethylbenzoic acid cannot as the *o*-methyl groups prevent the atoms of the carboxyl group from lying in the same plane as the benzene ring and π orbital interaction is thus much reduced or prevented. Formation of the acylium ion (LII) removes this restriction however and the substituent methyl groups can indeed further delocalise the positive charge by hyperconjugation:

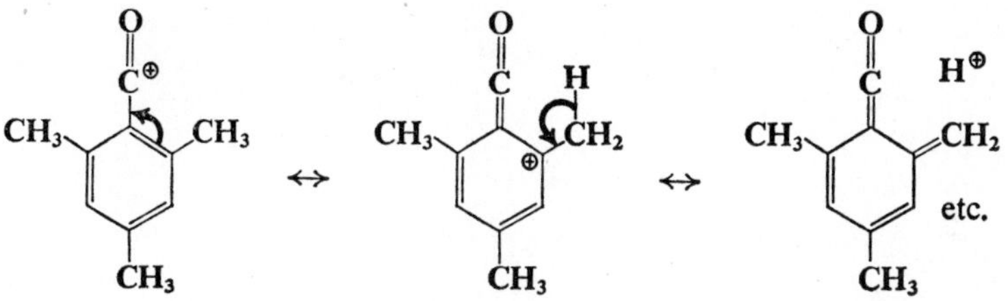

(LII)

(iii) Addition reactions of nitriles

Nitriles also undergo nucleophilic addition reactions due to:

$$-C\equiv N: \quad \leftrightarrow \quad -\overset{\oplus}{C}=\overset{\ominus}{\underset{\bullet\bullet}{N}}:$$

Thus they will undergo acid-catalysed addition of ethanol to yield salts of imino-ethers (LIV)

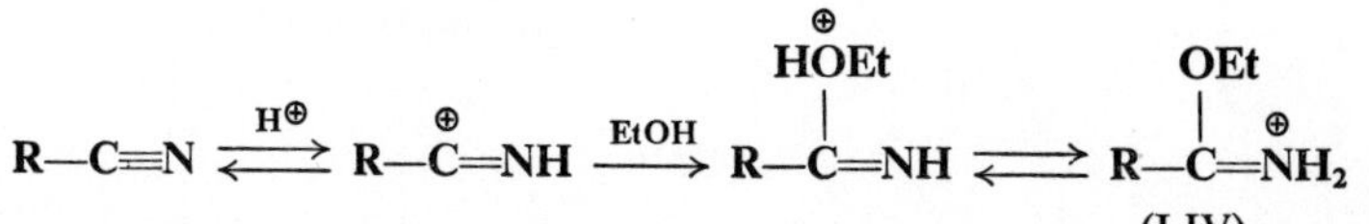

(LIV)

and also acid or base catalysed addition of water:

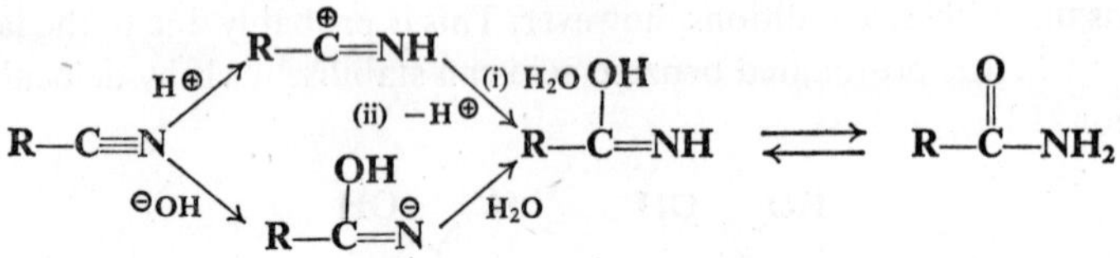

$$
R-C{\equiv}N
\begin{array}{c}
\nearrow \overset{\oplus}{R-C{=}NH} \searrow \\
H^{\oplus} \quad \quad \text{(ii)} -H^{\oplus} \\
\underset{\ominus OH}{\searrow} \quad \underset{R-\overset{|}{C}{\equiv}N^{\ominus}}{\overset{OH}{\nearrow}}
\end{array}
\begin{array}{c}
\text{(i) } H_2O \\
\xrightarrow{} \\
\xrightarrow{H_2O}
\end{array}
\quad R-\overset{OH}{\underset{|}{C}}{=}NH
\quad \rightleftharpoons \quad
R-\overset{O}{\overset{\|}{C}}-NH_2
$$

It is often difficult to isolate the amide, however, for this undergoes readier hydrolysis than the original nitrile yielding the acid or its anion.

Nitriles will, of course, also undergo addition of Grignard reagents to yield ketones and of hydrogen to yield primary amines.

9 ELIMINATION REACTIONS

ELIMINATION reactions are those in which two atoms or groups are removed from a molecule without being replaced by other atoms or groups. In the great majority of such reactions the groups are lost from adjacent carbon atoms, one of the groups eliminated very often being a proton and the other a nucleophile, $Y:$ or $Y^{\ominus}$, resulting in the formation of a multiple bond:

$$H-\overset{\beta|}{\underset{|}{C}}-\overset{|\alpha}{\underset{|}{C}}-Y \xrightarrow{-HY} \;\; \overset{}{\underset{}{>}}C=C\overset{}{\underset{}{<}} \quad \overset{H}{\underset{}{>}}C=C\overset{}{\underset{Y}{<}} \xrightarrow{-HY} \; -C\equiv C-$$

Among the most familiar examples are the base-induced elimination of hydrogen halide from alkyl halides

$$RCH_2CH_2Hal \xrightarrow{\ominus OH} RCH=CH_2 + H_2O + Hal^{\ominus}$$

the acid-catalysed dehydration of alcohols

$$RCH_2CH_2OH \xrightarrow{H^{\oplus}} RCH=CH_2 + H_3O^{\oplus}$$

and the Hofmann degradation of quaternary alkylammonium hydroxides:

$$RCH_2CH_2\overset{\oplus}{N}R_3 \xrightarrow{\ominus OH} RCH=CH_2 + H_2O + NR_3$$

β-ELIMINATION

The carbon atom from which Y is removed is generally referred to as the α-carbon and that losing a proton as the β-carbon, the overall process being designated as a 1,2- or β-elimination, though as this type of elimination reaction is by far the most common the β- is often omitted. Some α-elimination reactions are known, however, in which both groups are lost from the same carbon atom (p. 229), and elimination reactions are also known in which groups are lost from atoms

211

other than carbon: in the conversion of the acetates of aldoximes to nitriles, for example,

$$ArCH{=}NOCOMe \xrightarrow{-MeCO_2H} ArC{\equiv}N$$

or in the reversal of the addition reactions of carbonyl groups (p. 184)

$$\underset{\underset{CN}{\mid}}{\overset{\overset{H}{\mid}}{R-C-OH}} \xrightarrow{-HCN} \underset{}{\overset{\overset{H}{\mid}}{R-C{=}O}}$$

though these reactions have been studied in less detail.

Elimination reactions have been shown to take place by either a uni- or a bi-molecular mechanism, designated as E1 and E2 respectively, by analogy with the S_N1 and S_N2 mechanisms of nucleophilic substitution which they often accompany in, for example, the attack of base on an alkyl halide:

$$RCH_2CH_2Br \underset{\overset{\ominus}{OH}}{\overset{\overset{\ominus}{OH}}{\diagdown}}$$

$$RCH{=}CH_2 + H_2O + Br^{\ominus} \qquad \textit{Elimination}$$

$$RCH_2CH_2OH + Br^{\ominus} \qquad \textit{Substitution}$$

THE E1 MECHANISM

This mechanism, like the S_N1, envisages the rate of the reaction as being dependent on the substrate concentration only, the rate-determining stage involving this species alone. Thus with the halide, Me_3CBr

$$\text{Rate} \propto [Me_3CBr]$$

the reaction rate being measured is that of the formation of the transition state for which the carbonium ion (I), as a constituent of an ion pair, may be taken as a model:

$$Me_3CBr \rightarrow Me_3C^{\oplus} Br^{\ominus}$$
$$(I)$$

Rapid, non rate-determining attack by other species in the system, for example $^{\ominus}OH$ or $H_2O:$, can then take place. If these act as nucleophiles (i.e. electron pair donors towards carbon) the result is an overall substitution

$$
\begin{array}{c}
Me_3C{-}OH \\
\nearrow \, {}^{\ominus}OH \\
Me_3C^{\oplus} \\
\searrow \, H_2O \\
Me_3C{-}\overset{\oplus}{\underset{H}{O}}H \quad \longrightarrow \quad Me_3C{-}OH + H^{\oplus}
\end{array}
$$

while if they act as bases (electron pair donors towards hydrogen), the result is removal of a proton from a β-carbon atom to yield an alkene:

$$
\begin{array}{c}
CH_2{=}CMe_2 + H_2O \\
HO^{\ominus}{\downarrow}H \quad Me \quad \nearrow \\
(H_2O:) \quad \underset{Me}{CH_2{-}\overset{\oplus}{C}} \, {}^{\ominus}OH \\
\searrow \, H_2O \\
CH_2{=}CMe_2 + H_3O^{\oplus}
\end{array}
$$

Obviously conditions that promote S_N1 reactions (p. 75) will lead to E1 reactions also, for carbonium ion formation is the significant stage in both. Thus the ratio of unimolecular elimination to substitution has, in a number of cases, been shown to be fairly constant for a given alkyl group, no matter what the halogen atom or other group lost as an anion from it. This shows that E1 and S_N1 are *not* proceeding by quite separate, competing pathways and lends support to a carbonium ion as the common intermediate; for otherwise the nature of the leaving group would be expected to play a significant role leading to a change in the proportion of elimination to substitution products as it was varied. Variation of the structure of the alkyl group, however, has a considerable effect on the relative amounts of elimination and substitution that take place. It is found that branching at the β-carbon atom tends to favour E1 elimination; thus $MeCH_2CMe_2Cl$ yields only 34 % of alkene whereas Me_2CHCMe_2Cl yields 62 %. The reason for this may, in part at least, be steric: the more branched the halide, the more crowding is released when it is converted to the carbonium ion intermediate, but crowding is again introduced when the latter reacts with an entering group ($\rightarrow$ substitution); by contrast, loss of a proton (→ elimination) results, if anything, in further relief of strain and so is preferred. Study of a range of halides shows that this is not the whole of the story, however. Hyperconjugation may also play a part, as will be seen below (p. 219) in considering the preferential formation of one isomeric alkene rather than another from a carbonium ion in which there is more than one β-carbon atom which can lose a proton. The E1 mechanism is also encountered in the acid-induced dehydration of alcohols:

$$Me_3COH \xrightarrow{H^\oplus} \underset{H}{Me_3\overset{\oplus}{C}OH} \xrightarrow{-H_2O} Me_3C^\oplus \xrightarrow{-H^\oplus} Me_2C{=}CH_2$$

THE E2 MECHANISM

In the alternative E2 mechanism, the rate of elimination of, for example, hydrogen halide from an alkyl halide induced by $^\ominus OH$ is given by:

$$\text{Rate} \propto [RHal][^\ominus OH]$$

This rate law has been interpreted as involving the abstraction of a proton from the β-carbon atom by base, accompanied by a *simultaneous* loss of halide ion from the α-carbon atom:

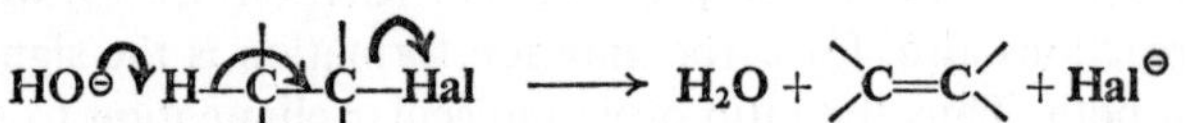

$$HO^\ominus\,H{-}C{-}C{-}Hal \longrightarrow H_2O + {>}C{=}C{<} + Hal^\ominus$$

Evidence for the involvement of C—H bond fission in the rate-limiting step of the reaction is provided by the occurrence of a primary kinetic isotope effect when H is replaced by D on the β-carbon atom.

It might be objected that there is no necessity for proton abstraction and halide elimination to be simultaneous, that initial removal of proton by base followed by the faster, non rate-determining elimination of halide ion from the resultant carbanion, *as a separate step*, would still conform to the above rate law:

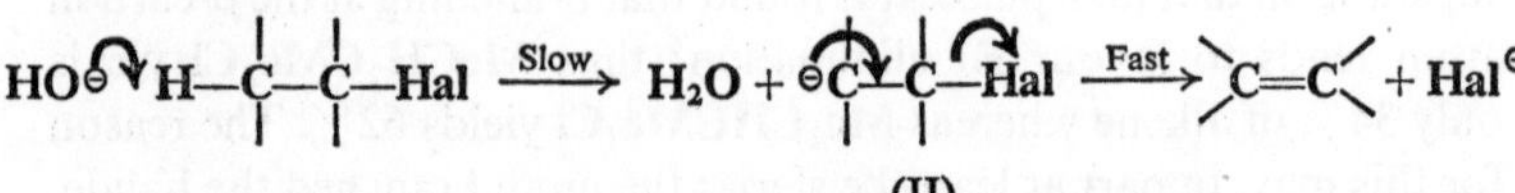

$$HO^\ominus\,H{-}C{-}C{-}Hal \xrightarrow{Slow} H_2O + {}^\ominus C{-}C{-}Hal \xrightarrow{Fast} {>}C{=}C{<} + Hal^\ominus$$

$$\text{(II)}$$

The formation of a carbanion such as (II) with so little possibility of stabilisation (*cf.* p. 234) seems inherently unlikely and evidence against the actual participation of carbanions is provided by a study of the reaction of β-phenylethyl bromide (III) with $^\ominus$OEt in EtOD. Carbanion formation would, *a priori*, be expected to be particularly easy with this halide because of the stabilisation that can occur by delocalisation of the negative charge via the π orbitals of the benzene nucleus (IIIa):

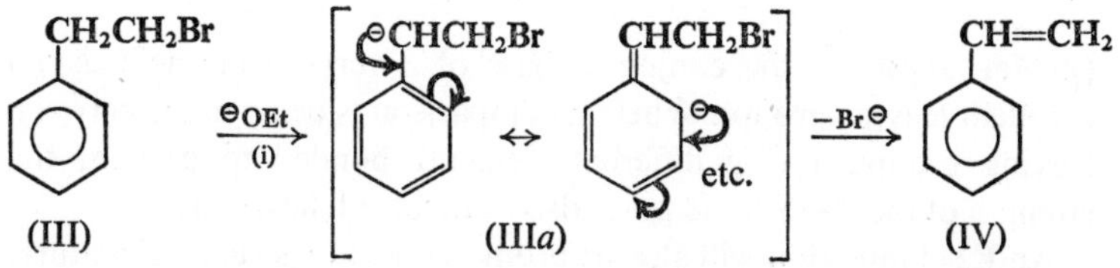

Carrying out the reaction in **EtOD** should, if such a carbanion is involved, lead to the formation of **PhCHDCH$_2$Br** by reversal of (i) and this in its turn should yield some **PhCD=CH$_2$** as well as **PhCH=CH$_2$** in the final product. In practice, if the reaction in **EtOD** is stopped short of completion, i.e. while some bromide is still left, it is found that neither this nor the styrene (IV) formed contain any deuterium. Thus a carbanion is not formed as an intermediate even in this especially favourable case, and it seems likely that in such E2 eliminations, abstraction of proton, formation of the double bond and elimination of the halide ion or other nucleophile normally occur simultaneously as a concerted process.

Though this is generally true there is, in the highly special case of the elimination reactions of trichloro- and a number of dihalo-ethylenes, some evidence that carbanions are involved:

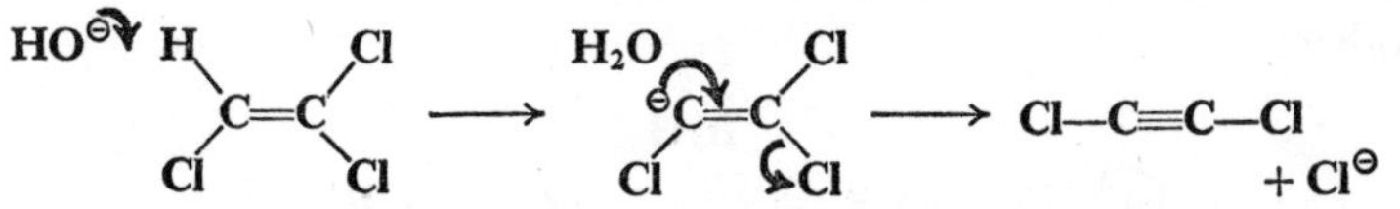

We have thus now seen elimination reactions in which the **H—C** bond is broken *before* (trichloroethylene above), *simultaneously* with (**E2**), and *after* (**E1**), the **C—Y** bond.

The rate of E2 elimination is, not surprisingly, found to be affected by the strength of the base employed; thus we find $^\ominus$**NH$_2$** > $^\ominus$**OEt** > $^\ominus$**OH**. The rate is also considerably influenced by the ability of the leaving group to depart with an electron pair: the anions (conjugate bases) of strong acids are found to be the best leaving groups. Thus alcohols do not normally (*cf.* p. 196, however) eliminate water under basic conditions ($^\ominus$**OH**, the conjugate base of a very weak acid, is a poor leaving group), whereas tosylates yield alkenes very readily

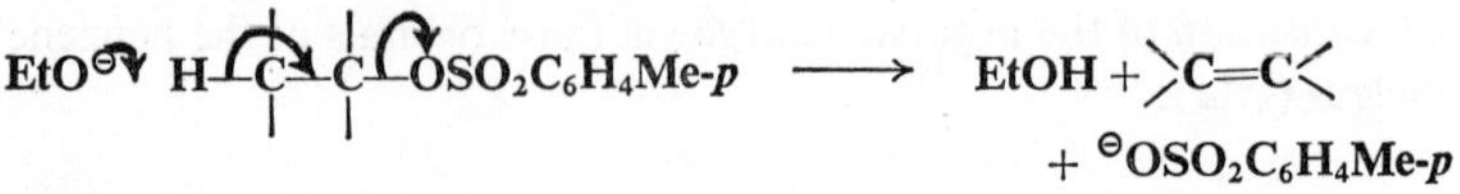

$$\text{EtO}^\ominus \curvearrowright \text{H} \overset{|}{\underset{|}{\text{C}}} \overset{|}{\underset{|}{\text{C}}} \text{OSO}_2\text{C}_6\text{H}_4\text{Me-}p \longrightarrow \text{EtOH} + {>}\text{C}{=}\text{C}{<}$$
$$+ \, ^\ominus\text{OSO}_2\text{C}_6\text{H}_4\text{Me-}p$$

(p-MeC$_6$H$_4$SO$_3^\ominus$, the conjugate base of a very strong acid, is an excellent leaving group). Where a comparison is being made between leaving groups having different elements bonded to carbon, the strength of the **C—Y** bond must also be involved, however.

An **E2** elimination will also be promoted by any structural feature, that serves, by conjugation or other means, to stabilise the resultant alkene or, more particularly, the transition state that leads to it, e.g. a phenyl group on the β-carbon atom (V):

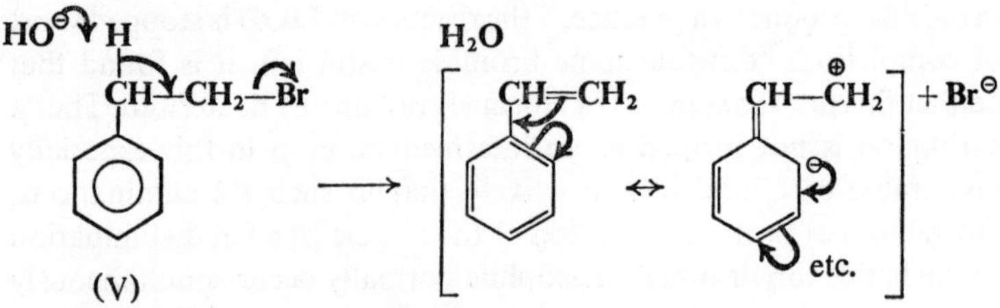

(V)

(i) Stereospecificity in E2 eliminations

It has been found that E2 elimination reactions exhibit a high degree of stereospecificity, proceeding very much more readily when the atoms or groups to be eliminated are—or can become—*trans* to each other:

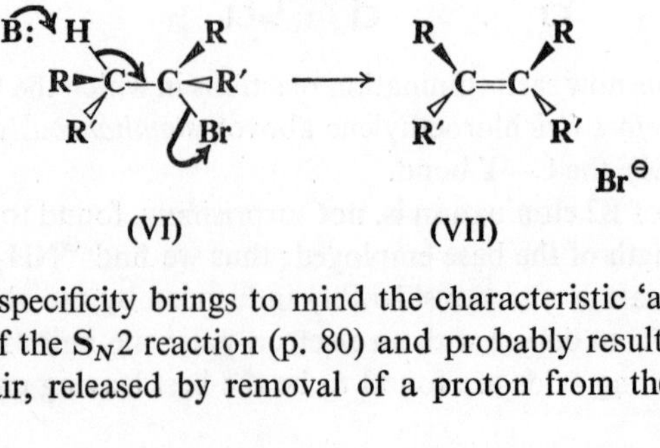

This stereospecificity brings to mind the characteristic 'attack from the back' of the S_N2 reaction (p. 80) and probably results from the electron pair, released by removal of a proton from the β-carbon atom, attacking the α-carbon atom 'from the back' with displacement of the leaving group. It has been suggested that it is necessary that the attacking atom of the base, the **H** to be eliminated, C_β, C_α and the other leaving group should all be coplanar in the transition state in an E2 elimination, but other factors may also be involved in securing stereospecific *trans* elimination.

Where restricted rotation about a bond prevents the leaving groups from getting into this preferred orientation, elimination is normally found to be very much more difficult. Thus of the stereoisomers of hexachlorocyclohexane, $C_6H_6Cl_6$, one isomer is found to eliminate **HCl** 10^4 times more slowly than any of the others, and it can be shown that this is the only one that has *no* chlorine and hydrogen atoms, on adjacent carbons, *trans* to each other. A similar situation is encountered when the groups to be eliminated are locked in position with respect to each other by a double bond; thus base-induced elimination of **HCl** to yield acetylene dicarboxylic acid (VIII) proceeds much more rapidly from chlorofumaric acid (IX) than from chloromaleic acid (X),

as does the elimination of acetic acid from *anti-* as compared with *syn*-benzaldoxime acetate (XI and XII, respectively) to yield benzonitrile (XIII):

A fact that may be made use of in assigning configurations to a pair of stereoisomeric aldoximes.

In cases such as the hexachlorocyclohexane isomer above where a stereospecific *trans* elimination cannot take place *cis* elimination can be made to occur, though normally only with considerable difficulty.

217

The reaction then probably proceeds via carbanion formation (*cf.* p. 214), a route that normally involves a considerably higher free energy of activation than that via a normal E2 'trans' transition state with consequent increase in the severity of the conditions necessary to effect it.

In compounds in which no restriction of rotation about a bond is imposed, the leaving groups will arrange themselves so as to be as far apart as possible when they are eliminated. Thus *meso* dibromostilbene (XIV) yields a *cis* unsaturated compound (XV), whereas the corresponding DL-compound (XVI) yields the *trans* form (XVII):

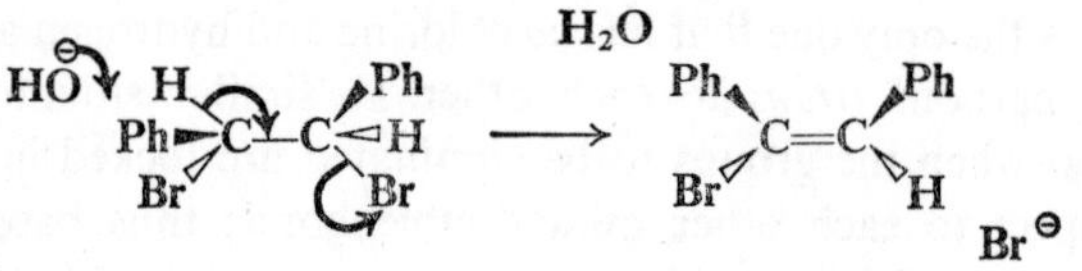

(XIV)   (XV)

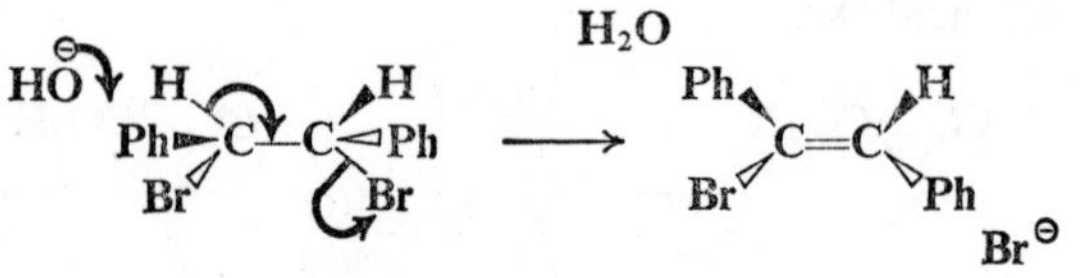

(XVI)   (XVII)

A 'ball-and-stick' or other model will be found useful for confirming the true stereochemical course of these eliminations.

(ii) Orientation in E2 eliminations: Saytzeff *v.* Hofmann

The situation frequently arises in base-induced elimination reactions of alkyl halides, **RHal**, and alkyl *onium* salts, such as $\overset{\oplus}{\mathbf{R}}\mathbf{NR'_3}$ and $\overset{\oplus}{\mathbf{R}}\mathbf{SR'_2}$, that more than one alkene can, in theory, be produced:

$RCH_2CH_2CH{=}CH_2$
(XIX)

$RCH_2CH{=}CHCH_3$
(XX)

$RCH_2CH_2{-}\underset{\underset{Y}{|}}{CH}{-}CH_3$
(XVIII)

$(Y = Hal, \overset{\oplus}{N}R'_3 \text{ or } \overset{\oplus}{S}R'_2)$

Three factors, essentially, influence the relative proportions of alkene that are actually obtained: (*a*) the relative ease with which a proton can be lost from the available, alternative β-positions, (*b*) the relative stability of the alkenes, once formed (more accurately, the relative stability of the transition states leading to them), and (*c*) steric effects (arising from substitution at the β-positions, the size of the leaving group **Y**, and the size of the base used to induce the elimination). The relative significance of, and conflict between, these factors has led in the past to the empirical recognition of two opposing modes of elimination: *Saytzeff elimination*, leading preferentially to the more stable alkene, normally the one carrying the *larger* number of alkyl groups, i.e. (XX) rather than (XIX), and *Hofmann elimination*, leading preferentially to the less stable alkene, normally the one carrying the *smaller* number of alkyl groups, i.e. (XIX) rather than (XX).

The Saytzeff mode, which is principally encountered in the elimination reactions of halides and of alcohols, is easy to justify in terms of (*b*) for the alkene carrying the larger number of alkyl groups can be shown by combustion experiments to be more stable than its less alkylated isomers, a fact that may be explained by hyperconjugation. Thus (XX) has *five* C—H linkages adjacent to the double bond compared with only *two* for (XIX) and a greater number of forms such as (XXI) can therefore contribute to its stabilisation by delocalisation (*cf.* p. 26):

$$RCH_2\overset{\ominus}{CH}-CH=CH_2 \overset{\oplus}{H}$$

(XXI)

It should be remembered, however, that it is the hyperconjugative effect of alkyl groups in the E2 transition state, rather than in the end product, that is of prime importance: alkyl hyperconjugation with the forming double bond lowers the energy of that transition state in which it occurs and hence favours its preferential formation.

At first sight, therefore, it might be concluded that the Saytzeff mode of elimination was the normal one and the Hofmann mode merely an occasional, abnormal departure therefrom. In fact, it is the latter that predominates with onium salts. Thus on heating (XXII), it is largely ethylene, rather than propylene, that is obtained despite the greater stability (due to hyperconjugation) of the latter:

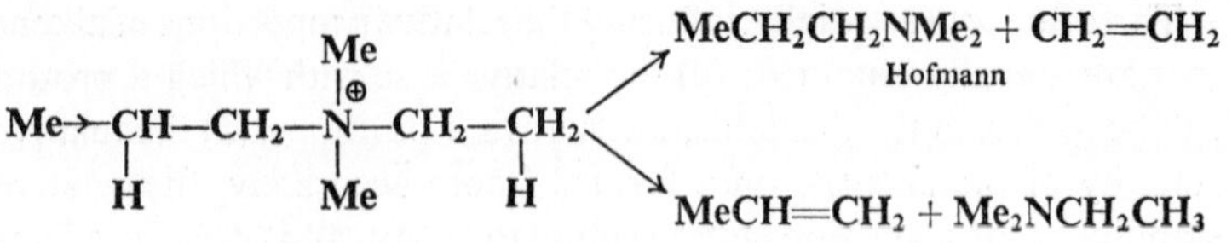

$$Me \rightarrow \underset{\underset{H}{|}}{CH} - CH_2 - \underset{\underset{Me}{|}}{\overset{\oplus}{N}} - CH_2 - \underset{\underset{H}{|}}{CH_2} \begin{array}{l} \nearrow \text{MeCH}_2\text{CH}_2\text{NMe}_2 + \text{CH}_2{=}\text{CH}_2 \\ \quad\quad\quad\text{Hofmann} \\ \\ \searrow \text{MeCH}{=}\text{CH}_2 + \text{Me}_2\text{NCH}_2\text{CH}_3 \end{array}$$

(XXII)

This can be explained by assuming that, in this case, (*a*) is of prime importance: the inductive effect of the methyl group in the propyl substituent causes a lowering of the acidity of the hydrogens attached to the β-carbon atom in this group and thus leads to preferential removal of a proton from the β-carbon atom of the ethyl substituent which is not so affected.

Thus it could be claimed that in Saytzeff elimination the hyperconjugative effects of alkyl groups are in control while in Hofmann elimination it is their inductive effects that predominate. But this leaves unanswered the question as to what causes the shift from the former mode to the latter. As has already been observed, Hofmann elimination is more common in the elimination reactions of onium salts and undoubtedly groups such as $\overset{\oplus}{R_3N}-$ and $\overset{\oplus}{R_2S}-$ will be much more potent in promoting acidity in β-hydrogens by their inductive effects than will halogen atoms so that the relative acidity of the β-hydrogen atoms could well come to be the controlling influence in the reaction. But this is not the whole of the story: another obvious difference is that the groups eliminated from onium compounds are usually considerably *larger* than those lost in the elimination reactions of halides; so much so that the preferential formation of that E2 transition state in which there is least crowding becomes imperative, even though this may not be the one favoured by hyperconjugative stabilisation.

The importance of the steric factor has been confirmed in a number of ways. Thus increase in the size of the leaving group in a compound of given structure leads to a corresponding increase in the proportion of Hofmann product produced, and the same result is observed when branching is introduced into the structure of a compound (with halides as well as onium salts) that might be expected to lead to increasing crowding in the E2 transition state. Perhaps most cogent of all, an increase in the proportion of Hofmann product is seen when the size of the initiating base is increased. Thus in the dehydrobromination of $MeCH_2CMe_2Br$ the change from $MeCH_2O^{\ominus}$ $\rightarrow Me_3CO^{\ominus} \rightarrow EtMe_2CO^{\ominus} \rightarrow Et_3CO^{\ominus}$ leads to formation of the

Hofmann product, $MeCH_2C(Me){=}CH_2$, in yields of 29, 72, 78 and 89% respectively. The importance that steric factors can play in deciding which type of elimination will result can, perhaps, best be seen by comparing the transition states (XXIII) and (XXIV) for the two modes of E2 elimination from RCH_2CMe_2X:

(XXIII)

(XXIV)

It can be seen that if **Y** is large, and especially if **R** is large as well, transition state (XXIV) will be favoured over (XXIII) as, in the former, **R** is much better able to get out of **Y**'s way; as indeed will be the case if $^\ominus OH$ is replaced by a bulkier base, **Y** being the same distance away in both cases but **R** being much less of a hindrance in (XXIV) than in (XXIII).

The classical Hofmann elimination reaction has, in the past, been of the utmost value in structure elucidation, particularly in the alkaloid field. Any basic nitrogen atom present is converted to the quaternary salt by exhaustive methylation and the corresponding quaternary hydroxide then heated. Removal of the nitrogen from the compound by one such treatment indicates that it was present in a side-chain, while elimination after two or three treatments indicates its presence in a saturated ring or at a ring junction, respectively. The resultant alkene is then investigated so as to shed further light on the structure of the original natural product.

The presence of a phenyl group on the α- or β-carbon atom very markedly promotes E2 eliminations because of its stabilisation of the resultant alkene by delocalisation (α-phenyl also promotes E1 eliminations for the carbonium ion formed by loss of, for example, $Hal^\ominus$

from the α-carbon is now of the benzylic type, *cf.* p. 79). The effect is more marked in the β- than in the α-position, however, because of the additional effect of phenyl in increasing the acidity of the β-hydrogens from this position and so facilitating their removal. The effect is sufficiently pronounced so as to control the orientation of elimination, resulting in the Saytzeff mode even with onium salts:

$$\text{PhCH}_2\text{CH}_2-\overset{\overset{\displaystyle \text{Me}}{|}}{\underset{\underset{\displaystyle \text{Me}}{|}}{\overset{\oplus}{\text{N}}}}-\text{CH}_2\text{CH}_3 \xrightarrow{\ominus\text{OH}} \text{PhCH}=\text{CH}_2 + \text{Me}_2\text{NCH}_2\text{CH}_3$$

A vinyl group will have much the same effect.

A steric limitation on elimination reactions is codified in Bredt's rule that reactions which would introduce a double bond on to a bridgehead carbon in fused ring systems do not normally take place. This is presumably due to the rigidity of the ring systems being great enough to prevent the attainment of a degree of coplanarity by the *p* orbitals on adjacent carbon atoms sufficient to allow of the formation of a double bond between them. Thus (XXV) does not yield the bicycloheptene (XXVI) which has, indeed, never been prepared:

$$\text{(XXV)} \xrightarrow[\times]{-\text{HBr}} \text{(XXVI)} \quad \text{i.e.}$$

The Bredt rule no longer applies if the rings fused at the bridgehead are large enough, and hence flexible enough, for significant *p* orbital overlap to occur:

(XXVII)

Nor does it apply if the necessary overlap can be attained by mutual twisting of the ring structures, as happens in the formation of the most interesting bridgehead alkene yet synthesised, a bicyclononene:

Finally, it should be emphasised that in E1 eliminations where alternative alkenes may be obtained, the more stable product will almost always predominate no matter what the nature of the leaving group. This is hardly surprising, for the leaving group will have already departed by the time the carbonium ion intermediate has been formed, and it can thus exert no influence on which β-carbon undergoes loss of proton: this will be determined by the relative stability of the alternative alkenes (more properly, the transition states leading to them), i.e. (XXVIII) will be obtained rather than (XXIX):

$$RCH_2CMe_2Br \longrightarrow RCH_2\overset{\oplus}{C} \overset{\displaystyle Me}{\underset{\displaystyle Me}{<}} \begin{array}{l} \nearrow \ RCH{=}CMe_2 \\ \ \ \ \ (XXVIII) \\ \\ \searrow \ RCH_2C(Me){=}CH_2 \\ \ \ \ \ (XXIX) \end{array}$$

ELIMINATION v. SUBSTITUTION

Broadly speaking changes in reaction conditions that would be expected to promote an S_N2 reaction at the expense of an S_N1 (p. 75) will promote the often competing E2 reaction at the expense of an E1 and, of course, vice-versa. The features that will favour overall elimination at the expense of substitution are a little more subtle, though some passing attention has already been paid to them; thus in the E1 reaction reference has already been made to steric features. The more crowded a halide, for example, the greater is the release of strain when the carbonium ion intermediate is formed. This strain is reintroduced on attack by a nucleophile but is not increased, and may even be further reduced, on removal of a proton to yield the alkene. The sheer steric effect here becomes merged with other features, however, for increasing alkyl substitution may also

223

lead to the possible formation of alkenes that are increasingly stabilised by hyperconjugation, thus favouring their formation at the expense of substitution. This, of course, is the reason for the greater tendency of tertiary and secondary, as compared with primary, halides to undergo unimolecular elimination rather than substitution reactions whatever the reagent employed:

$$\underset{\downarrow}{RCH_2CH_2Hal} \quad \underset{\downarrow}{RCH_2CHMeHal} \quad \underset{\downarrow}{RCH_2CMe_2Hal}$$
$$RCH{=}CH_2 \enspace < \enspace RCH{=}CHMe \enspace < \enspace RCH{=}CMe_2$$

In bimolecular reactions also it is found that increasing alkyl substitution favours elimination at the expense of substitution, for while it retards S_N2 because of overcrowding in the transition state that would lead to substitution, it promotes E2 because of the hyperconjugative stabilisation of the incipient alkene in the alternative (and probably less crowded) transition state that would lead to elimination.

One of the most potent factors influencing the elimination/substitution ratio with a given substrate, however, is change of mechanism from uni- to bi-molecular. Thus solvolysis of Me_3CBr with **EtOH** (largely $E1/S_N1$) yields only 19 % of $Me_2C{=}CH_2$, but if $^{\ominus}OEt$ is present (now largely $E2/S_N2$) the yield of alkene is essentially quantitative. The $E1/S_N1$ product ratio will be fixed, as will the $E2/S_N2$ ratio, and, provided the reaction is proceeding by a purely uni- *or* bi-molecular mechanism, the ratio will thus be independent of the concentration of, for example, $^{\ominus}OH$. As the concentration of $^{\ominus}OH$ is increased, however, there will come a changeover from an initially unimolecular to a bimolecular mechanism, a changeover that takes place quite suddenly with strong bases such as $^{\ominus}OH$ and which leads to a different, usually higher, proportion of the elimination product. This reflects the well-known use of high concentrations of strong bases for the actual preparation of alkenes.

It might be expected that the reagent employed would be of great significance in influencing the relative amounts of $E2/S_N2$ in a particular system, for basicity (i.e. electron pair donation to hydrogen) and nucleophilicity (i.e. electron pair donation to carbon) do not run wholly in parallel in a series of reagents, $Y^{\ominus}$ or $Y{:}$. Thus the use of tertiary amines, e.g. triethylamine, rather than $^{\ominus}OH$ or $^{\ominus}OEt$ for converting primary halides to alkenes depends on the amines being moderately strong bases but weak nucleophiles while the latter reagents are powerful nucleophiles as well as being strong bases (ethyl bromide yields only 1% of ethylene with $^\ominus$**OEt**, by far the major product being diethyl ether). The particular preparative value of pyridine for this purpose, despite its being a considerably weaker base than simple tertiary amines, **R$_3$N**: (*cf.* p. 71), arises in part at least from the stability of the pyridinium cation once formed; reversal of the abstraction of **H$^\oplus$** by pyridine is thus unlikely. The use of an amine that is relatively high boiling is also advantageous (see below). Reagents such as $^\ominus$**SR** which show the widest divergences between their basicity and nucleophilicity, are in general too weak bases to be of much value in inducing elimination reactions.

Careful investigation has shown that where substitution and elimination reactions compete in a given system, elimination normally has the higher activation energy and is thus the more favoured, relatively, of the two by rise in temperature: a fact that has long been recognised in preparative chemistry.

EFFECT OF ACTIVATING GROUPS

Thus far we have only considered the effect of alkyl, and occasionally aryl, substituents in influencing elimination reactions, but a far more potent influence is exerted by strongly electron-withdrawing groups such as $-NO_2$, $>SO_2$, $-CN$, $>C=O$, $-CO_2Et$, etc., in facilitating eliminations. Their influence is primarily on increasing the acidity of the β-hydrogen atoms:

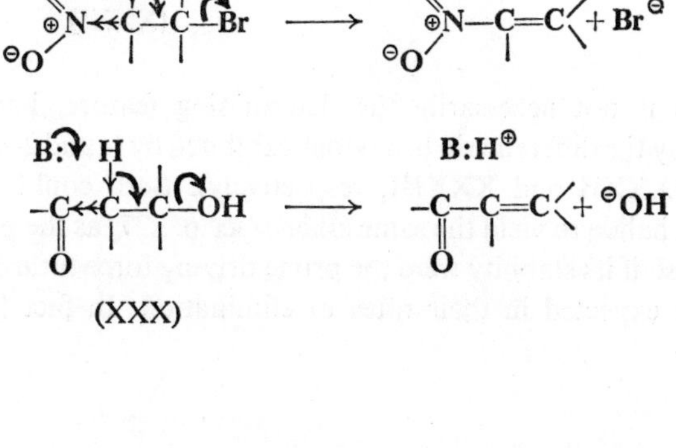

(XXX)

225

The reactions can proceed by a one- or a two-stage mechanism depending on whether the removal of proton and the other leaving group is concerted or whether an intermediate anion is actually formed. An added effect of substituents like the above is, of course, to stabilise such an anion by delocalisation

$$-\overset{\overset{\displaystyle O}{\|}}{C}-\overset{\ominus}{\underset{|}{C}}-\underset{|}{C}-Y \quad \leftrightarrow \quad -\overset{\overset{\displaystyle \ominus O}{}}{C}=\underset{|}{C}-\underset{|}{C}-Y$$

$$\overset{\overset{\displaystyle O}{\diagdown}}{\underset{\underset{\displaystyle \ominus O}{\diagup}}{\oplus N}}-\overset{\ominus}{\underset{|}{C}}-\underset{|}{C}-Y \quad \leftrightarrow \quad \overset{\overset{\displaystyle \ominus O}{\diagdown}}{\underset{\underset{\displaystyle \ominus O}{\diagup}}{\oplus N}}=\underset{|}{C}-\underset{|}{C}-Y$$

but it is not certain that such intermediates are formed as a matter of course in all these reactions, however. An interesting example above is the way in which the $\diagup C{=}O$ group makes possible a *base*-induced elimination of water from the aldol (XXX), whereas the elimination of water from a compound not so activated is nearly always acid-induced.

It has been suggested that a good deal of the driving force for the elimination reactions of suitably substituted carbonyl compounds is due to the product (and to some extent the transition state leading to it) being conjugated and so able to stabilise itself by delocalisation (XXXI):

$$\overset{\overset{\displaystyle OH}{|}}{\underset{\underset{\displaystyle B:\, H}{}}{MeCH{-}CH{-}CH}}{=}O \longrightarrow \begin{bmatrix} MeCH{=}CH{-}CH{=}O \\ \updownarrow \\ \overset{\oplus}{MeCH}{-}CH{=}CH{-}O^{\ominus} \end{bmatrix}$$

$$(XXXI)$$

That this is not necessarily the determining feature, however, is revealed by the difference in behaviour exhibited by 1- and 2-halogeno-ketones (XXXII and XXXIII, respectively). Both could eliminate hydrogen halide to yield the same alkene (see p. 227) as the product of reaction, so if its stability were the prime driving force little difference would be expected in their rates of elimination. In fact (XXXIII)

$$\underset{\text{(XXXII)}}{\overset{\displaystyle O}{\overset{\|}{\text{MeC}}}-\underset{\underset{\text{Br}}{|}}{\text{CH}}-\text{CH}_3}$$

$$\overset{\displaystyle O}{\overset{\|}{\text{MeC}}}-\text{CH}{=}\text{CH}_2$$

$$\underset{\text{(XXXIII)}}{\overset{\displaystyle O}{\overset{\|}{\text{MeC}}}-\text{CH}_2-\underset{\underset{\text{Br}}{|}}{\text{CH}_2}}$$

eliminates very much more rapidly than (XXXII) suggesting that the main effect of the carbonyl substituent is in increasing the acidity of the hydrogens on the adjacent carbon atom: this is the one that loses proton in (XXXIII) but *not* in (XXXII). It is indeed found to be generally true that the elimination-promoting effect of a particular electron-withdrawing substituent is much greater in the β- than in the α-position. The influence of such activating groups is often sufficiently great to lead to the elimination of more unusual leaving groups such as **OR** and **NH$_2$**.

A rather interesting intramolecular reaction of this type is the loss of **CO$_2$** and bromide ion from the anion of a β-bromoacid, for example cinnamic acid dibromide (XXXIV), in solvents such as acetone:

(XXXIV)

DEBROMINATION

Attention has been confined so far almost wholly to reactions in which one of the leaving groups has been hydrogen, and although these are the most common and important eliminations, the dehalogenation of 1,2-dihalides, particularly bromides, with metals such as zinc and magnesium also has some mechanistic and preparative interest:

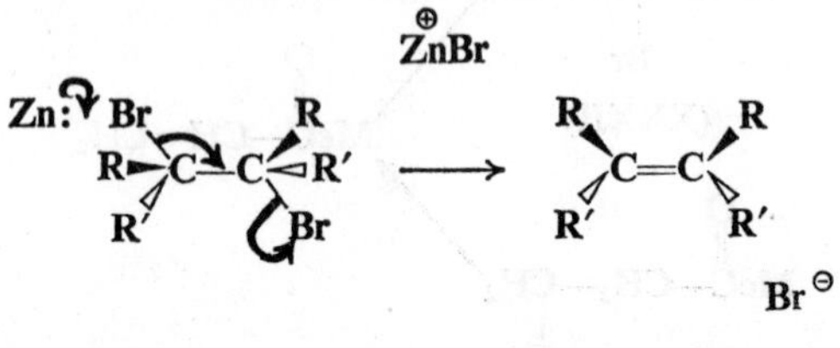

The reactions are carried out in solvents such as acetic acid whose function is probably to remove the halide salt from the reactive surface of the metal. The fact that the reactions are heterogeneous makes them somewhat difficult to study but there is evidence that in some cases they may proceed by a two step process involving a carbanion intermediate (resulting from initial removal of **Br** by **Zn**) rather than by the concerted E2 mode shown above. Apart from any preparative or diagnostic value the reaction may have, it is of course the above state of affairs that normally makes it impossible to prepare Grignard and similar organo-metallic compounds from 1,2-dihalides. Similar eliminations can be made to proceed with 1,2-haloesters and 1,2-haloethers. The reactions normally proceed stereospecifically *trans*.

A similar stereospecificity has been observed in the preparatively more useful debromination of 1,2-dibromides by iodide ion, particularly by use of **KI** in acetone. The kinetic law followed is of the form

$$\text{Rate} \propto [\text{I}^{\ominus}]\,[1,2\text{-dibromide}]$$

and the reaction probably follows a similar course to that with zinc:

(1)

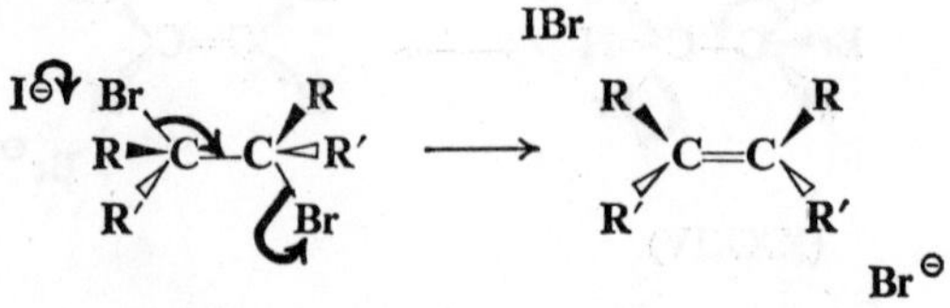

(2) $\text{I}^{\ominus} + \text{IBr} \longrightarrow \text{I}_2 + \text{Br}^{\ominus}$

With ethylene dibromide, however, there is evidence that the initial, and rate-determining, attack of $\text{I}^{\ominus}$ is on carbon (S_N2) to yield the 1,2-iodobromide as an intermediate which then undergoes further attack as above, but on **I** rather than **Br**, to yield ethylene.

228

Practical use is made of this reaction in the purification of alkenes. The usually crystalline dibromides are purified by recrystallisation and the pure alkene then regenerated, as above, under extremely mild conditions.

α-ELIMINATION

A small number of cases are known of the elimination of hydrogen halide where both atoms are lost from the same carbon: these are known as α-elimination reactions. The best known example occurs in the hydrolysis of chloroform with strong bases:

$$HO^{\ominus} H-C\overset{Cl}{\underset{Cl}{\overset{|}{-}}}Cl \quad \overset{Fast}{\rightleftharpoons} \quad H_2O + {}^{\ominus}CCl_3 \quad \overset{Slow}{\longrightarrow} \quad Cl^{\ominus} + :CCl_2$$

$$\text{Fast} \downarrow H_2O$$

$$HCO_2^{\ominus} \quad \overset{Slow}{\underset{{}^{\ominus}OH}{\longleftarrow}} \quad CO$$

Hydrogen halide is lost in a two-stage process to yield dichlorocarbene (*cf.* p. 49), as an intermediate in the hydrolysis. The latter then reacts with water to yield **CO** as the primary product and this then undergoes further slow attack by ${}^{\ominus}$**OH** to yield formate anion. The initial attack on **H** rather than **C** (with expulsion of $Cl^{\ominus}$) by ${}^{\ominus}$**OH** is due to the electron-withdrawing effect of the chlorine atoms increasing the acidity of the hydrogen atom, a property which is reflected in its ready base-catalysed exchange with deuterium in D_2O. Confirmation of the existence of CCl_2, i.e. of an α-elimination, is provided by the introduction of substrates into the system that would be expected to react readily with such a species; thus alkenes have been converted into cyclopropane derivatives

$$CHCl_3 \quad \overset{t\text{-BuO}^{\ominus}}{\longrightarrow} \quad CCl_2 \quad \overset{\text{Me} \quad \text{Me}}{\longrightarrow} \quad$$

and dichlorocarbene is also involved in the **Reimer-Tiemann reaction** (p. 247).

Another example of an α-elimination is seen in the action of potassium amide on 2,2-diphenylvinyl bromide (XXXV):

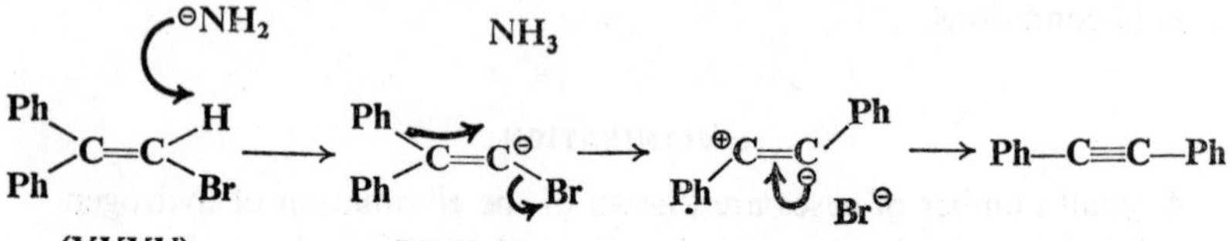

(XXXV) (XXXVI) → Ph—C≡C—Ph

Whether the whole process proceeds as a concerted operation or whether the carbanion (XXXVI) is actually formed, as such, is not known however.

CIS-ELIMINATION

A number of esters, particularly acetates, are known to undergo elimination reactions on heating, alone or in an inert solvent, to yield alkenes:

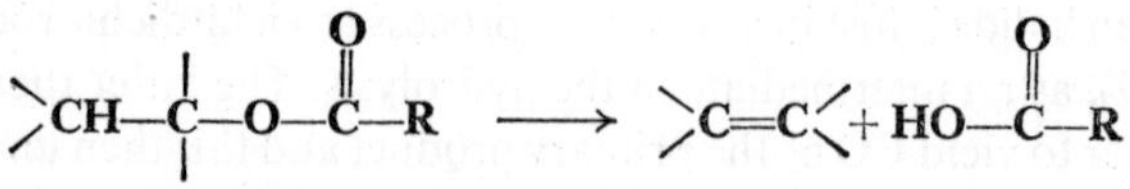

The kinetic law followed is of the form

$$\text{Rate} \propto [\text{ester}]$$

and the reaction is believed to proceed through a cyclic transition state (XXXVII):

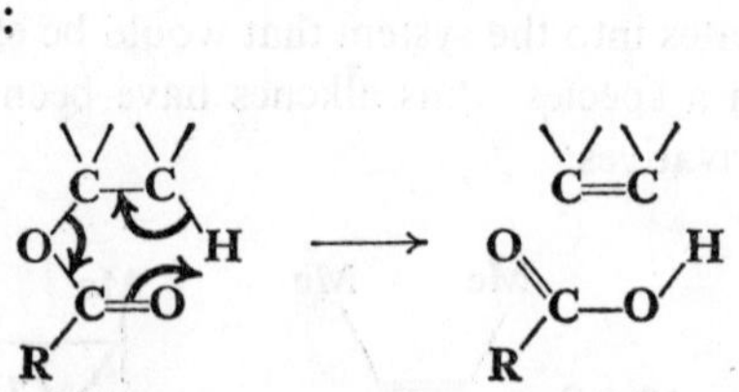

(XXXVII)

Such eliminations are generally referred as proceeding by the E*i* mode.

The strongest evidence in favour of such a route is that these eliminations are found to proceed stereospecifically *cis* in contrast to the very much commoner *trans* eliminations that we have encountered to date. It is thus probable, in most cases, that the breakage of the C—H and C—O bonds occurs simultaneously, but the fact that small, though detectable, amounts of *trans* elimination have been observed to take place suggests that in some cases the latter breaks before the former, leading to the transient formation of an ion pair, which can then undergo mutual reorientation. It seems unlikely in such cases that radicals, rather than ion pairs, are ever formed as the reactions appear to be unaffected by either radical initiators or inhibitors (*cf*. p. 257).

A very closely analogous elimination is the Chugaev reaction which involves the indirect conversion of an alcohol to the corresponding alkene via pyrolysis of the methyl xanthate (XXXVIII) obtained by the action of $CS_2/^{\ominus}OH/MeI$ on the original alcohol:

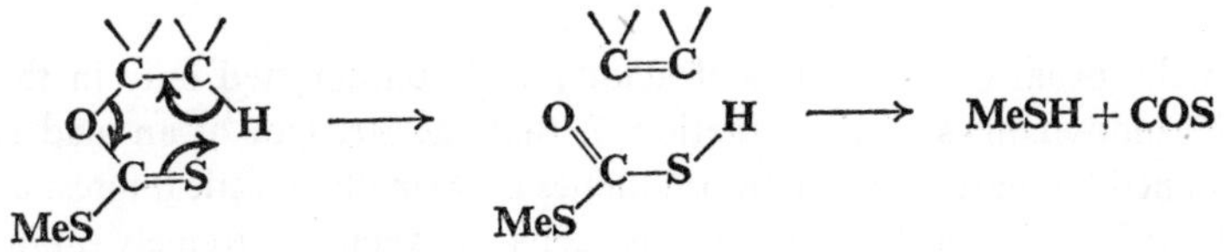

(XXXVIII)

This reaction has the greater preparative value as the pyrolysis may be successfully carried out at lower temperatures (*ca*. 150° as compared with *ca*. 400° for acetates); the advantage of both, compared with other methods of alkene formation in complicated structures, is their high yield and relative freedom from simultaneous molecular rearrangement.

A more recent elimination reaction of preparative value analogous to the above is that of tertiary amine oxides (XXXIX), the Cope reaction. This elimination proceeds smoothly at even lower temperatures, as low as 25° in some cases using dimethyl sulphoxide as solvent, with the elimination of a dialkylhydroxylamine (XL):

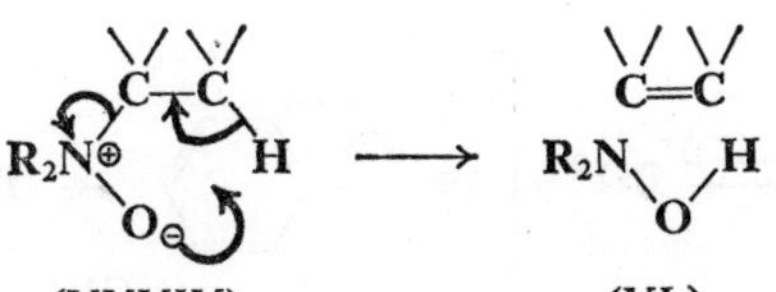

(XXXIX)　　　　　(XL)

231

10 CARBANIONS AND THEIR REACTIONS

SOME organic compounds are known which function as acids, in the classical sense, in that a proton may be liberated from a **C—H** bond by a base, the resultant conjugate base (I) being known as a *carbanion*:

$$\text{>C—H} \quad \curvearrowright :B \quad \rightleftharpoons \quad \text{>C}^{\ominus} + HB^{\oplus}$$
$$\text{(I)}$$

In the usual consideration of acidity we are interested only in the thermodynamics of the situation in that the strength of an acid is defined by the above equation; kinetics is normally of little interest as the transfer of proton from atoms such as oxygen is extremely rapid. With carbon acids, however, the *rate* at which proton is transferred may be sufficiently slow as to constitute the limiting factor, acidity then being controlled kinetically rather than thermodynamically (*cf.* p. 241).

FORMATION OF CARBANIONS

The above tendency is, not surprisingly, but little marked with aliphatic hydrocarbons for the **C—H** bond is a fairly strong one and there is normally no structural feature that either promotes acidity in the hydrogen atom or that leads to significant stabilisation of the carbanion with respect to the original undissociated molecule (*cf.* p. 54); thus methane has been estimated to have a pK_a of ≈ 43, compared with 4·76 for acetic acid. Triphenylmethane (II), however, whose related carbanion (III) can be stabilised by delocalisation

$$Ph_3C—H + :B \rightleftharpoons BH^{\oplus} + \left[Ph_2\overset{\ominus}{C} \underset{}{\bigcirc} \quad \leftrightarrow \quad Ph_2C = \overset{\ominus}{\bigcirc} \right]$$
$$\text{(II)} \qquad\qquad\qquad\qquad \text{(III)}$$

is a much stronger acid ($pK_a \approx 33$) by relative, if not by absolute, standards. The presence of one or more strongly electron-withdrawing groups has an even more marked effect; thus tricyanomethane (IV)

$$(NC)_3C\!-\!H + :B \;\rightleftharpoons\; BH^\oplus + \left[(NC)_2\overset{\ominus}{C}\!-\!C\!\equiv\!N \;\leftrightarrow\; (NC)_2C\!=\!C\!=\!\overset{\ominus}{N} \right]$$

(IV)

approaches the mineral acids in strength. The cumulative effect of introducing successive cyano groups is of interest here for acetonitrile (V), containing only one such group, still has a pK_a as high as 25:

$$CH_3\!-\!C\!\equiv\!N + :B \;\rightleftharpoons\; BH^\oplus + \left[\overset{\ominus}{C}H_2\!-\!C\!\equiv\!N \;\leftrightarrow\; CH_2\!=\!C\!=\!\overset{\ominus}{N} \right]$$

The presence of electron-withdrawing groups can therefore lead to sufficient acidity in such hydrogen atoms as to permit their ready removal by bases; thus the following compounds

EtO$_2$CCH$_2$CO$_2$Et	CH$_3$NO$_2$
(VI) $pK_a = 13\cdot3$	$pK_a = 10\cdot2$
MeCOCH$_2$CO$_2$Et	MeCOCH$_2$COMe
(VII) $pK_a = 10\cdot7$	(VIII) $pK_a = 8\cdot8$

readily yield carbanions in this way which are of the utmost importance as intermediates in a wide variety of reactions. Other compounds which lose a proton less readily may sometimes be converted into carbanions by treatment with very strong bases in anhydrous media, as is the case with acetylene ($pK_a = 25$):

$$HC\!\equiv\!CH \xrightarrow[\text{liq.}\ NH_3]{\ \ominus NH_2\ } HC\!\equiv\!C^\ominus + NH_3$$

Here the acidity arises from the increasing electronegativity of carbon as its hybridisation changes from $sp^3 \to sp^2 \to sp^1$ (*cf.* p. 57). Carbanions that cannot be obtained by direct removal of proton by base may often be formed by the action of metals on the corresponding halides

$$R—Br + 2M \rightarrow R—M + MBr$$

or a variant thereon; the less electronegative the metal the more polar the **R—M** bond and the greater the carbanion character of **R**; thus alkyls and aryls of the most electropositive metals exist as an ion pair, $R^{\ominus} M^{\oplus}$, e.g. $Me(CH_2)_2CH_2^{\ominus} Na^{\oplus}$. Many other compounds containing carbon-metal bonds, though not ionised, are sufficiently polarised to serve as sources of negative carbon, e.g. Grignard reagents (p. 192), lithium alkyls etc.

Where the actual isolation or ready detection of a carbanion as such is not feasible, the transient formation of such a species may be inferred from dissolving the compound, in the presence of base, in D_2O or **EtOD** and observing if deuterium becomes incorporated. Thus chloroform (*cf.* p. 229) is found to undergo such an incorporation:

$$HCCl_3 \xrightarrow[H_2O]{-H^{\oplus}} {}^{\ominus}CCl_3 \xrightarrow[-D^{\oplus}]{D_2O} DCCl_3$$

The rate of exchange can also be used to investigate the kinetic, as opposed to the thermodynamic, acidity of carbon acids (*cf.* p. 232).

STABILITY OF CARBANIONS

Carbanions can be stabilised by high s character in the orbital accommodating the electron pair left by loss of proton; we have already mentioned this effect above in the case of acetylene and it is responsible for increasing acidity on going $H_3C—CH_3 \rightarrow H_2C=CH_2 \rightarrow HC\equiv CH$. Carbanions can also be stabilised purely by inductive effects if sufficiently powerful electron-withdrawing atoms or groups are present. This is probably only really significant with $R_3N^{\oplus}$ and F but is responsible for the pK_a's of HCF_3 and $HC(CF_3)_3$ being ≈ 28 and 11, respectively compared with ≈ 43 for CH_4; other effects may also be operating in $HC(CF_3)_3$, however. As might be expected, the effect of electron-withdrawing groups is most pronounced when they can exert a mesomeric as well as an inductive effect; thus a carbonyl or a nitro group will be more effective than $R_3N^{\oplus}$ or CF_3:

234

The smaller pK_a, and hence greater stability of the carbanion, of acetylacetone (VIII) compared with ethyl acetoacetate (VII), and of the latter compared with diethyl malonate (VI) arises from the carbonyl group in a ketone being more effective at electron-withdrawal and delocalisation than the carbonyl group in an ester. This springs from the occurrence in the latter of

which lowers the effectiveness of the $\diagup\!\!C\!\!=\!\!O$ group at withdrawing electrons from the rest of the molecule. The decreasing activating effect of $-COY$ on going $-COH \rightarrow -COR \rightarrow -CO_2R \rightarrow -CONH_2 \rightarrow -CO_2^{\ominus}$ is thus due to Y becoming more electron-donating as the series is traversed.

An interesting carbanion, the cyclopentadienyl anion (IX),

(IX)

owes its considerable stability to the fact that, in the system, a total of six π electrons ($n = 1$ in $2 + 4n$, *cf.* p. 17) is available and these can distribute themselves so as to form delocalised π orbitals covering all five carbon atoms, leading to the quasi-aromatic structure (IXa):

(IXa)

In this stabilisation by aromatisation the stability of the ion is reflected in the acidity of cyclopentadiene itself ($\mathbf{p}K_a \approx 15$), demonstrating the readiness with which the latter can be made to lose a proton in order to attain the stable carbanion state. The quasi-aromaticity cannot be demonstrated by electrophilic substitution, for attack by $\mathbf{X}^{\oplus}$ would merely lead to direct combination with the ion, but true aromatic character (Friedel-Crafts reactions, etc.) is shown by the remarkable series of extremely stable, neutral compounds such as *ferrocene* (X) (obtained by attack of (IXa) on metallic halides such as $FeCl_2$),

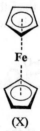

(X)

in which the metal atom is held by π bonds in a kind of molecular 'sandwich' between two cyclopentadienyl structures.

It has also proved possible to isolate, as a crystalline solid, the dipotassium salt of cyclooctatraene dianion:

This doubly charged carbanion is made by the action of potassium metal on cyclooctatetraene and it too fulfils the necessary requirement of $2 + 4n$ π electrons delocalised cyclically (here $n = 2$) to ensure great stabilisation by aromatisation.

The stabilising effects of some electron-withdrawing groups on carbanions is found to be considerably greater than might be expected on the basis of the inductive effects they can exert, e.g. $R_3P^\oplus$, RSO_2, etc. It is significant that these groups contain second row elements which have d orbitals available, and the extra stabilisation is believed to arise from overlap of these d orbitals with the orbital containing the electron pair on the adjacent carbanion carbon atom.

STEREOCHEMISTRY OF CARBANIONS

In theory a simple carbanion of the form $R_3C^\ominus$ could assume a pyramidal (sp^3) or a planar (sp^2) configuration, or perhaps somewhere in between. The former (pyramidal) is the most likely by analogy with tertiary amines, $R_3N:$, with which simple carbanions are isoelectronic. Rapid interconversion of the mirror image forms of such amines, via a planar intermediate, has been demonstrated spectroscopically and carbanions would be expected to behave similarly:

In reactions in which carbanions are developed as intermediates from optically active starting materials, predominant retention, racemisation, and inversion of configuration have all been demonstrated depending on the conditions employed. The solvent is certainly much involved and it has been suggested that the differing behaviour is the result of unsymmetrical solvation of a carbanion which can oscillate between its opposite pyramidal configurations, if its lifetime is long enough.

As soon as one or more of the groups attached to the carbanion carbon atom are capable of stabilising the ion by delocalisation, then limitations are imposed on its configuration because of the near coplanarity necessary if significant delocalisation via the overlapping of parallel p orbitals is to take place. This will apply to the three bonds to the carbanion carbon in the triphenylmethyl anion (XI)

(XI)

237

and also in cases where the carbanion carbon is adjacent to $>C=O$
NO_2, etc. A good example of this is seen in the compound (XII):

(XII)   (XIII)

Here the hydrogen, despite being flanked by *two* carbonyl groups,
shows little sign of acidity because the carbanion (XIII) resulting from
its removal would be unable to stabilise itself by delocalisation owing
to the rigid ring structure preventing the *p* orbitals on the two carbon
atoms involved from becoming parallel; significant overlapping thus
could not take place and the carbanion does not form. This is in
contrast to cyclohexan-1,3-dione (lacking the trans-annular bridge)
which is quite a strong carbon acid.

It is well known that asymmetric centres carrying a hydrogen atom
adjacent to carbonyl groups (e.g. XIV) are very readily racemised in
the presence of base. This can, nevertheless, *not* be taken as entirely
unambiguous confirmation of the planar nature of any carbanion
intermediate involved (e.g. XV), despite its likelihood on other
grounds for the enol form (XVI) which *must* be planar will also be in
equilibrium with it:

(XIV)   (XV)

(XV)   (XVI)

CARBANIONS AND TAUTOMERISM

The enolisation mentioned above, is, of course, an example of the
larger phenomenon of *tautomerism*. This term, strictly defined, can be

used to describe the reversible interconversion of isomers, in all cases and under all conditions; in practice, its use has become increasingly limited to the case of isomers that are fairly readily interconvertible and that differ from each other only in electron distribution and in the position of a relatively mobile atom or group. The mobile atom or group involved is hydrogen in the great majority of examples, in which case the phenomenon is known as *prototropy*; familiar examples are ethyl acetoacetate and aliphatic nitro-compounds:

$$\overset{O}{\overset{\|}{MeC}}-CH_2CO_2Et \; \rightleftharpoons \; \overset{OH}{\overset{|}{MeC}}=CHCO_2Et$$

$$\qquad \textit{Keto} \qquad\qquad\qquad \textit{Enol}$$

$$RCH_2-N\overset{\oplus}{\underset{O^{\ominus}}{\overset{O}{\diagup}}} \; \rightleftharpoons \; RCH=N\overset{\oplus}{\underset{O^{\ominus}}{\overset{OH}{\diagup}}}$$

$$\textit{Pseudo-acid form} \qquad\qquad \textit{Aci-form}$$

(i) Concerted *v.* stepwise mechanism

These interconversions could, in theory, take place by either a stepwise or a concerted mechanism depending on whether the abstraction of hydrogen from one atom takes place prior to, or simultaneously with, the addition of hydrogen to the other; both types are encountered in practice. If the conversion is base-induced, the former mechanism would, in suitable cases, involve the participation of a carbanion, while the second would not. Many common examples, including β-ketoesters, β-diketones, aliphatic nitro-compounds (i.e. the common ketol/enol systems and their relatives), are believed to involve the stepwise mechanism

$$\text{(XVII)} \qquad\qquad \text{(XVIII)} \qquad\qquad \text{(XIX)}$$

239

and, as might be expected, the more stable the carbanion intermediate (XVIII), i.e. the more acidic the substrate from which it may be derived, the more is the stepwise mechanism favoured with respect to the concerted one. The above example enables emphasis to be laid on the distinction between tautomerism and mesomerism which so often apparently leads to confusion. Thus (XVII) and (XIX) are tautomers, the so-called keto and enol forms respectively, and are quite distinct chemical entities. Although often readily interconvertible *both* can, in suitable cases (e.g. ethyl acetoacetate), actually be isolated and characterised. By contrast, the intermediate involved in their interconversion, the carbanion (XVIII), is a *single species,* a mesomeric hybrid of the two hypothetical structures written, neither of which has any real existence. It is, of course, a commonplace to find a pair of tautomers underlain, as it were, by a carbanion stabilised by delocalisation in this way.

By contrast, the tautomerisation of a number of compounds of the form $R_2CH-N=CR'_2$, the azomethines, has been shown to proceed via the concerted mechanism. Thus tautomerisation of (XX) → (XXI) has been carried out in **EtOD** with **EtO**$^\ominus$ as catalyst, and been found to result in deuterium exchange as well as tautomerisation. A concerted mechanism must lead to deuteration and tautomerisation proceeding at exactly the same rate for the latter cannot take place without the former

$$(XX) \rightleftharpoons (XXI)$$

whereas with a stepwise mechanism (see p. 241) deuteration can take place *without* tautomerisation by the reversal of (i) and so should proceed the faster. In practice, both *are* found to take place at the same rate so, knowing the reaction is reversible, it can be said with confidence that it proceeds via the concerted mechanism.

Which mechanism is at work in a particular case is also influenced by the medium involved, polar solvents, not surprisingly, favouring the stepwise mode. Although not relevant to carbanions, it should perhaps be emphasised that keto-enol interconversions can be

$$p\text{-PhC}_6\text{H}_4 \overset{\ominus}{\underset{\text{Ph}}{\overset{\text{EtO}}{\underset{}{}}}} \overset{\text{H}}{\underset{}{\overset{*}{\text{C}}}} - \text{N} = \text{C} \overset{\text{Ph}}{\underset{\text{H}}{}}$$

(XX)

EtOD ↑↓ $\ominus$OEt
(i)

$$\left[\begin{array}{c} p\text{-PhC}_6\text{H}_4 \\ \text{Ph} \end{array} \overset{\ominus}{\text{C}} - \text{N} = \text{C} \overset{\text{Ph}}{\underset{\text{H}}{}} \quad \leftrightarrow \quad \begin{array}{c} p\text{-PhC}_6\text{H}_4 \\ \text{Ph} \end{array} \text{C} = \text{N} - \overset{\ominus}{\text{C}} \overset{\text{Ph}}{\underset{\text{H}}{}} \right]$$

$\ominus$OEt ↑↓ EtOD
(ii)

$$\begin{array}{c} p\text{-PhC}_6\text{H}_4 \\ \text{Ph} \end{array} \text{C} = \text{N} - \overset{*}{\text{C}} \overset{\text{Ph}}{\underset{\text{D}}{\overset{\text{H}}{}}}$$

(XXI)

catalysed by acids, e.g. **HA**, as well as by bases. This is another example of *general* acid catalysis (*cf.* p. 182):

$$\text{A}^{\ominus} \text{H} - \text{C} - \text{C} = \text{O} \cdots \text{H} - \text{A} \quad \longrightarrow \quad \text{AH} + \underset{|}{>}\text{C} = \text{C} - \text{OH} + \text{A}^{\ominus}$$

The role played above by the acid anion, $\text{A}^{\ominus}$, can equally well be played by a solvent molecule in many cases.

(ii) Rate of tautomerisation

It should perhaps be emphasised that in the tautomerisations about which we have been speaking a **C—H** bond must undergo dissociation; so that although a number of the conversions are fairly rapid they are not like ionic reactions where no such bond-breaking is involved. The actual rates of a number of these conversions can, as implied in the last section, be followed by measuring the rate at which the compounds involved will incorporate deuterium from D_2O, EtOD, etc. When base is added to a pure tautomer it is, in the more familiar examples, usually the rate of formation of the anionic intermediate that is being observed and, on subsequent acidification, the rate of re-formation of one or both tautomers from the ion:

$$\text{MeC}\overset{\text{O}}{\underset{}{\|}}\text{—CH}_2\text{CO}_2\text{Et} \xrightarrow{\ominus\text{OH}} \left[\begin{array}{c} \overset{\text{O}}{\underset{}{\text{MeC}\overset{\|}{\text{—}}\overset{\ominus}{\text{CHCO}_2\text{Et}}}} \\ \updownarrow \\ \overset{\ominus\text{O}}{\underset{}{\text{MeC}\overset{}{=}\text{CHCO}_2\text{Et}}} \end{array} \right] \begin{array}{c} \nearrow \\ \text{H}^{\oplus} \\ \searrow \end{array} \begin{array}{c} \overset{\text{O}}{\underset{}{\|}} \\ \text{MeC—CH}_2\text{CO}_2\text{Et} \\ \\ \overset{\text{OH}}{\underset{}{|}} \\ \text{MeC}{=}\text{CHCO}_2\text{Et} \end{array}$$

The correlation of relative rates of tautomerisation with structure is not quite so simple as might have been expected and while it is broadly true that structural changes that lead to greater acidity lead also to more rapid reactions, this is by no means universally the case. Thus MeNO_2 ($pK_a = 10.2$) is more acidic than $\text{CH}_2(\text{CO}_2\text{Et})_2$ ($pK_a = 13.3$) but ionises more slowly (*cf.* p. 232).

(iii) Structure and position of equilibrium

Information on this subject is most readily available for keto/enol tautomers and their near relatives and we shall confine ourselves to them. The relative proportions of keto and enol forms in equilibrium with each other may often be estimated chemically but is usually more conveniently done spectroscopically. Normally speaking, the occurrence of a significant amount of the enol form in the keto/enol equilibrium mixture requires the presence of a group or groups capable of stabilising the enol by delocalisation of the π electrons of its carbon–carbon double bond:

$$\overset{\curvearrowleft}{\text{O}}{=}\text{C}\overset{\curvearrowleft}{\underset{|}{—}}\text{C}{=}\text{C}\overset{\curvearrowleft}{\underset{|}{—}}\overset{..}{\text{O}}\text{H} \quad \leftrightarrow \quad \overset{\ominus}{\text{O}}\overset{}{\underset{|}{—}}\text{C}{=}\text{C}\overset{}{\underset{|}{—}}\text{C}{=}\overset{\oplus}{\text{O}}\text{H}$$

Thus the proportion of enol in acetone is almost negligible while with a β-diketone it is present as the major species:

Compound		% enol in liquid
MeCOCH_3		0.00025
$\text{MeCOCH}_2\text{CO}_2\text{Et}$	(XXII)	7.5
$\text{MeCOCHPhCO}_2\text{Et}$	(XXIII)	30
$\text{MeCOCH}_2\text{COMe}$	(XXIV)	80

The relative effectiveness to this end of a second keto group as opposed to the $>\text{C}{=}\text{O}$ of an ester group is seen in comparing acetylacetone (XXIV) with ethyl acetoacetate (XXII), and the further delocalisation effected, by the introduction of a phenyl group on the methylene carbon atom by comparing ethyl acetoacetate (XXII) with its α-phenyl derivative (XXIII).

Another factor is also contributing to the stabilisation of the enol compared with the keto form in such species, however, namely internal hydrogen bonding:

(XXIV) (XXII)

Apart from the further stabilisation thereby effected, this hydrogen bonding leads to a decrease in the polar character of the enol and a compact, 'folded up' conformation for the molecule (by contrast to the extended conformation of the keto form) with the rather surprising result that, where the keto and enol forms can actually be separated, the latter usually has the lower boiling point of the two despite its hydroxyl group. The part played by such hydrogen bonding in stabilising the enol form with respect to the keto is seen by comparing the proportion of the enol form of acetylacetone (XXIV) in water and in a non-hydroxylic solvent such as hexane. In the latter there is 92% enol (more than the 80% in the pure liquid) whereas in the former where the keto form can also form hydrogen bonds—with the solvent water molecules—the proportion of the enol is down to 15%.

In the above examples, the composition of the equilibrium mixture is governed by the relative thermodynamic stabilities of the two forms under the particular circumstances being considered and this will, of course, ultimately always be the case. An interesting situation arises with aliphatic nitro-compounds such as phenylnitromethane (XXV), however:

Pseudo-acid form (XXVII) *Aci-form*
(XXV) (XXVI)

243

Here the normal nitro or pseudo-acid form (XXV) is thermo-dynamically the more stable of the two and at true equilibrium this form is present to the almost total exclusion of the aci-form (XXVI). Nevertheless, on acidification of the sodium salt of the system, i.e. (XXVII), it is very largely (XXVI) that is obtained. This results from the more stable pseudo-acid (XXV) being formed *more slowly* from the mesomeric ion than is the aci-form (XXVI), for the transition state between (XXVII) and (XXV) is at a considerably *higher* energy level than that between (XXVII) and (XXVI), reflecting the greater difficulty of breaking a C—H bond rather than an O—H bond. The reaction is thus kinetically rather than thermodynamically controlled, leading to (XXVI) rather than (XXV).

Although the *immediate* result is the preferential formation of the aci-form (XXVI), slow, spontaneous re-ionisation of (XXVI) leads inexorably to the formation of (XXV), so that the *ultimate* composition of the product is thermodynamically controlled.

REACTIONS OF CARBANIONS

(i) Addition reactions

Carbanions and sources of negative carbon take part in a wide variety of addition reactions, many of which involve additions to carbonyl systems. Thus Grignard (p. 192) and acetylide ion (p. 193) additions have already been considered, as have aldol (p. 194), Perkin (p. 197), Claisen ester (p. 198), and benzoin (p. 200) reactions and carbanion addition to $\alpha\beta$-unsaturated carbonyl systems in the Michael reaction (p. 175).

(a) **Carbonation:** Alkyls, aryls or acetylides of metals more electro-positive than magnesium, or the corresponding Grignard reagents, will all add on to the very weak electrophile carbon dioxide to yield the salt of the corresponding carboxylic acid:

$$M^{\oplus} \; R^{\ominus} \curvearrowright \underset{\underset{O}{\overset{\|}{\|}}}{\overset{\overset{O}{\|}}{C}} \; \longrightarrow \; R-C \underset{O^{\ominus} \; M^{\oplus}}{\overset{O}{\diagup}}$$

The reaction is often carried out by adding the organometallic compound, dissolved in a suitable inert solvent, to a large excess of powdered solid CO_2, and is particularly useful for the preparation of acetylenic acids. It may also be used to detect the formation of a carbanion in a particular case, the resultant carboxylic acid being particularly easily isolated and characterised. The Kolbe-Schmidt reaction (p. 249) is another example.

(ii) Displacement reactions

Carbanions are also involved in a number of displacement reactions.

(a) **Alkylation**: Suitable carbanions, such as those derived from malonic ester, β-ketoesters, β-dicarbonyl compounds, etc., will effect ready displacement reactions with alkylating agents such as alkyl halides and other reactive halogen-containing compounds, this being a useful preparative method for the formation of carbon–carbon bonds. In most cases the carbanion is generated in non-aqueous solution by bases such as $EtO^{\ominus}$ and the reaction then follows a normal S_N2 course:

$$(EtO_2C)_2HC^{\ominus} + \overset{H}{\underset{H}{\overset{\diagdown}{H\!\!\Rightarrow\!\!C}}}\!\!-Br \longrightarrow (EtO_2C)_2HC\overset{\delta-}{\cdots}\overset{H\diagup H}{\underset{|}{C}}\overset{\delta-}{\cdots}Br$$

$$\Big\uparrow {}^{\ominus}OEt \qquad\qquad \Big\downarrow$$

$$(EtO_2C)_2HC\!-\!H \qquad\qquad (EtO_2C)_2HC\!-\!\overset{H}{\underset{H}{C}}\!\!\nwarrow\!H + Br^{\ominus}$$

The S_N2 character has been confirmed kinetically in some examples and inversion of configuration has been shown to take place at the carbon atom attacked. A similar reaction of preparative value occurs with acetylide ion:

$$HC\!\equiv\!CH \underset{}{\overset{{}^{\ominus}NH_2}{\rightleftarrows}} HC\!\equiv\!C^{\ominus} + RBr \rightarrow HC\!\equiv\!CR + Br^{\ominus}$$

It should be remembered, however, that the above carbanions, and particularly the acetylide ion, are derived from very weak acids, and are, therefore, themselves strong bases with the result that they can induce elimination (p. 223) as well as displacement reactions; reaction with tertiary halides thus commonly results in alkene formation to the exclusion of alkylation.

(b) **Grignard and other organo-metallic reagents**: Grignard reagents can be looked upon, formally, as providing a source of negative

9

carbon though it is unlikely that they act as *direct* sources of carbanions as such, and more than one molecule of the reagent may actually be involved in their reactions as has been seen in discussing their addition to carbonyl groups (p. 192). They can also take part in displacement reactions as in the preparatively useful formation of aldehydes, via their acetals, from ethyl orthoformate (XXVIII):

$$\overset{\delta-}{R}\overset{\delta+}{Mg}Br + CH(OEt)_3 \rightarrow RCH(OEt)_2 \xrightarrow[H_2O]{H^{\oplus}} RCHO$$

(XXVIII)

Of other organo-metallic compounds, there is some evidence that sodium and potassium derivatives do provide actual carbanions as the active species in the reactions in which they take part, though even here the carbanions may not always be present as the simplest possible species but as dimers, trimers and higher aggregates. Lithium derivatives, which are synthetically more useful, correspond more closely to Grignard reagents in possessing a greater degree of covalent character not normally leading to the formation of free carbanions. They are not entirely analogous to Grignard reagents, however, as a number of cases are known where the two lead to different products from the same substrate.

(c) Wurtz reaction: A number of claims have been advanced that the Wurtz reaction proceeds via a radical mechanism:

$$RCH_2CH_2Br + Na\cdot \rightarrow RCH_2CH_2\cdot + Br^{\ominus} + Na^{\oplus}$$
$$2RCH_2CH_2\cdot \rightarrow R(CH_2)_4R$$

One of the pieces of supporting evidence quoted to corroborate this is that disproportionation is observed to take place as well as the expected dimerisation, an alternative reaction well known in radical chemistry (p. 261):

$$2RCH_2CH_2\cdot \rightarrow RCH_2CH_3 + RCH{=}CH_2$$

While such a mechanism is probably valid in the vapour phase it is somewhat less likely in solution and reaction of alkyl chlorides with sodium in hydrocarbon solvents has been shown to lead to the formation of sodium alkyls, which can then be made to react with a second alkyl halide to give quite high yields of the mixed hydrocarbon:

$$RCH_2CH_2Cl \xrightarrow{2Na} Na^\oplus + Cl^\ominus + RCH_2CH_2{}^\ominus Na^\oplus \xrightarrow{R'Cl} RCH_2CH_2R'$$

Further, the disproportionation, which has often been taken as confirmation of a radical mechanism, can equally be explained, on a carbanion basis, by elimination. The carbanion acts as a base, abstracting a proton from the β-carbon atom of the halide, while chlorine is lost from the α-carbon atom under the influence of $Na^\oplus$:

RCH₂CH₂⊖ H   RCH₂CH₃

R—CH—CH₂   ⟶   RCH=CH₂

Cl Na⊕   Cl⊖ Na⊕

Whether the process is actually initiated by $Na^\oplus$, $RCH_2CH_2{}^\ominus$ or both is not certain, however.

Further support for such an ionic mechanism for the Wurtz reaction is provided by the behaviour of some optically active chlorides, the first formed carbanion attacking a second molecule of chloride by the S_N2 mechanism and leading to inversion of configuration at the carbon atom attacked. Such coupling reactions, while often not preparatively useful, do introduce difficulties in the preparation of a number of simple metal alkyls from the corresponding halides, particularly iodides.

(d) **Reimer-Tiemann reaction:** An example of a species which can lay claim to be a carbanion only because of the π delocalisation that results in the transfer of negative charge from oxygen to carbon is the phenoxide ion (XXIX):

O⊖   O

⟷   ⊖   etc.

(XXIX)

This reacts with chloroform in the presence of strong bases to yield the *o*-aldehyde, salicaldehyde (XXX), plus a small amount of *p*-hydroxybenzaldehyde in the Reimer-Tiemann reaction, the quasi-carbanion (XXIX) attacking the electron-deficient carbene CCl_2 obtained by partial hydrolysis of chloroform in the alkaline solution (*cf.* p. 229):

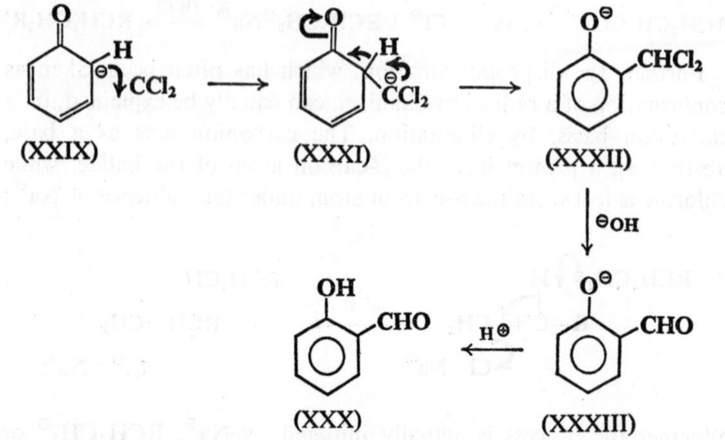

(XXIX) → (XXXI) → (XXXII)

(XXX) ← (XXXIII)

Thus attack on CCl_2 yields (XXXI) which undergoes a proton shift to form (XXXII). This undergoes hydrolysis in the system to yield (XXXIII), the free aldehyde not being obtained until the reaction mixture is acidified after the reaction proper is over. Some *p*-hydroxy-benzaldehyde is also obtained for the negative charge on the nucleus of (XXIX) is not of course confined to the *o*-position.

Some support for the above mechanism involving attack by dichlorocarbene is provided by the fact that when the reaction is carried out on the anion of *p*-cresol (XXXIV), in addition to the expected *o*-aldehyde (XXXV), it is also possible to isolate the un-

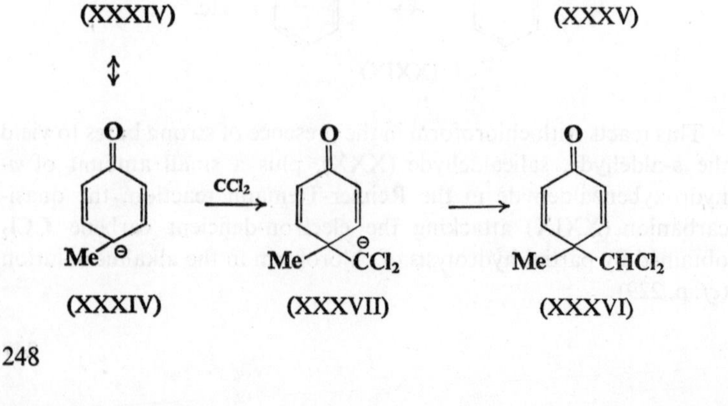

(XXXIV) → (XXXV)

(XXXIV) → (XXXVII) → (XXXVI)

hydrolysed dichloro-compound (XXXVI) arising from attack by CCl_2 at the *p*-position (see p. 248). The initial product of the attack (XXXVII) has no hydrogen atom that can be eliminated as $H^{\oplus}$ from the relevant carbon atom to allow reformation of an aromatic structure, so the introduced $^{\ominus}CCl_2$ group acquires $H^{\oplus}$ from the solvent. The dichloro-compound (XXXVI) is somewhat resistant to further hydrolysis, partly due to its insolubility in aqueous alkali but also because the two chlorine atoms are in a neopentyl type environment (*cf.* p. 80).

Somewhat analogous attack of CO_2 on powdered sodium phenate (the Kolbe-Schmidt reaction) is used to prepare sodium salicylate:

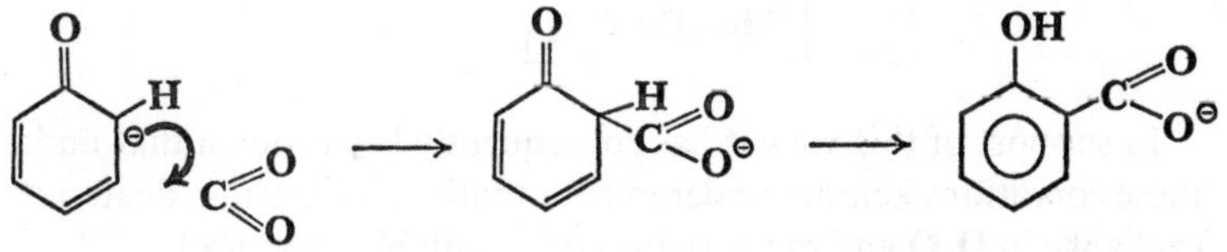

(XXIX)

Only traces of the salt of *p*-hydroxybenzoic acid are obtained, but if the reaction is carried out on potassium phenate, the salt of the *p*-acid becomes the major product. It has been suggested that this probably arises from stabilisation of the transition state for *o*-attack through chelation by $Na^{\oplus}$ in the ion pair (*cf.* p. 143):

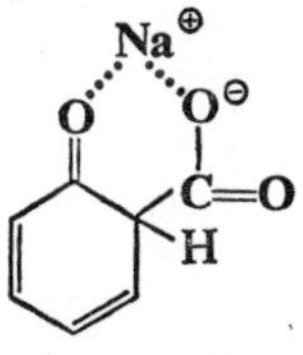

The larger $K^{\oplus}$ ion is likely to be less effective in this role and attack on the *p*-position will thus become more competitive.

(iii) Halogenation of ketones

One of the earliest observations having a bearing on the subject of the occurrence of carbanions as reaction intermediates was that the bromination of acetone in the presence of base followed the kinetic equation

$$\text{Rate} \propto [\text{MeCOMe}]$$

and was independent of [Br_2]. The rate was, however, found to be proportional to [base] and this was subsequently interpreted as involving the formation of a carbanion as the rate-determining step, followed by rapid bromination:

$$Me-\overset{\overset{O}{\parallel}}{C}-CH_3 \underset{slow}{\overset{\ominus OH}{\rightleftharpoons}} \left[\begin{array}{c} Me-\overset{\overset{O}{\parallel}}{C}-\overset{\ominus}{C}H_2 \\ \updownarrow \\ Me-\overset{\overset{\ominus O}{|}}{C}=CH_2 \end{array} \right] \overset{Br_2}{\underset{fast}{\longrightarrow}} Me-\overset{\overset{O}{\parallel}}{C}-CH_2Br + Br^{\ominus}$$

In support of this view it has subsequently been shown that under these conditions ketones undergo chlorination, iodination, deuterium exchange in D_2O and racemisation (if of suitable structure)

$$\overset{R}{\underset{H}{\overset{R'}{>}}}\overset{\overset{O}{\parallel}}{C}-CR \underset{\underset{slow}{H_2O}}{\overset{\ominus OH}{\rightleftharpoons}} \left[\overset{R}{\underset{R'}{>}}\overset{\ominus}{C}-\overset{\overset{O}{\parallel}}{C}R \leftrightarrow \overset{R}{\underset{R'}{>}}C=\overset{\overset{O^{\ominus}}{|}}{C}R \right]$$

$$(+) \qquad\qquad (XXXVIII)$$

$$\ominus OH \updownarrow H_2O$$

$$\overset{R}{\underset{R'}{\overset{H}{>}}}\overset{\overset{O}{\parallel}}{C}-CR$$

$$(-)$$

at exactly the same rate that they undergo bromination, suggesting the participation of a common intermediate in all. The reconversion to the ketone of the planar intermediate (XXXVIII), or the enol derived from it (*cf.* p. 238), leads as readily to the (+) as to the (−) form and thus results in racemisation.

Further base-induced halogenation of a mono-halogenated ketone (XXXIX) will take place preferentially at the carbon atom that has already been substituted, provided that it still carries a hydrogen atom; for not only will the inductive effect of the halogen atom make the hydrogen atoms attached to the halogen-substituted carbon atom more acidic, and therefore more readily abstracted by base, but it will also help to stabilise the resultant carbanion, leading to the formation of (XL) rather than (XLI):

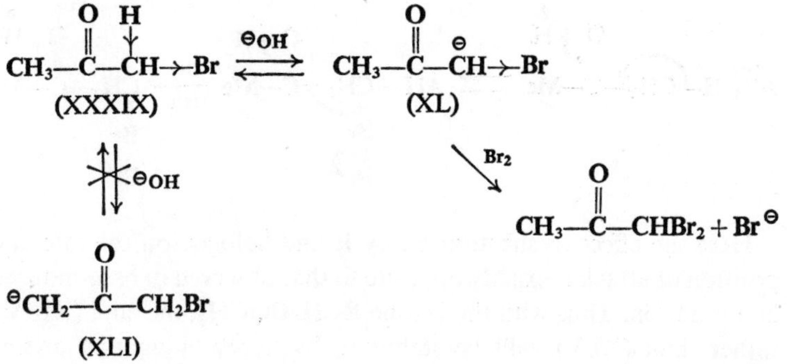

This is, of course, the reason for the exclusive production of $MeCOCX_3$ in the base-induced halogenation of acetone, the introduction of the second and third halogen atoms proceeding much more rapidly than that of the first. As a final stage in the *haloform* reaction, this species then undergoes attack by base, e.g. $^\ominus OH$, on the carbonyl carbon because of the highly positive character that that atom has now acquired:

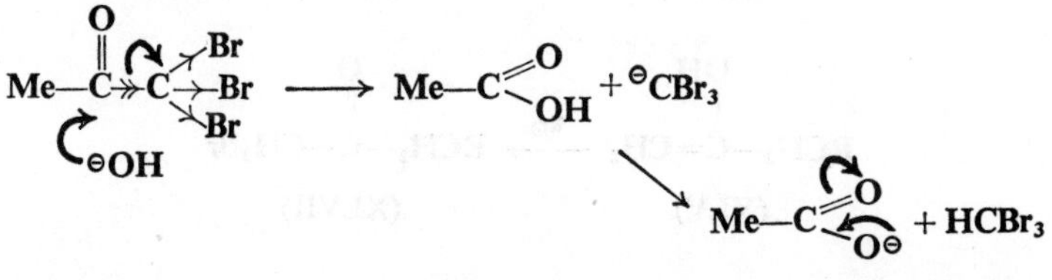

In the base-induced halogenation of the ketone, RCH_2COCH_3, it is the methyl rather than the methylene group that is attacked, for the inductive effect of the R group will serve to decrease the acidity of the hydrogen atoms attached to the methylene group, while those of the methyl group are unaffected, thus leading to preferential formation of the carbanion (XLII) which will be more stable than (XLIII):

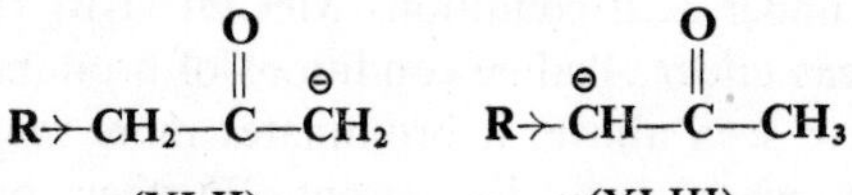

251

The halogenation of ketones is also catalysed by acids, the reaction probably proceeding through the enol form of the ketone (*cf.* p. 240) whose formation is the rate-determining step of the reaction:

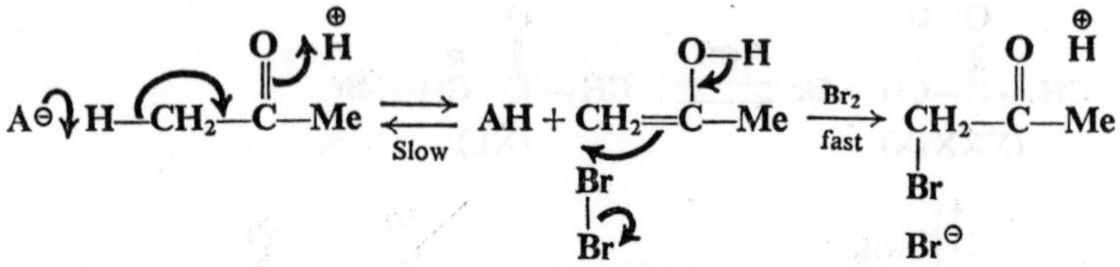

$$A^{\ominus} H - CH_2 - \overset{\overset{\displaystyle O}{\|}}{C} - Me \underset{\text{Slow}}{\rightleftharpoons} AH + CH_2 = \overset{\overset{\displaystyle O-H}{|}}{C} - Me \xrightarrow[\text{fast}]{Br_2} CH_2 - \overset{\overset{\displaystyle O}{\|}}{C} - Me$$

Here the effect of substitution by **R** and halogen on the rate and position of attack is exactly opposite to that observed in base-induced halogenation. Thus with the ketone **RCH$_2$COCH$_3$**, the enol (XLIV), rather than (XLV), will be stabilised by hyperconjugation arising from the α-hydrogens of the methyl group in addition to any in the **R** group, whereas only the methylene group will be operative in (XLV):

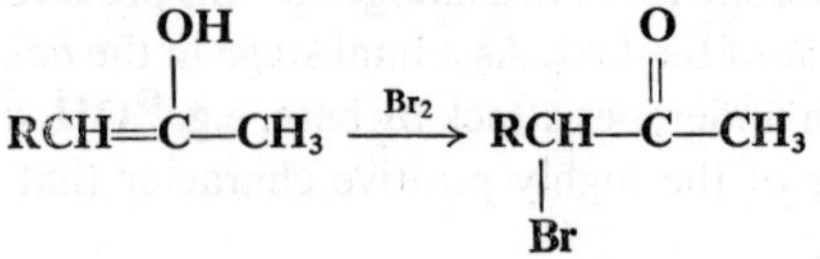

$$\underset{\text{(XLIV)}}{RCH = \overset{\overset{\displaystyle OH}{|}}{C} - CH_3} \xrightarrow{Br_2} \underset{\text{(XLVI)}}{RCH - \overset{\overset{\displaystyle O}{\|}}{C} - CH_3}$$

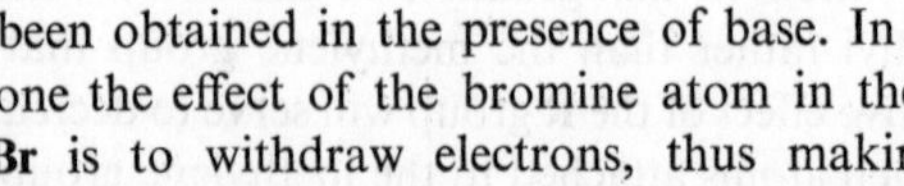

$$\underset{\text{(XLV)}}{RCH_2 - \overset{\overset{\displaystyle OH}{|}}{C} = CH_2} \xrightarrow{Br_2} \underset{\text{(XLVII)}}{RCH_2 - \overset{\overset{\displaystyle O}{\|}}{C} - CH_2Br}$$

This leads to the formation of (XLVI) rather than (XLVII), which would have been obtained in the presence of base. In the bromination of acetone the effect of the bromine atom in the first-formed **MeCOCH$_2$Br** is to withdraw electrons, thus making the initial uptake of proton by the $>$C$=$O, in forming the enol, *less* ready in bromoacetone than in acetone itself, resulting in preferential attack on the acetone rather than the bromoacetone in the system. The net effect is that under acid conditions **MeCOCH$_2$Br** can actually be isolated whereas under alkaline conditions of bromination it cannot for, as we have seen above, it brominates more readily than does acetone itself when base is present. Further bromination of

MeCOCH$_2$Br under acid conditions results in preferential attack on the methyl rather than the methylene group.

(iv) Decarboxylation

Another reaction involving carbanions is the decarboxylation of a number of carboxylic acids via their anions

$$\overset{\ominus}{O}\overset{\overset{\displaystyle O}{\|}}{-C}-R \longrightarrow CO_2 + R^{\ominus} \overset{H^{\oplus}}{\longrightarrow} R-H$$

the resultant carbanion $R^{\ominus}$ subsequently acquiring a proton from solvent or other source. It would thus be expected that this mode of decarboxylation would be assisted by the presence of electron-withdrawing groups in **R** because of the stabilisation they would then confer on the carbanion intermediate. This is borne out by the extremely ready decomposition of nitroacetate

$$\overset{\ominus}{O}-\overset{\overset{\displaystyle O}{\|}}{C}-CH_2-\overset{\oplus}{\underset{\underset{\displaystyle O^{\ominus}}{|}}{N}}{=}O \longrightarrow CO_2 + \left[\begin{array}{c} \overset{\ominus}{C}H_2-\overset{\oplus}{\underset{\underset{\displaystyle O^{\ominus}}{|}}{N}}{=}O \\ \updownarrow \\ CH_2{=}\overset{\oplus}{\underset{\underset{\displaystyle O^{\ominus}}{|}}{N}}-O^{\ominus} \end{array} \right] \overset{H^{\oplus}}{\longrightarrow} CH_3NO_2$$

and by the relative ease with which the decarboxylation of trihaloacetates and 2,4,6-trinitrobenzoates may be accomplished. That carbanions are involved may be demonstrated by carrying out the decarboxylation in the presence of bromine. This is found to have no effect on the rate of decomposition of, for example, $O_2NCH_2CO_2{}^{\ominus}$ but O_2NCH_2Br is then obtained, instead of O_2NCH_3, in conditions under which the latter is known not to be brominated. The new product thus arises by rapid, not rate-determining, attack of bromine of the carbanion, $O_2NCH_2{}^{\ominus}$, resulting from decarboxylation, and the latter is thereby 'trapped'.

The decarboxylation of β-keto acids may also proceed via their anions leading to stabilised carbanions such as (XLVIII):

$$\overset{\ominus}{O} - \overset{O}{\overset{\|}{C}} - CH_2 - \overset{O}{\overset{\|}{C}} - Me \longrightarrow CO_2 + \begin{bmatrix} \overset{O}{\overset{\|}{C}} \\ \overset{\ominus}{C}H_2 - \overset{\|}{C} - Me \\ \updownarrow \\ \overset{O^\ominus}{\overset{|}{C}} \\ CH_2 = \overset{|}{C} - Me \\ (XLVIII) \end{bmatrix} \overset{H^\oplus}{\longrightarrow} CH_3 - \overset{O}{\overset{\|}{C}} - Me$$

The overall rate law for the decarboxylation is, however, found to contain a term referring to [keto acid] itself as well as to the concentration of its anion; this is believed to be due to incipient transfer of proton to the keto group through hydrogen bonding:

$$Me - \overset{O^{\cdot\cdot}H}{\overset{\|}{C}}\overset{O}{\underset{CH_2}{C}} = O \longrightarrow Me - \overset{O}{\underset{CH_2}{C}}\overset{H}{\underset{CH_2}{\overset{O}{\|}}}\overset{O}{C} = O \longrightarrow Me - \overset{O}{\overset{\|}{C}} - CH_3 + CO_2$$

(XLIX)

Confirmation of this mode of decarboxylation of the free acid has been obtained by 'trapping' the acetone-enol intermediate (XLIX). $\beta\gamma$-unsaturated acids can also decarboxylate by a rather similar pathway:

$$R - \overset{R_2C}{\underset{CH_2}{\overset{H}{\overset{\|}{C}}}}\overset{O}{\underset{}{C}} = O \longrightarrow R - \overset{R_2C}{\underset{CH_2}{\overset{H}{\overset{/}{C}}}}\overset{O}{\underset{}{C}} = O$$

(v) Rearrangement

Rearrangements involving carbanions are very much less common than those of carbonium ions, but examples are known of 1,2-aryl shifts (see p. 255) as indicated by the structure of the products resulting from protonation and carbonation. If lithium is used instead of sodium and the temperature kept low similar products of the unrearranged carbanion can be obtained, but only rearranged ones if the ion is first warmed. Investigation of the relative migratory aptitudes of *p*-substituted aryl groups confirms that carbanions rather than

$$\text{Ph}_3\text{C}-\text{CH}_2\text{Cl} \xrightarrow{\text{Na}} \left[\text{Ph}_3\text{C}-\text{CH}_2^{\ominus} \overset{\text{Na}^{\oplus}}{} \right]$$

$$\downarrow$$

$$\underset{\text{Ph}_2\overset{\ominus}{\text{C}}-\text{CH}_2\text{Ph}}{\text{Na}^{\oplus}}$$

$$\xrightarrow{\text{ROH}} \underset{\text{H}}{\text{Ph}_2\text{C}-\text{CH}_2\text{Ph}}$$

$$\xrightarrow{\text{CO}_2} \underset{\text{CO}_2^{\ominus}\text{Na}^{\oplus}}{\text{Ph}_2\text{C}-\text{CH}_2\text{Ph}}$$

radicals are involved. The driving force for the shift is presumably supplied by the extensive delocalisation possible in the rearranged carbanion, the slightly more vigorous conditions necessary for rearrangement of the lithium derivatives reflecting the element of covalent character in the C—Li bond. The shift may well involve a bridged intermediate or transition state (*cf.* p. 99).

Simple 1,2-alkyl shifts, which are very common in carbonium ions, are unknown in carbanions, reflecting the very much higher energy level of the relevant three-membered transition state that would here involve four electrons rather than the two in a carbonium ion shift (*cf.* p. 286). A number of 1,2-shifts of alkyl from nitrogen, oxygen and sulphur to a carbanion carbon atom are known, however.

11 RADICALS AND THEIR REACTIONS

MOST of the reactions that have been considered to date have involved the participation, however transiently, of charged intermediates, i.e. carbonium ions or carbanions as constituents of ion pairs, produced by the *heterolytic fission* of covalent bonds:

$$\underset{/}{\overset{|}{\gtrless}}C:X \quad \underset{\searrow}{\overset{\nearrow}{}} \quad \begin{array}{l} \underset{/}{\overset{|}{\gtrless}}C^{\oplus} \quad :X^{\ominus} \\[2ex] \underset{/}{\overset{|}{\gtrless}}C:^{\ominus} \quad X^{\oplus} \end{array}$$

But reactive intermediates possessing an unpaired electron, i.e. *radicals*, can also be generated if a covalent bond undergoes *homolytic fission*:

$$\underset{/}{\overset{|}{\gtrless}}C:X \quad \longrightarrow \quad \underset{/}{\overset{|}{\gtrless}}C\cdot \quad \cdot X$$

The activation energy for such homolysis is usually little, if any, greater than the energy of the bond that is broken, for the activation energy of the reverse reaction—the recombination of the two radicals—is usually negligible. The *rate* of homolysis thus tends, hardly surprisingly, to increase as the strength of the bond to be broken decreases.

Reactions involving such radicals occur widely in the gas phase, but they also occur in solution, particularly if the reaction is carried out in non-polar solvents and if it is catalysed by light or the simultaneous decomposition of substances known to produce radicals, e.g. peroxides. Radicals, once formed, are in general found to be far less selective in their attack on other species than are carbonium ions and carbanions.

Once a radical reaction has been started, if often proceeds with very great rapidity owing to the establishment of fast chain reactions (see below). These arise from the ability of the first formed radical to generate another on reaction with a neutral molecule, the new radical being able to repeat the process, and so the reaction is carried on. Thus in the bromination of a hydrocarbon, **R—H**, the reaction may need starting by introduction of the radical, **Ra·**, but once started it is self-perpetuating:

$$Ra\cdot + Br{-}Br \longrightarrow Ra{-}Br + \cdot Br$$

$$\downarrow$$

$$R{-}H + \cdot Br \longrightarrow R\cdot + H{-}Br$$

$$\uparrow \qquad \downarrow Br_2$$

$$\cdot Br + R{-}Br$$

The chief characteristics of radical reactions are their rapidity, their initiation by radicals themselves or substances known to produce them (*initiators*), and their inhibition or termination by substances which are themselves known to react readily with radicals (*inhibitors*), e.g. phenols, quinones, diphenylamine, iodine etc. Apart from the short-lived radicals that occur largely as reaction intermediates, some others are known which are more stable and which consequently have a longer life; these will be considered first.

LONG-LIVED RADICALS

The colourless solid hexaphenylethane, Ph_3CCPh_3, is found to yield a yellow solution in non-polar solvents such as benzene. This solution reacts very readily with the oxygen of the air to form triphenyl-methyl peroxide, $Ph_3COOCPh_3$, or with iodine to yield triphenyl-methyl iodide, Ph_3CI. In addition, the solution is found to be *paramagnetic*, i.e. to be attracted by a magnetic field, indicating the presence of unpaired electrons (compounds having only paired electrons are *diamagnetic*, i.e. are repelled by a magnetic field). These observations have been interpreted as indicating that hexaphenyl-ethane undergoes reversible dissociation into triphenylmethyl radicals:

$$Ph_3C{:}CPh_3 \rightleftarrows Ph_3C\cdot + \cdot CPh_3$$

In support of this hypothesis, it is significant that the C—C bond energy in hexaphenylethane is only 11 kcals/mole compared with 83 kcals/mole for this bond in ethane itself.

The degree of dissociation of a 3% solution in benzene has been estimated as about 0·02 at 20° and about 0·1 at 80°. The reason for this behaviour, in contrast to hexamethylethane which does not exhibit it, has been ascribed to the stabilisation of the triphenylmethyl radical, with respect to undissociated hexaphenylethane, that can arise from the delocalisation of the unpaired electron through the π orbitals of the benzene nuclei:

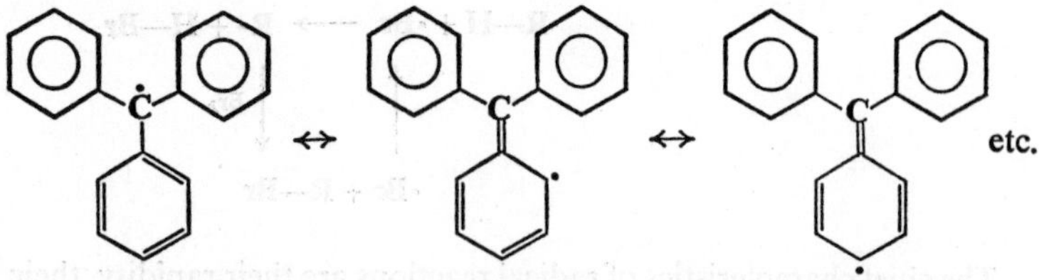

A number of contributing structures of this kind can be written, but the stabilisation thereby promoted is not so great as might, at first sight, be expected, for interaction between the hydrogen atoms in the *o*-positions prevents the nuclei attaining coplanarity. The radical is thus not flat, but probably more like a three-bladed propeller with angled blades, so that delocalisation of the unpaired electron, with consequent stabilisation, is slightly inhibited. Nevertheless, in the hexaphenylethane derivative with six *p*-NO_2 substituents, the extra delocalisation effected by the electron-withdrawing nitro groups is sufficient to cause almost complete dissociation.

The ready formation and stability of the radicals are, indeed, due in no small measure to the steric crowding in hexaphenylethane that can be relieved (tetrahedral → planar about the substituted carbon atom) by dissociation. In support of this explanation, it is found that the C—C distance in this compound is significantly longer (by $\approx 0·04$ Å) than in ethane. Also, while the introduction of a variety of substituents into the nuclei promotes dissociation, this is particularly marked when substituents are in the *o*-positions where they would be expected to contribute most to steric crowding. Further, it is found that the compound (I, see p. 259), in which two of the benzene nuclei on each carbon atom are bonded to each other and so held back from 'crowding' near the C—C bond, is not dissociated at room temperature though the possibilities of stabilising the radical, that could be

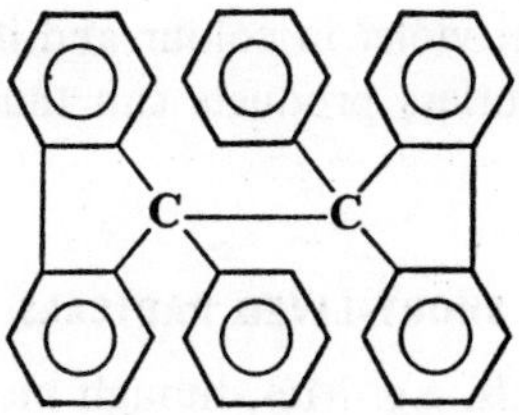

(I)

obtained from (I), by delocalisation are at least as great as those for triphenylmethyl.

Somewhat less stable radicals may be obtained by warming tetra-arylhydrazines in non-polar solvents, green solutions being obtained:

$$2Ph_2NH \xrightarrow{KMnO_4} Ph_2N:NPh_2 \rightleftarrows Ph_2N\cdot + \cdot NPh_2$$

Here, promotion of dissociation by steric crowding is clearly less important than with hexaphenylethane; stabilisation of the radical due to delocalisation may be more significant, but dissociation is certainly favoured by the lower energy of the N—N bond.

Similarly, solutions of diphenyl disulphide become yellow on heating

$$PhS:SPh \rightleftarrows PhS\cdot + \cdot SPh$$

and the radicals formed may be detected by the classical device of adding a second radical and isolating a mixed product:

$$PhS\cdot + \cdot CPh_3 \rightarrow PhS:CPh_3$$

The sulphide obtained is, however, rapidly decomposed in the presence of air. The best radical to use for such detection is 1,1-diphenyl-2-picrylhydrazyl (II)

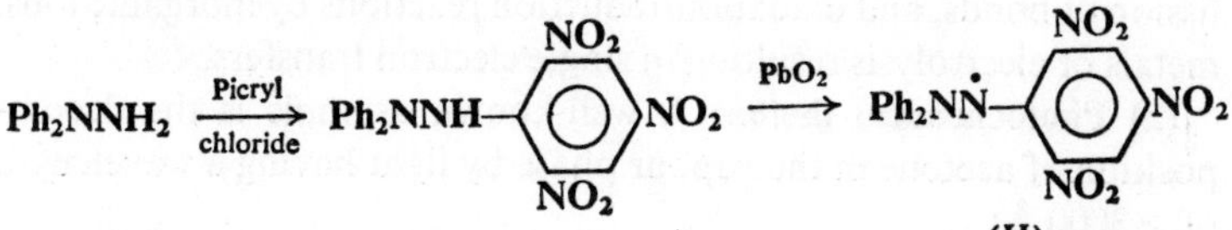

(II)

for this is very stable (due to delocalisation of the unpaired electron) and, further, will form stable products with many radicals. In addition,

259

its solutions are bright violet in colour and its reaction with other radicals to yield colourless products can thus be readily followed colorimetrically.

SHORT-LIVED RADICALS

The short-lived radicals, e.g. **Me·**, though more difficult to handle, are of much greater importance as participants in chemical reactions; as their short life suggests, they are extremely reactive.

The relative stability of simple alkyl radicals is reflected in the ease with which the C—H bond of the corresponding hydrocarbon may be broken homolytically, and is found to be in the same order as that of the corresponding carbonium ions (p. 78),

$$R_3C\cdot > R_2CH\cdot > RCH_2\cdot > CH_3\cdot$$

the sequence reflecting decreasing stabilisation by hyperconjugation and relief of steric strain as the series is traversed. As might be expected, however, the differences in stability between the radicals is less marked than between corresponding carbonium ions. Radicals involving allylic or benzylic positions show greatly enhanced stability arising from the delocalisation via π orbitals that is then possible:

$$CH_2{=}CH{-}\overset{\centerdot}{C}H_2 \quad \leftrightarrow \quad \overset{\centerdot}{C}H_2{-}CH{=}CH_2$$

etc.

(i) Methods of formation

There are numerous methods by which short-lived radicals may be formed, of which the most important are photochemical and thermal fission of bonds, and oxidation/reduction reactions by inorganic ions, metals or electrolysis resulting in single electron transfers.

(a) **Photochemical fission:** A well-known example is the decomposition of acetone in the vapour phase by light having a wavelength of ≈ 3000 Å:

$$\underset{\displaystyle Me{-}\overset{\displaystyle O}{\overset{\|}{C}}{-}Me}{} \;\rightarrow\; Me\cdot + \cdot\overset{\displaystyle O}{\overset{\|}{C}}{-}Me \;\rightarrow\; CO + \cdot Me$$

Another classic example is the conversion of molecular chlorine to chlorine atoms by sunlight

$$Cl—Cl \rightarrow Cl\cdot + \cdot Cl$$

that occurs as the first step in a number of photo-catalysed chlorinations (p. 269). Normally speaking, such photochemical decomposition may only be effected by visible or ultraviolet light of definite wavelengths corresponding—hardly surprisingly—to absorption maxima in the spectrum of the compound, and to bond dissociation energies of ≈ 35–120 kcal/mole. Reactions of this type also occur in solution, but the life of the radical is then usually shorter owing to the opportunities afforded for reaction with solvent molecules (*cf*. p. 267).

(b) **Thermal fission:** Much of the early work on short-lived radicals, including studies of their half-lives, was carried out on the products obtained from the thermal decomposition of metal alkyls:

$$Pb(CH_2CH_3)_4 \rightarrow Pb + 4 \cdot CH_2CH_3$$

Further reference is made to this work when the methods for detecting short-lived radicals are discussed below. In the vapour phase, the life of such radicals can be ended not only by dimerisation

$$CH_3CH_2\cdot + \cdot CH_2CH_3 \rightarrow CH_3CH_2CH_2CH_3$$

but also by disproportionation:

$$CH_3CH_2\cdot + CH_3CH_2\cdot \rightarrow CH_3CH_3 + CH_2{=}CH_2$$

The use of lead tetraethyl as an anti-knock agent depends in part on the ability of the ethyl radicals that it produces to combine with radicals resulting from the over-rapid combustion of petrol, thus terminating chain reactions which are building up towards explosion, but the complete details of the mechanism by which it prevents knocking are still unknown.

In solution, of course, the relative abundance of solvent molecules means that the initial radicals most commonly meet their end by reaction with solvent

$$CH_3CH_2\cdot + H—R \rightarrow CH_3CH_3 + \cdot R$$

but a new radical is then obtained in exchange and this may possibly be capable of establishing a new reaction chain.

The thermal fission of carbon–carbon bonds is seen in the radical-induced cracking (at $\approx 600°$) of long-chain hydrocarbons, in which the initial radicals introduced into the system act by abstracting a hydrogen atom from a CH_2 group of the chain. The radical so formed then undergoes fission at the β-position yielding an alkene of lower molecular weight and also a new radical to maintain the reaction chain:

$$\underset{|}{Ra \cdot H} \qquad\qquad RaH$$

$$RCH_2\overset{|}{C}HCH_2CH_2R' \;\rightarrow\; RCH_2\dot{C}HCH_2CH_2R'$$

$$\downarrow$$

$$RCH_2CH{=}CH_2 + \cdot CH_2R'$$

This is, of course, the reversal of the addition of radicals to alkenes in vinyl polymerisation (p. 274). Termination of the reaction by mutual interaction of radicals will tend not to take place to any marked extent until the concentration of long-chain hydrocarbons has dropped to a low level.

Bonds involving some elements other than carbon may undergo easier thermal fission. Thus azomethane is decomposed at 300°

$$CH_3{-}N{=}N{-}CH_3 \;\rightarrow\; 2CH_3 \cdot + N{\equiv}N$$

homolytic $C{-}N$ bond fission being promoted, as it results in formation of the extremely stable nitrogen molecule. Organic peroxides which contain the very weak $O{-}O$ bond (bond dissociation energy ≈ 30 kcal/mole) are particularly easily decomposed at quite low temperatures and, because of the ease with which they will form radicals, are much used as initiators:

$$\underset{Ph{-}C{-}O{-}O{-}C{-}Ph}{\overset{O\quad\;\; O\qquad\;\; O\quad\;\; O}{\|\qquad\|}} \;\rightarrow\; Ph{-}\overset{O}{\overset{\|}{C}}{-}O\cdot + \cdot O{-}\overset{O}{\overset{\|}{C}}{-}Ph$$

The decomposition of benzoyl peroxide, in which the resultant benzoate radicals are stabilised by delocalisation, is discussed in more detail below (p. 266).

(c) **Oxidation/reduction reactions:** Perhaps the best-known example is the use of ferrous ion to catalyse oxidations with hydrogen peroxide, the mixture being known as Fenton's reagent:

$$H_2O_2 + Fe^{2\oplus} \;\rightarrow\; HO\cdot + {}^{\ominus}OH + Fe^{3\oplus}$$

The ferrous ion goes to the ferric state and a hydroxyl radical is liberated. The latter acts as the effective oxidising agent in the system, usually by abstracting a hydrogen atom from the substrate that is to be oxidised:

$$HO \cdot + H{-}X \rightarrow H_2O + \cdot X$$

A rather similar reaction, but involving reduction of the inorganic ion, may take place as the first step in the autoxidation of benzaldehyde (p. 281), which is catalysed by a number of heavy metal ions capable of one-electron transfers:

$$\underset{\displaystyle \overset{\text{O}}{\|}}{Ph{-}C{-}H} + Fe^{3\oplus} \rightarrow \underset{\displaystyle \overset{\text{O}}{\|}}{Ph{-}C\cdot} + H^{\oplus} + Fe^{2\oplus}$$

In the Kolbe electrolytic synthesis of hydrocarbons

$$2R{-}\overset{\text{O}}{\underset{\|}{C}}{-}O^{\ominus} \xrightarrow{-2e^{\ominus}} 2R{-}\overset{\text{O}}{\underset{\|}{C}}{-}O\cdot \xrightarrow{-2CO_2} 2R\cdot \xrightarrow{\text{Dimerisation}} R{-}R$$
$$\text{(III)} \qquad\qquad \text{(IV)} \qquad\qquad \text{(V)}$$

the carboxyl anion gives up an electron on discharge at the anode to yield the carboxyl radical (III) which rapidly decarboxylates to form the alkyl radical (IV). These alkyl radicals then dimerise, in part at any rate, to yield the expected hydrocarbon (V).

Electrolysis of ketones in aqueous acid solution results in their reduction at the cathode to pinacols (VII) via the radical anion (VI)

$$2R_2C{=}O \xrightarrow{+2e^{\ominus}} 2R_2\underset{\cdot}{C}{-}O^{\ominus} \xrightarrow{\text{Dimerisation}} \begin{matrix} R_2C{-}O^{\ominus} \\ | \\ R_2C{-}O^{\ominus} \end{matrix} \xrightarrow{H^{\oplus}} \begin{matrix} R_2C{-}OH \\ | \\ R_2C{-}OH \end{matrix}$$
$$\text{(VI)} \qquad\qquad\qquad\qquad \text{(VII)}$$

which has already been encountered in a reaction of aromatic ketones in which sodium contributes an electron in the absence of air (p. 190); it also resembles the radical anion obtained in the first stage of the acyloin reaction (p. 190). The above are but two cases of electrolytic reaction, several examples of which have considerable synthetic importance.

(ii) Methods of detection

The classical work on the detection of short-lived radicals was done by Paneth using thin metal, e.g. lead, mirrors deposited on the inside wall of glass tubes. These mirrors disappeared when attacked by radicals, so by varying (*a*) the distance of the mirrors from the point where the radicals were generated (by thermal decomposition of metal alkyls), and (*b*) the velocity of the inert carrier gas by which the radicals were transported, it was possible to estimate their half-lives. That of methyl, under these conditions proved to be *ca.* 8×10^{-3} sec.

Some, more stable, radicals, e.g. **Ph$_3$C·**, may be detected by molecular weight determinations, but it is only rarely that this can be accomplished satisfactorily. Several radicals are coloured, though the compounds from which they are derived are not, so that colorimetric estimation may be possible; and even though the radicals themselves may not be coloured, the rate at which they discharge the colour of the stable radical, diphenylpicrylhydrazyl (II), may serve to determine their concentration. This is an example of 'set a radical to catch a radical' that has already been mentioned (p. 259), the evidence being strengthened by the isolation of the mixed product formed by mutual interaction of the two radicals, if that is possible. Another chemical method of detection involves the ability of radicals to initiate polymerisation of, for example, alkenes; reference is made to this below (p. 274).

The use of magnetic fields to detect the paramagnetism arising from the presence of unpaired electrons in radicals has already been referred to (p. 257). Though simple in essence, it can be fraught with much difficulty in practice and is relatively insensitive, only being able to detect relatively large concentrations of radicals: other more sensitive methods of detection are thus commonly preferred. The most useful of these to date is electron paramagnetic (or spin) resonance (epr or esr) spectroscopy, which depends on the magnetic moment (arising from the spin) of an unpaired electron being able to orient itself with or against an applied magnetic field, corresponding to magnetic quantum numbers of $+\frac{1}{2}$ and $-\frac{1}{2}$. The electron can thus exist in either of two energy levels, and transitions between them leads to a characteristic absorption line in the microwave region of the spectrum. It is thus possible to detect transient radical intermediates present only in low concentration ($\approx 10^{-12}$ molar).

Where it is desired merely to try and discover whether a particular reaction proceeds via radical intermediates or not, one of the simplest procedures is to observe the effect on the rate of the reaction of adding (*a*) compounds that readily form radicals e.g. organic peroxides, and (*b*) compounds known to react readily with radicals, i.e. inhibitors such as hydroquinone.

(iii) Stereochemistry

A good deal of attention has been devoted to the question of whether radicals in which the unpaired electron is on carbon have a planar structure (VIII), a pyramidal structure (IX),

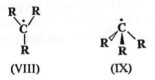

(VIII)   (IX)

or something in between. There is little doubt that in radicals that may be considerably stabilised by delocalisation of the unpaired electron, the three bonds attached to the carbon atom will be coplanar. Thus in triphenylmethyl, although interference between the *o*-hydrogen atoms of the benzene nuclei prevents the latter from lying in a common plane (*cf.* p. 258), the bonds joining the radical carbon atom to the three phenyl groups are almost certainly coplanar, for movement of one of these bonds out of the common plane would lower delocalisation possibilities without any compensating relief of steric strain. The benzene nuclei are angled to this common plane like the blades of a propeller so as to relieve as much steric strain as possible, while losing the minimum amount of delocalisation stabilisation due to their non-coplanarity.

By contrast, radicals in which the radical carbon atom constitutes the bridgehead of a rigid cyclic system will have the pyramidal configuration forced upon them, e.g. the apocamphyl radical (X):

Me   Me

(X)

There is found to be no difficulty in forming such non-planar radicals, unlike the corresponding carbonium ions, but there is

265

evidence that such radicals are somewhat less stable than simple tertiary aliphatic radicals upon which no such stereochemical restraint is imposed.

For radicals which do not have their configuration thrust upon them in this way, or which are not notably stabilised by delocalisation, the evidence available to date while not conclusive is certainly suggestive. Thus spectroscopic evidence indicates that the methyl radical, its deutero-derivative and $^{13}CH_3$ are planar or very nearly so, and with other simple alkyl radicals any stabilisation that may be possible, either by hyperconjugation or on steric grounds, will tend to favour the planar configuration; though this tendency is presumably less marked than with the corresponding carbonium ions as the resultant stabilisation of the radicals is much less pronounced than that of the ions.

(iv) Reactions

As with the carbonium ions and carbanions that have already been considered, radicals, once formed, can take part in three principal types of reaction: addition, displacement, and rearrangement, the latter normally being followed by one or other of the former. Before these reaction types are considered in detail, however, reference will be made to the formation and behaviour of a typical radical to illustrate the complexity of the secondary reactions that may result and, consequently, the wide variety of products that may be formed.

(a) **The thermal fission of benzoyl peroxide:** Benzoyl peroxide (a crystalline solid obtained by the reaction of benzoyl chloride with hydrogen peroxide in alkaline solution under Schotten-Baumann conditions) undergoes extremely ready thermal decomposition (*cf.* p. 262) to yield benzoate radicals:

$$Ph-\underset{\substack{\|\\O}}{C}-O-O-\underset{\substack{\|\\O}}{C}-Ph \longrightarrow Ph-\underset{\substack{\|\\O}}{C}-O\cdot + \cdot O-\underset{\substack{\|\\O}}{C}-Ph$$

It can be looked upon as consisting of two dipoles joined negative end to negative end as indicated above, and part, at least, of its inherent instability may stem from this cause. It would thus be expected that substitution of the benzene nucleus with electron-donating groups would enhance this instability leading to even more ready decomposition and this is, in fact, found to be the case. Electron-withdrawing groups are, correspondingly, found to exert a stabilising influence as compared with the unsubstituted compound.

Solutions of the peroxide are observed to liberate CO_2 as the temperature is raised, and even to a slight extent at room temperature, due to:

$$\underset{\underset{\displaystyle }{\overset{\textstyle O}{\overset{\|}{Ph-C-O\cdot}}}}{} \rightarrow Ph\cdot + CO_2$$

Thus, phenyl radicals will be present as well as benzoate radicals and often in quite considerable concentration; this is, indeed, one of the most useful sources of phenyl radicals. Production of the radicals in solution, as is the normal practice, can lead to further complications. Thus with benzene as solvent, the following initial reactions can, in theory, take place:

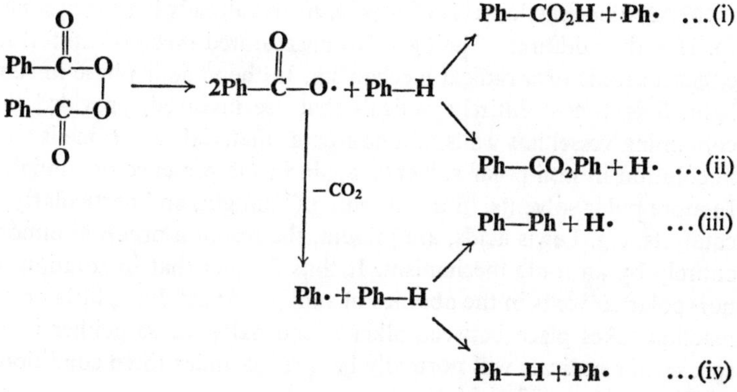

Reaction (iv) will, of course, not be directly detectable, but would serve to prolong the apparent life of phenyl radicals in the solution. In fact, (i) is found to be the main reaction taking place. It should be emphasised, however, that the above is only the *first stage*, for either

$\overset{\textstyle O}{\overset{\|}{Ph-C-O\cdot}}$ or Ph· can then attack the products derived from (i), (ii), and (iii). Thus, further attack on diphenyl from (iii) by Ph· leads to the formation of ter- and quater-phenyl, etc. It should, however, be pointed out that reactions (ii) and (iii) are almost certainly not direct displacements as shown, but proceed by addition followed by removal of a hydrogen atom from the addition product by another radical (*cf.* p. 284):

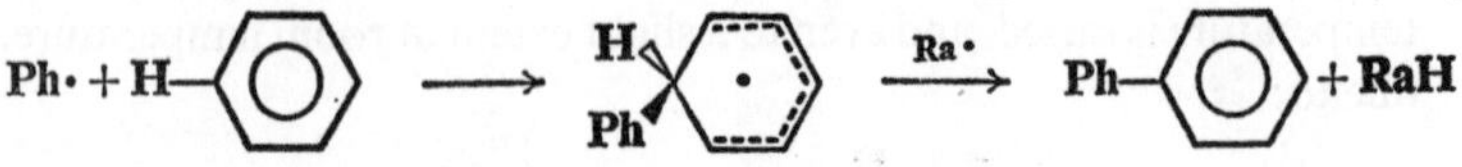

A further possibility is the attack of benzoate or phenyl radicals on as yet undecomposed benzoyl peroxide leading to the formation in the system of new radicals, $RaC_6H_4\overset{\text{O}}{\underset{\|}{C}}$—$O\cdot$ and $RaC_6H_4\cdot$, which can give rise to a further range of possible products. As this is only a simple case, the possible complexity of the mixture of products that may result from radical reactions in general will readily be realised.

The most important group of radical reactions are probably those involving addition.

(b) **Addition reactions:** (*i*) *Halogens.* As has already been mentioned (p. 156) the addition of halogens to unsaturated systems can follow either an ionic or a radical mechanism. In the vapour phase in sunlight, it is almost entirely radicals that are involved, provided the containing vessel has walls of a non-polar material. The same is true in solution in non-polar solvents, again in the presence of sunlight. In more polar solvents, in the absence of sunlight, and particularly if catalysts, e.g. Lewis acids, are present, the reaction proceeds almost entirely by an ionic mechanism. It thus follows that in solution in non-polar solvents in the absence of sunlight or catalysts, little or no reaction takes place between alkenes and halogens as neither ionic species nor radicals will normally be formed under these conditions without some specific initiating process.

The photochemically catalysed addition of chlorine to tetrachloroethylene (XI), for example, may be formulated as:

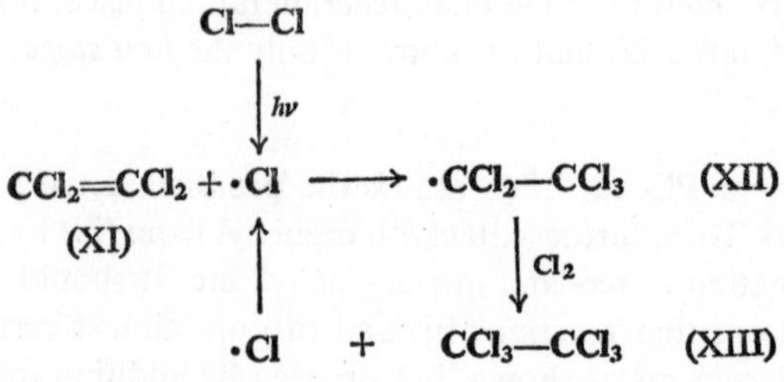

It will be seen that the initiating step, the photochemical fission of a molecule of chlorine, will lead to the formation of *two* reactive entities, i.e. free chlorine atoms, which are, of course, radicals. In support of this it is found that

$$\text{Rate} \propto \sqrt{\text{Intensity of absorbed light}}$$

i.e. each quantum of energy absorbed did, in fact, lead to the initiation of *two* reaction chains. The addition of a free chlorine atom to the unsaturated compound results in the formation of a second radical (XII) which is capable of undergoing a radical displacement reaction with a molecule of chlorine to yield the final addition product (XIII) and a free chlorine atom. This is capable of initiating a similar reaction cycle with a second molecule of unsaturated compound and so the process goes on, i.e. an extremely rapid, continuing chain reaction is set up by *each* initiating chlorine atom produced photochemically.

Such a continuing chain reaction, self-perpetuating once initiated, is perhaps the most characteristic feature of reactions proceeding via a radical mechanism. In support of the above reaction scheme, it is found that each quantum of energy absorbed leads to the conversion of several thousand molecules of (XI) $\rightarrow$ (XIII). Until the later stages of the reaction, i.e. when nearly all the unsaturated compound, (XI), is used up, the concentration of Cl· will always be very low compared to that of $Cl_2C{=}CCl_2$, so that termination of reaction chains owing to interaction of active intermediates, e.g. of (XII) or Cl· with themselves or each other, will be a very uncommon happening, and hence chain termination relatively infrequent. The reaction is inhibited by oxygen as the latter's molecule contains two unpaired electrons, ·O—O·, causing it to behave as a diradical, albeit a not very reactive one. It can thus act as an effective inhibitor by converting highly active radicals to the much less reactive peroxy radicals, RaO—O·. That the oxygen is reacting largely with pentachloroethyl radicals (XII) is shown by the

$$\text{formation of trichloroacetyl chloride, } Cl_3C-\overset{\displaystyle O}{\overset{\displaystyle \|}{C}}-Cl,$$

formation of trichloroacetyl chloride, $Cl_3C-\overset{O}{\overset{\|}{C}}-Cl$, when the addition of chlorine to tetrachloroethylene is inhibited by oxygen. The addition of halogens, X_2, to ethylene itself follows a closely parallel course. The initial stage (after photochemical formation of X·), i.e. addition of X· to $CH_2{=}CH_2$, is highly exothermic for Cl·, less so for Br·, and endothermic for I·; by contrast, the final stage, i.e. attack of XCH_2—$CH_2\cdot$ on X_2, is exothermic in all cases, with relatively little difference in magnitude from one to another. Thus the addition of chlorine is found to be rapid and to have long reaction chains, that of bromine to be slower with somewhat shorter reaction chains, while the addition of iodine does not normally take place at all.

The addition of chlorine to many unsaturated compounds is found to be irreversible at room temperature and for some way above (*cf.* p. 277), whereas the addition of bromine is often readily reversible. This results in the use of bromine radicals for the *cis* → *trans* isomerisation of geometrical isomers:

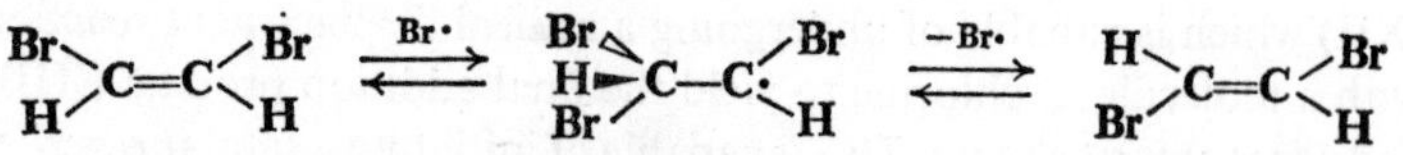

The radical (XIV) formed initially can then eliminate $Br\cdot$ very rapidly and so be reconverted to the *cis* starting material, (XV), or rotation about the C—C bond can take place first followed by subsequent elimination of $Br\cdot$ to yield the *trans* isomer, (XVI). As the latter is the more stable, it will come to preponderate, leading to an overall conversion of (XV) → (XVI).

The addition of chlorine to benzene—one of the few addition reactions of an unactivated benzene nucleus—has also been shown to proceed via a radical mechanism, i.e. it is catalysed by light and the presence of peroxides, and is slowed or prevented by the usual inhibitors. This presumably proceeds:

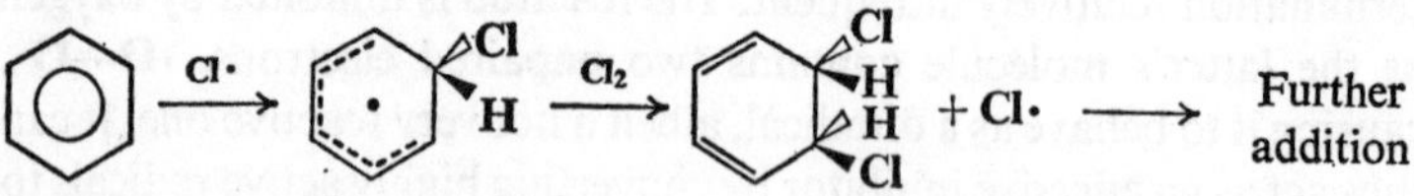

A mixture of several of the eight possible geometrical isomers of hexachlorocyclohexane is obtained. In the absence of sunlight and radicals, no addition of chlorine can take place, while if Lewis acid catalysts or their precursors are present (p. 125) electrophilic substitution occurs. With toluene, under radical conditions, attack on the methyl group offers an easier reaction path leading to predominant side-chain chlorination (substitution), rather than addition to the nucleus as with benzene, because of the stability and consequent ease of formation of the initial product, the benzyl radical, $PhCH_2\cdot$.

(*ii*) *Hydrogen halide.* The addition of hydrogen bromide to propylene via ionic intermediates to yield 2-bromopropane, has already been referred to (p. 160). In the presence of peroxides or other radical sources, however, the addition proceeds via a rapid chain reaction to yield 1-bromopropane (XVII) (i.e. the so-called 'anti-Markownikov' addition or *peroxide effect*). The difference in product under differing conditions is due to the fact that in the former case the addition is initiated by $H^{\oplus}$, while in the latter it is initiated by **Br·** (the alternative attack of **Ra·** on HBr to yield **H·** + **Ra—Br** is energetically much less favourable):

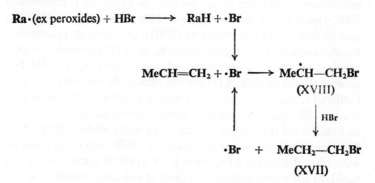

In the addition of **Br·** to propylene, the radical (XVIII) is formed rather than the possible alternative, **MeCHBrCH₂·**, since a secondary radical is more stable than a primary one (p. 260).

The addition reaction may not need the presence of added radicals to initiate it, however, for alkenes absorb oxygen from the air forming peroxides which can then themselves sometimes act as initiators. Such auto-initiation can be avoided by rigorous purification of the alkene prior to reaction, but this is not easy to achieve in practice, and formation of 2-bromopropane, i.e. predominance of the ionic reaction leading to so-called normal or Markownikov addition, is more easily secured by the addition of radical *acceptors* (inhibitors) such as phenols, quinones, etc., to absorb any radicals that may be present in the system and so prevent the occurrence of the rapid chain reaction.

It should not be thought that the presence of radicals in any way inhibits the ionic mechanism; it is merely that the radical reaction which they initiate, being a chain reaction, is so very much more rapid that it results in the vast majority of the propylene being converted to 1-bromopropane, (XVII), despite the fact that the ionic reaction is proceeding simultaneously. The virtually complete control of orientation of addition of **HBr** that can be effected by introducing radicals *or* radical acceptors into the reaction is very useful preparatively; it is not confined to propylene and applies to a number of other unsymmetrical unsaturated compounds, e.g. allyl bromide, $CH_2\!=\!CHCH_2Br$, which can be converted into 1,2- or 1,3-dibromopropane at will. In some cases, however, the ionic mechanism of addition is sufficiently fast to compete effectively with that induced by radicals and clear-cut control cannot then, of course, be effected.

It should, moreover, be emphasised that the reversal of the normal orientation of addition in the presence of peroxides is confined to **HBr**. This is due to the fact that with **HBr** the formation of (XVIII) and its subsequent conversion to (XVII), i.e. the steps propagating the chain-reaction, are both exothermic. With **HF** too much energy is required to produce **F·** in the second stage, and though with **HI, I·** is formed readily enough, it is then not sufficiently reactive to proceed further, i.e. the energy gained in forming a carbon–iodine bond is so much smaller than that lost in breaking a carbon–carbon double bond as to make the reaction energetically not worth while. With **HCl** the energetics more closely resemble those with **HBr** and radical additions have been observed in a few cases, but the radical reaction is not very rapid, as the reaction chains are short at ordinary temperatures (the second stage is often slightly endothermic), and it competes somewhat ineffectively with the ionic mechanism.

Nothing has so far been said about the stereochemistry of radical-induced additions to unsaturated compounds. The addition that has been the most intensively studied is that of **HBr** to simple acyclic unsaturated compounds, and it has been found that at low temperatures ($\approx -80°$), and with high relative concentrations of **HBr**, the addition proceeds stereospecifically *trans*: thus *cis* 2-bromobut-2-ene (XIX) yielded 92 % of the expected *meso* dibromide (XX) under these conditions:

$$\begin{array}{ccccc}
\underset{Me}{\overset{H}{>}}C\!=\!C\underset{Me}{\overset{Br}{<}} & \xrightarrow{\ Br\,\cdot\ } & \underset{Me}{\overset{H}{>}}\underset{}{Br\!>\!C\!-\!C\underset{Me}{\overset{Br}{:}}} & \xrightarrow{\ HBr\ } & \underset{Me}{\overset{H}{>}}Br\!>\!C\!-\!C\underset{Me}{\overset{Br}{<}}H + Br\,\cdot \\
(XIX) & & (XXIa) & & (XX)
\end{array}$$

At higher (i.e. room) temperature and with lower relative concentrations of **HBr**, however, a mixed product is obtained (XXII (78 %), as well as XX (22 %)). This has been explained as being due to rotation about the **C—C** bond in the intermediate radical (**XXI**a) occurring sufficiently rapidly (*cf.* p. 270), under these conditions, for both conformations (**XXI**a and **XXI**b) to be available by the time subsequent attack by **HBr** takes place:

(**XXI**a)   (**XXI**b)

(**XX**)   (**XXII**)

It is significant that exactly the same mixture of products is obtained from **HBr** and *trans* 2-bromobut-2-ene under these conditions. Significant also is the fact that cyclic alkenes, in which no such rotation about a **C—C** bond is possible in the intermediate radical, are found to undergo overall *trans* addition of **HBr** no matter what the conditions. An alternative suggestion has been made that a bridged radical

(corresponding to a bromonium ion (*cf.* p. 158) but involving an extra electron) may be involved in promoting overall *trans* addition of **HBr**, but the evidence is not precise enough to decide between the two alternative explanations. Where a double bond is sterically hindered in such a way as to impede an overall *trans* addition of **HBr**, virtually exclusive overall *cis* addition has been observed.

The addition of thiols, RSH, to alkenes closely resembles that of **HBr** in many ways. Heterolytic addition (of $RS^{\ominus}$) can take place but radical additions may be initiated by the presence of peroxides and, as with **HBr**, the two mechanisms generally lead to opposite orientations of addition. H_2S normally adds by an ionic mechanism.

Radical-induced addition of **CBr₄** and **CCl₄** may also be effected through attack of the first-formed $\cdot CX_3$ on an alkene, but the resultant radical, $\cdot CH_2CH_2CX_3$, may now compete with $\cdot CX_3$ in attack on a further molecule of alkene so that low molecular weight polymers may also be obtained.

(*iii*) **Vinyl polymerisation.** This reaction has probably received more attention than any other involving radicals, not least because of its commercial implications in the manufacture of polymers. It can be said to involve three phases:

(a) *Initiation:* Formation of **Ra·** from peroxides, etc.

(b) *Propagation:*

$$\textbf{Ra·} + \textbf{CH}_2\!\!=\!\!\textbf{CH}_2 \;\rightarrow\; \textbf{RaCH}_2\!\!-\!\!\textbf{CH}_2\textbf{·} \xrightarrow{\;CH_2=CH_2\;} \textbf{Ra(CH}_2)_4\textbf{·} \quad \text{etc.}$$

(c) *Termination:*

(i) $\textbf{Ra(CH}_2)_{n-1}\,\textbf{CH}_2\textbf{·} + \textbf{·Ra} \;\rightarrow\; \textbf{Ra(CH}_2)_n\textbf{Ra}$

(ii) $\textbf{Ra(CH}_2)_{n-1}\,\textbf{CH}_2\textbf{·} + \textbf{·CH}_2\textbf{(CH}_2)_{n-1}\textbf{Ra} \;\rightarrow\; \textbf{Ra(CH}_2)_{2n}\textbf{Ra}$

The propagation stage is usually extremely rapid.

As the alkene monomers readily absorb oxygen from the air, forming peroxides which can themselves form radicals and so act as initiators of polymerisation, it is usual to add some inhibitor, e.g. quinone, to the monomer if it is to be stored. When, subsequently, the monomer comes to be polymerised sufficient radicals must be produced to 'saturate' this added inhibitor before any become available to initiate polymerisation; thus an induction period is often observed before polymerisation begins to take place.

The radicals acting as initiators cannot properly be looked upon as catalysts—though often referred to as such—for each one that initiates a polymerisation chain becomes irreversibly attached to the chain and, if of suitable chemical structure, may be detected in the final polymer. The efficiency of some radicals as initiators may be so great that, after any induction period, every radical formed leads to a polymer chain; the concentration of initiator radicals may thus be kept very low.

Termination of a growing chain can result from reaction with either an initiator radical or a second growing chain, but of these the latter is normally the more important as the initiator radicals will have been largely used up in setting the chains going in the first place. It should be emphasised that such mutual interaction of radicals can result not only in dimerisation as above but also in disproportionation (p. 261). The chain length of the polymer, which is often of the order of

1000 units or more, may be controlled by addition of terminators or of *chain transfer agents*. These are usually compounds, **XH**, which react with a growing chain by loss of a hydrogen atom, so terminating the chain:

$$Ra(CH_2)_nCH_2 \cdot + HX \rightarrow Ra(CH_2)_nCH_3 + \cdot X$$

A new radical, **X·**, is formed and in the case of terminators **X** is chosen so that this radical is of low reactivity and hence not capable of initiating addition polymerisation in more monomer. In the case of chain transfer agents, however, **X** is chosen so that **X·** is reactive enough to initiate a new reaction chain so that the length (molecular weight) of individual chains is then controlled without at the same time slowing down the overall rate at which monomer undergoes polymerisation. Thiols, **RSH** are often used as chain transfer agents yielding **RS·** radicals as the initiators of the new chains.

The radical-induced polymerisation of simple alkenes, e.g. ethylene, propylene, is difficult and requires extreme conditions, but many other substituted alkene monomers may be polymerised readily. These include $CH_2{=}CHCl \rightarrow$ polyvinyl chloride for making pipes etc., $CH_2{=}CHMeCO_2Me \rightarrow$ perspex, $PhCH{=}CH_2 \rightarrow$ polystyrene, and $CF_2{=}CF_2 \rightarrow$ teflon which has an extremely low coefficient of friction, and unusually high chemical inertness and m.p. The properties of the resultant polymer may be even further varied—virtually as required—by the *copolymerisation* of two different types of monomer so that both become incorporated in the polymer molecule.

Radical-induced polymerisation of alkenes has some drawbacks, however. Thus branched as well as straight chain polymers may be formed as a result of a developing radical chain abstracting a hydrogen atom from another such chain, or from an already formed polymer molecule, and so providing a new growing point. Also, in the polymerisation of $CH_2{=}CHX$ the X groups will end up arranged stereochemically at random about the carbon chain backbone of the polymer molecules in the solid, and such *atactic* polymers are generally found to be non-crystalline, low melting, and mechanically weak. However, polymerisation of, for example, $MeCH{=}CH_2$ may be effected under very mild conditions with a $TiCl_4 \cdot AlEt_3$ catalyst and the resultant polypropylene is then found to be crystalline, and of high mechanical strength, due to the **Me** groups being oriented regularly— all on the same side—about the carbon chain backbone of the now *isotactic* polymer; branching is avoided as well. The mechanism of this

275

coordination polymerisation is not wholly clear but appears to involve oriented transfer of monomer molecules to the growing chain while both are transiently bonded to **Ti**. Vinyl polymerisation, proceeding by ionic mechanisms, may also be initiated by acids and bases, and by Lewis acids.

(c) **Displacement reactions:** (*i*) *Halogenation.* The displacement reactions on carbon that proceed via a radical mechanism are not in fact *direct* displacements or substitutions but involve two separate stages. This may be seen in the photochemically catalysed chlorination of a hydrocarbon:

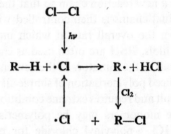

The reaction may also be initiated in the dark by heating but considerably elevated temperatures are required to effect Cl—Cl → Cl· ·Cl; thus the rate of chlorination of ethane in the dark at 120° is virtually indetectable. The reaction becomes extremely rapid, however, on the introduction of small quantities of **Pb(Et)₄** which undergoes decomposition at this temperature to yield ethyl radicals (*cf.* p. 261) capable of acting as initiators:

$$\text{Et}\cdot + \text{Cl—Cl} \rightarrow \text{Et—Cl} + \cdot\text{Cl}$$

The hydrogen atom on a tertiary carbon is more readily displaced than those on a secondary carbon and these, in their turn, more readily than those on a primary carbon; this reflects the relative stability of the radical, **R·**, that will be formed in the first instance (p. 260). The difference is often not sufficiently great, however, to avoid the formation of mixtures of products from hydrocarbons containing more than one position that may undergo attack; further, what preferential attack there is may be in large part negatived by a statistical effect. Thus, in isobutane, $(CH_3)_3CH$, although the hydrogen atom on the tertiary carbon is more readily attacked than those on the primary, there are no less than *nine* of the latter to attack compared with only *one* of the former, thus further limiting the preparative, i.e. selective, use of photochemical chlorination. If the chlorination is carried out in solution the product distribution is found to depend also on the nature of the solvent particularly on its ability to complex with a chlorine atom, thereby stabilising it and so making it more selective in its action.

The reaction is, however, also influenced by polar factors, for the electronegative **Hal·** as well as being a radical is at the same time an electrophilic reagent and will tend therefore to attack preferentially at a site where the electron density is high. Radical halogenation thus tends to be inhibited by the presence of electron-withdrawing atoms or groups, as is seen in the relative amounts of substitution at the four carbon atoms of 1-chlorobutane on photochemically catalysed chlorination at 35°:

$$CH_3—CH_2—CH_2—CH_2—Cl$$
$$25\% \quad 50\% \quad 17\% \quad 3\%$$

The variation for the three **CH$_2$** groups demonstrates the diminishing operation with distance of the inductive effect of **Cl**, the γ-**CH$_2$** behaving essentially analogously to that in **CH$_3$CH$_2$CH$_2$CH$_3$**; the lowered amount of attack on **CH$_3$** reflects the greater difficulty of breaking the **C—H** bond in **CH$_3$** rather than **CH$_2$**.

If the carbon indirectly attacked is asymmetric, e.g. **RR'R"CH**, then a racemic chloride is obtained. This racemisation does not constitute proof of the planar nature of the radical formed, however (*cf.* p. 265), for the same result would be obtained with a radical having a pyramidal configuration provided it could rapidly and reversibly turn itself 'inside out' as can the pyramidal molecules of ammonia and amines:

$$R \overset{\dot{C}}{\underset{R'}{\big|}} R'' \rightleftharpoons R \overset{R'}{\underset{\underset{\cdot}{C}}{\big|}} R''$$

At elevated temperatures (*ca.* 450°) propylene, **MeCH=CH$_2$**, is found to undergo chlorination to allyl chloride rather than addition of chlorine, for as the temperature rises the addition reaction becomes reversible (*cf.* p. 270) whereas the displacement reaction via a stabilised allyl radical does not:

$$Cl_2 + CH_3CH{=}CH_2 \rightarrow ClCH_2CH{=}CH_2 + HCl$$

The ΔH (kcal/mole) values for the two steps (after photochemical formation of $X\cdot$) of the chain reaction (p. 276) involved in the halogenation of methane are as follows:

Step	F_2	Cl_2	Br_2	I_2
(1) $CH_3{-}H + \cdot X \rightarrow CH_3\cdot + H{-}X$	-32	-1	$+15$	$+33$
(2) $CH_3\cdot + X{-}X \rightarrow CH_3{-}X + \cdot X$	-70	-23	-21	-18

Thus fluorination takes place with great readiness and though it appears to proceed via a radical mechanism, the reaction will often take place in the absence of light or initiators. Fluorine atoms are then believed to be produced, in the first instance, by the reaction:

$$\overset{}{\underset{}{{>}C{-}H}} + F{-}F \longrightarrow \overset{}{\underset{}{{>}C\cdot}} + H{-}F + \cdot F$$

The driving force of the reaction is provided by the 100 kcals by which the bond energy of $H{-}F$ exceeds that of $F{-}F$. If $F\cdot$ is produced in this way then small traces of F_2 should catalyse the chlorination of alkanes in the absence of photochemical activation: this does indeed occur at room temperature, or even lower, in the dark. Bromination is generally slower and less easy than chlorination, requiring above room temperatures except for reactive $C{-}H$ bonds as the stage in which a hydrogen atom is abstracted

$$\overset{}{\underset{}{{>}C{-}H}} + \cdot Br \longrightarrow \overset{}{\underset{}{{>}C\cdot}} + H{-}Br$$

is usually endothermic, whereas in chlorination this stage is slightly exothermic due to the greater bond energy of $H{-}Cl$ as compared with $H{-}Br$. The lower reactivity of bromine compared with chlorine is associated, as often happens, with greater selectivity in the position of attack, so that the difference in reactivity of tertiary, secondary and primary hydrogen is considerably more marked in bromination than chlorination. Thus for the series CH_4, $PhCH_3$, Ph_2CH_2, and Ph_3CH, relative rates of bromination are found to differ over a range of 10^9 but only over a range of 10^3 for chlorination; selectivity decreases, however, as the temperature is raised. Direct iodination is not normally practicable, for though $I\cdot$ can be readily formed it is not reactive enough to abstract a hydrogen atom from an alkane, the bond energy of the H—I that would be formed being low.

Radical halogenation by reagents other than the halogens themselves is often of greater synthetic importance. Thus chlorination may be effected by *t*-BuOCl in the presence of an initiator, the latter acting so as to liberate *t*-BuO· which has been shown to be the effective hydrogen abstractor from R—H; preparatively useful chlorination may also be effected by SO_2Cl_2 with an initiator, here ·$SOCl_2$ is obtained which readily loses SO_2 to yield Cl·. Especially useful is N-bromosuccinimide (XXIII) particularly for its highly selective attack on a position α- to a double bond or benzene ring—an allylic or benzylic carbon; thus cyclohexene (XXIV) yields largely the 3-bromoderivative (XXV):

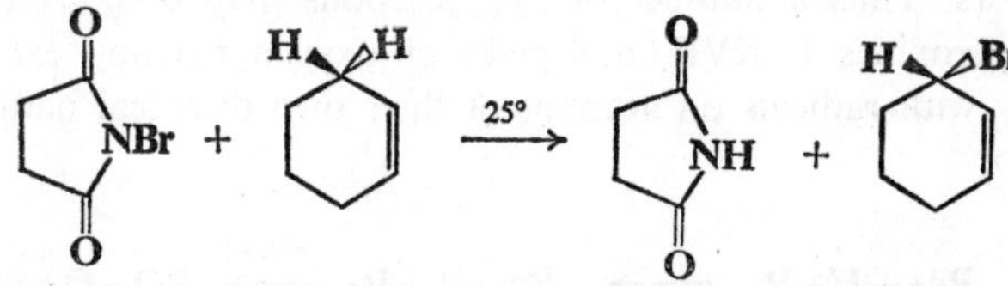

(XXIII)   (XXIV)   (XXV)

The reaction requires initiation by peroxides or light but a trace of HBr is also apparently necessary. It has been shown that (XXIII) undergoes a fast ionic reaction with HBr,

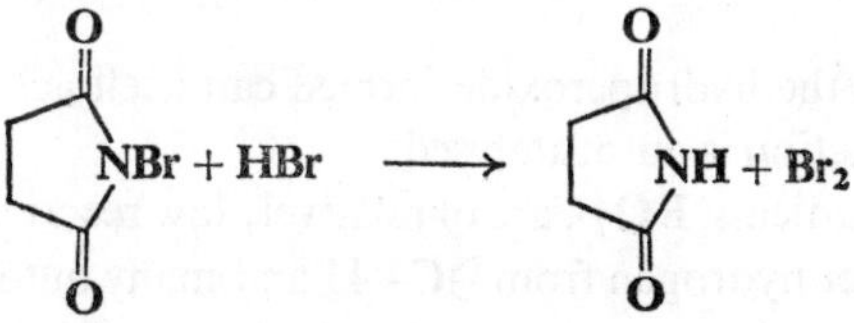

(XXIII)

and can thus act as the source of a small, steady concentration of Br_2 which will effect radical-induced bromination of the cyclohexene, the HBr thereby evolved liberating more Br_2 from (XXIII) by the above reaction. The selectivity of attack (little addition to the double bond takes place) is influenced by the stability of the intermediate allyl and benzyl radicals, but more particularly by the [Br_2] being kept low, for since addition of Br_2 is reversible, low bromine concentration favours hydrogen abstraction, which leads to overall substitution. Support for this interpretation is provided by the fact

279

that attack of Br_2, as such, in high concentration on cyclohexene is found to lead largely to addition, while attack in low ambient concentration coupled with efficient removal of the **HBr** evolved is found to lead to results exactly parallel to those with N-bromosuccinimide.

(*ii*) **Autoxidation.** Another displacement reaction involving radicals is autoxidation, the reaction of organic compounds with oxygen under mild conditions. The decomposition of most organic compounds exposed to air and sunlight is due to photosensitised oxidation; this is of importance in, for example, the deterioration of rubber and a number of plastics, and also in the hardening of oil paints and varnishes. Substances often contain impurities, e.g. trace metals, that can act as initiators so that the reaction then proceeds spontaneously; but, as always, added peroxides act as very powerful initiators. Thus a number of hydrocarbons may be converted to hydroperoxides (XXVI), molecules of oxygen reacting extremely readily with radicals on account of their own diradical nature (*cf.* p. 269):

$$Ra\cdot + H{-}R \longrightarrow Ra{-}H + R\cdot \xrightarrow{\cdot O_2\cdot} RO{-}O\cdot$$

$$\uparrow \qquad\qquad \downarrow {\scriptstyle R-H}$$

$$R\cdot \;+\; RO{-}OH$$

$$(XXVI)$$

In some cases the hydroperoxide formed can itself act as an initiator so that the reaction is autocatalysed.

As peroxy radicals, $RO_2\cdot$, are of relatively low reactivity they do not readily abstract hydrogen from $\geqslant C{-}H$ and many autoxidation reactions are highly selective. Thus tertiary hydrogens are usually the only ones attacked in simple saturated hydrocarbons but allylic, benzylic and other positions that can yield stabilised radicals are attacked relatively easily. Thus decalin (decahydronaphthalene) yields (XXVII), cyclohexene (XXVIII) and diphenylmethane, (XXIX), respectively:

OOH

OOH

OOH

CH

(XXVII)   (XXVIII)   (XXIX)

Ethers are particularly prone to autoxidation, attack taking place α- to the oxygen atom; the resultant hydroperoxides are dangerously explosive. Peroxidation reactions often proceed further, however, to yield alcohols, ketones, acids and more complex products, and with suitable unsaturated compounds the first formed hydroperoxy radical, $RO_2\cdot$ may initiate polymerisation: something like this is involved in the hardening of drying oils.

Reference has already been made (p. 118) to the large-scale conversion of cumene (isopropylbenzene) into phenol and acetone via the hydroperoxide; further study has shown that the presence of electron-withdrawing substituents (e.g. p-NO_2) in the benzene ring slows down the rate of formation of the hydroperoxide, while electron-donating substituents speed it up. The air oxidation of tetralin (tetrahydronaphthalene) to the ketone α-tetralone may also be accomplished preparatively via the action of alkali on the first-formed hydroperoxide:

In addition, the corresponding alcohol α-tetralol may be obtained by reductive fissions of the hydroperoxide.

Aldehydes also readily autoxidise: thus benzaldehyde, in air, is extremely easily converted into benzoic acid (see p. 282). This reaction is catalysed by light and also by a number of metal ions, provided these are capable of a one electron oxidation/reduction transition (e.g. $Fe^{3+} \rightarrow Fe^{2+}$). The perbenzoate radical (XXXI), obtained by addition of $\cdot O_2\cdot$ to the first-formed benzoyl radical (XXX), removes a hydrogen atom from a second molecule of benzaldehyde to form perbenzoic acid (XXXII) plus a benzoyl radical (XXX) to continue the reaction chain.

281

$$
\underset{\text{(XXX)}}{\overset{\displaystyle O}{Ph-\overset{\|}{C}-H}} + Fe^{3\oplus} \longrightarrow Fe^{2\oplus} + H^{\oplus} + \underset{\text{(XXX)}}{\overset{\displaystyle O}{Ph-\overset{\|}{C}\cdot}} \overset{\cdot O_2\cdot}{\longrightarrow} \underset{\text{(XXXI)}}{\overset{\displaystyle O}{Ph-\overset{\|}{C}-O-O\cdot}}
$$

$$
\underset{\text{(XXX)}}{\overset{\displaystyle O}{Ph-\overset{\|}{C}\cdot}} + \underset{\text{(XXXII)}}{\overset{\displaystyle O}{Ph-\overset{\|}{C}-O-OH}}
$$

$$
\underset{\text{(XXXII)}}{\overset{\displaystyle O}{Ph-\overset{\|}{C}-O-OH}} + \overset{\displaystyle O}{Ph-\overset{\|}{C}-H} \overset{H^{\oplus}}{\longrightarrow} 2\overset{\displaystyle O}{Ph-\overset{\|}{C}-OH}
$$

The perbenzoic acid reacts with a further molecule of benzaldehyde, however, to yield two molecules of benzoic acid. This reaction is catalysed by hydrogen ions and so is accelerated as the amount of benzoic acid formed increases. The presence of electron-donating groups in the benzene nucleus, as might be expected, facilitates removal of a hydrogen atom from the aldehyde to yield the initial radical, corresponding to (XXX).

The autoxidation of aldehydes may be lessened by very careful purification but more readily by the addition of anti-oxidants, such as phenols and aromatic amines, that react preferentially with any radicals that may be present.

An interesting autoxidation is the photo-oxidation of hydrocarbons such as anthracene (XXXV), especially in solution in CS_2:

      (XXXV)   (XXXIII)   (XXXIV)

The light absorbed converts the anthracene to an excited state, such as the di-radical (XXXIII) or something like it (*cf.* p. 286), which then adds on a molecule of oxygen to form the trans-annular peroxide (XXXIV) in a non-chain process. This photo-oxidation proceeds so readily with higher *lin* aromatic hydrocarbons such as hexacene (XXXVI)

(XXXVI)

that it is impossible to work with them in sunlight.

(*iii*) **Arylation.** Phenyl or other aryl radicals can take part in a number of reactions, one of the most common being the Sandmeyer reaction of a diazonium salt with cuprous chloride or bromide as a complex ion, e.g. $CuCl_2{}^{\ominus}$. Here the aryl radical is believed to be generated by an electron transfer, coupled with loss of nitrogen

$$ArN_2{}^{\oplus} + CuCl_2{}^{\ominus} \rightarrow Ar\cdot + N_2 + CuCl_2$$

which is then followed by a displacement reaction on a chlorine atom of the cupric chloride:

$$Ar\cdot + Cl\!-\!Cu\!-\!Cl \rightarrow Ar\!-\!Cl + Cu\!-\!Cl$$

Again no reaction chain is set up as the reaction in which the radical is consumed does not lead to the production of a second one in its place. It is found that

$$\text{Rate} \propto [ArN_2{}^{\oplus}][CuCl_2{}^{\ominus}]$$

indicating that the first reaction is the slower, i.e. rate-determining, one. The actual intervention of radicals can be confirmed, under suitable conditions, by their initiation of the chain polymerisation of acrylonitrile. Reactions of diazonium solutions can also proceed through an ionic mechanism as has already been mentioned (p. 131).

Phenyl or other aryl radicals, generated in a number of ways, can also react with aromatic species, e.g. the Gomberg-Bachmann reaction in which aryl radicals are generated by the decomposition of the anhydrides (XXXVII) derived from diazohydroxides, in contact with the aromatic compound that is to be arylated:

$$ArN\!\!=\!\!N\!-\!O\!-\!N\!\!=\!\!NAr \rightarrow Ar\cdot + N_2 + \cdot O\!-\!N\!\!=\!\!NAr$$
(XXXVII)

It might be expected that the attack on the aromatic species, C_6H_5X, would then proceed via a direct displacement:

$$Ar\cdot + H—C_6H_4X \rightarrow Ar—C_6H_4X + H\cdot$$

This would, however, involve the breaking of a carbon–hydrogen bond and the formation of a carbon–carbon bond and, as the former is usually considerably stronger than the latter, would thus lead to a high activation energy and so to a slow reaction: this is not what is actually observed. Also the addition of substances that would readily be reduced by free hydrogen atoms have never been found to result in such reduction, indicating that it is unlikely that hydrogen atoms ever do in fact become free. It seems more likely, therefore, that the reaction proceeds as a two-stage process:

The hydrogen atom is removed by the diazotate, or other, radical or by attack on the original source of aryl radicals:

Evidence for such a mechanism is provided by the fact that such arylations show no isotope effect, i.e. deuterium and tritium are displaced as readily as hydrogen, indicating that arylation cannot be initiated by fission of a carbon–hydrogen (deuterium, or tritium) bond as the rate-determining step of the reaction.

The spread in relative reactivity of mono-substituted benzenes towards homolytic attack is very much less marked than towards electrophiles (p. 135) or nucleophiles (p. 148), showing only a ten-fold variation over the whole range of substituents; further, most substituted benzenes are more readily attacked than benzene itself, irrespective of the nature of **X**. Directive effects of **X** are also very much less marked than in electrophilic attack and all substituents, whether electron-donating or -withdrawing, appear slightly to favour *o/p*-attack probably because of the delocalisation possibilities:

Such stabilisation of intermediates is much less marked than in ionic reactions, and there is an unexplained preference for predominant attack on the *o*-position unless **X** is very large (e.g. *t*-**Bu**): other details of the mechanism of homolytic aromatic substitution are not yet clear either.

Yields in these reactions tend to be low, partly owing to the two-phase (aqueous/organic) system involved. They may be improved by using substances such as N-nitrosoacetanilide, **PhN(NO)COMe**, or dibenzoyl peroxide (heated) as sources of phenyl radicals, for the reaction may then be carried out in homogeneous solution using the substance to be phenylated as solvent, but mixed products and tarry by-products still result.

(**d**) **Rearrangements:** The few known 1,2-rearrangement reactions of radicals nearly always involve aryl residues as migrants, and even then only from an atom on which the attached groups are strained by crowding. Thus the radical (**XXXVIII**) derived from the aldehyde, **Ph₂C(Me)CH₂CHO**, may be made to undergo loss of **CO**, and the products ultimately formed are found to be derived largely from radical (**XXXIX**), rather than (**XL**):

$$\text{Ph}_2\text{MeC—CH}_2\dot{\text{C}}\text{=O} \xrightarrow{-\text{CO}} \text{Ph}_2\text{MeC—}\dot{\text{C}}\text{H}_2 \rightarrow \text{PhMe}\dot{\text{C}}\text{—CH}_2\text{Ph}$$

$$\text{(XXXVIII)} \qquad\qquad \text{(XL)} \qquad\qquad \text{(XXXIX)}$$

It has been suggested that the migration of phenyl rather than an alkyl group is encouraged by the possibility, with the former, of proceeding via an intermediate (**XLI**) that is stabilised by delocalisation:

$$\text{PhMe}\dot{\text{C}}\text{—}\dot{\text{C}}\text{H}_2 \quad \text{PhMeC—CH}_2 \quad \text{PhMe}\dot{\text{C}}\text{—}\dot{\text{C}}\text{H}_2$$

(**XLI**)

285

For in the above case it is migration of methyl rather than phenyl that might be expected to occur, thereby yielding the more stable radical, **Ph$_2\overset{\bullet}{\text{C}}CH_2$Me**; further the non-phenylated radical, **EtMe$_2$C$\overset{\bullet}{\text{C}}H_2$**, corresponding to (XL) is found not to rearrange at all. Simple 1,2-alkyl shifts are indeed unknown in radicals, and in carbanions too (p. 255), which is in marked contrast to carbonium ions in which they are very common (p. 104). This reflects the fact that in the transition states that would be involved in the three shifts

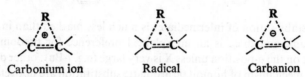

| Carbonium ion | Radical | Carbanion |

overlap of atomic orbitals of the three atoms concerned would lead to three molecular orbitals and these would have to accommodate two, three, and four electrons respectively:

Energy

Anti-bonding

Bonding

Carbonium ion   Radical   Carbanion

The carbonium ion transition state can accommodate its two electrons in the relatively low energy, bonding orbital and so forms fairly readily, but the other two transition states would, in addition, have to accommodate one and two electrons, respectively, in the much higher energy, non-bonding orbitals, and do not therefore form. 1,2-shifts of chlorine and bromine atoms in radicals have been observed, however.

But migration is not confined to shifts on to carbon; thus triphenylmethyl peroxide (p. 257) undergoes the following changes on heating:

$$\text{Ph}_3\text{C}-\text{O}-\text{O}-\text{CPh}_3 \;\rightarrow\; 2\text{Ph}_3\text{C}-\text{O}\cdot \;\rightarrow\; 2\overset{\bullet}{\text{Ph}_2\text{C}}-\text{OPh} \;\rightarrow\; \begin{array}{l}\text{Ph}_2\text{C}-\text{OPh} \\ | \\ \text{Ph}_2\text{C}-\text{OPh}\end{array}$$

(v) Diradicals

The oxygen molecule with an unpaired electron on each atom has already been referred to as a diradical, albeit an unreactive one, and the photochemical excitation of anthracene to a diradical, or some-

286

thing rather like it, has also been mentioned (p. 282); if the excitation is carried out in the absence of air or oxygen a photo-dimer (XLII) is formed:

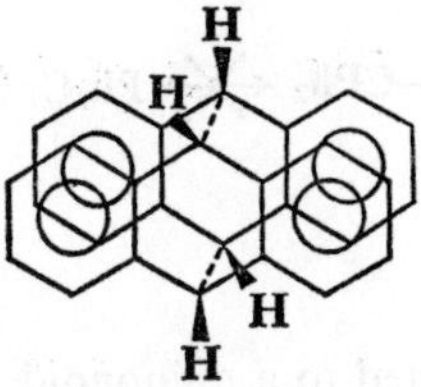

(XLII)

Diradicals, in the form of radical anions, have also been encountered as intermediates in the reduction of ketones to pinacols (p. 190) and in the acyloin reaction of esters (p. 190). The pyrolytic isomerisation of cyclopropane to propylene probably proceeds through an intermediate diradical also:

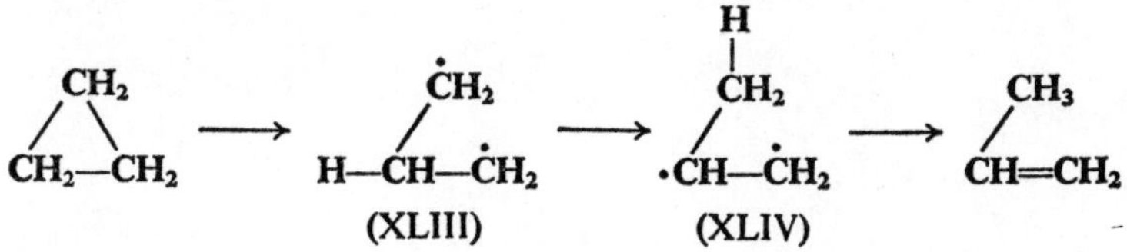

The driving force for the 1,2-hydrogen shift in (XLIII) → (XLIV) is provided by the possibility of consequent electron pairing in bond formation that is thereby afforded.

All these diradicals, with the exception of the oxygen molecule, are highly unstable but, surprisingly, a number of diradicals are known which are quite stable. Thus the hydrocarbon (XLV) exists in the diradical form:

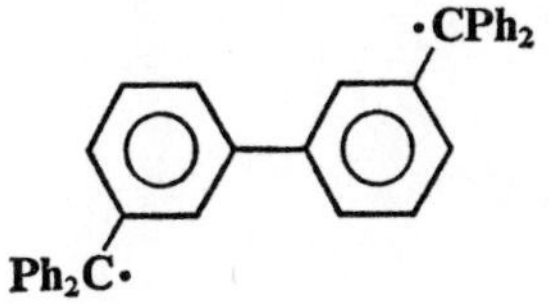

(XLV)

This is due to the fact that the diradical is greatly stabilised by delocalisation and also that a quinonoid structure embracing both nuclei, that would result in electron pairing, cannot be formed. The diradical (XLVI)

$$Ph_2\overset{\cdot}{C}-\langle O \rangle-\langle O \rangle-\overset{\cdot}{C}Ph_2 \quad \overset{Cl\ Cl}{\underset{Cl\ Cl}{\longleftrightarrow\!\!\!\!\times\!\!\!\!\longrightarrow}} \quad Ph_2C=\langle \rangle=\langle \rangle=CPh_2$$

(XLVI)	(XLVII)

can, in theory, be converted to a quinonoid form (XLVII) in which its electrons are paired, but formation of the latter is inhibited as the bulky chlorine atoms prevent the two benzene nuclei from becoming coplanar, a necessary condition if there is to be the effective over-lapping between their π orbitals that formation of (XLVII) requires. Both these diradicals undergo reversible association in solution, however.

SELECT BIBLIOGRAPHY

Valence and Chemical Bonding

CARTMELL, E., and FOWLES, G. W. A. *Valency and Molecular Structure* (Butterworths, 3rd Edition, 1966).

COULSON, C. A. *Valence* (O.U.P., 2nd Edition, 1961).

GRAY, H. B. *Electrons and Chemical Bonding* (Benjamin, 1964).

MORTIMER, C. T. *Reaction Heats and Bond Strengths* (Pergamon, 1962).

MURRELL, J. N., KETTLE, S. F. A., and TEDDER, J. M. *Valence Theory* (Wiley, 1965).

PIMENTEL, G. C., and MCCLELLAN, A. L. *The Hydrogen Bond* (Freeman, 1960).

STREITWIESER, A. *Molecular Orbital Theory for Organic Chemists* (Wiley, 1961).

Structure and Reaction Mechanism

AMIS, E. S. *Solvent Effects on Reaction Rates and Mechanisms* (Academic Press, 1966).

Aromaticity. Spec. Pub. No. 21 (Chemical Society, 1967).

BADGER, G. M. *Aromatic Character and Aromaticity* (C.U.P., 1969).

BANTHORPE, D. V. *Elimination Reactions* (Elsevier, 1963).

BARTLETT, P. D. *Nonclassical Ions* (Benjamin, 1965).

BETHELL, D., and GOLD, V. *Carbonium Ions* (Academic Press, 1967).

CRAM, D. J. *Fundamentals of Carbanion Chemistry* (Academic Press, 1965).

DE MAYO, P. (Ed.). *Molecular Rearrangements* (Interscience, Vol. I, 1963; Vol. II, 1964).

DEWAR, M. J. S. *Hyperconjugation* (Ronald Press, 1962).

ELIEL, E. L. *Stereochemistry of Carbon Compounds* (McGraw-Hill, 1962).

GILCHRIST, T. L. and REES, C. W. *Carbenes, Nitrenes and Arynes* (Nelson, 1969).

GOULD, E. S. *Mechanism and Structure in Organic Chemistry* (Holt-Dryden, 1959).

HINE, J. *Divalent Carbon* (Ronald Press, 1964).

HINE, J. *Physical Organic Chemistry* (McGraw-Hill, 2nd Edition, 1962).

HOFFMAN, R. W. *Dehydrobenzene and Cycloalkynes* (Academic Press, 1967).

Select Bibliography

INGOLD, SIR C. K. *Structure and Mechanism in Organic Chemistry* (Bell, 1953).

KIRMSE, W. *Carbene Chemistry* (Academic Press, 1964).

KOSOWER, E. M. *An Introduction to Physical Organic Chemistry* (Wiley, 1968).

LEFFLER, J. E., and GRUNWALD, E. *Rates and Equilibria of Organic Reactions* (Wiley, 1963).

MARE, P. B. D. DE LA, and BOLTON, J. *Electrophilic Addition to Unsaturated Systems* (Elsevier, 1966).

MARE, P. B. D. DE LA, and RIDD, J. H. *Aromatic Substitution: Nitration and Halogenation* (Butterworths, 1959).

MELANDER, L. *Isotope Effects on Reaction Rates* (Ronald Press, 1960).

MILLER, J. *Aromatic Nucleophilic Substitution* (Elsevier, 1968).

NEWMAN, M. S. (Ed.). *Steric Effects in Organic Chemistry* (Wiley, 1956).

NORMAN, R. O. C., and TAYLOR, R. *Electrophilic Substitution in Benzenoid Compounds* (Elsevier, 1965).

PRYOR, W. A. *Free Radicals* (McGraw-Hill, 1966).

STEWART, R. *Oxidation Mechanisms* (Benjamin, 1964).

STREITWIESER, A. *Solvolytic Displacement Reactions* (McGraw-Hill, 1962).

WALLING, C. *Free Radicals in Solution* (Wiley, 1957).

WATERS, W. A. *Mechanisms of Oxidation of Organic Compounds* (Methuen, 1964).

WELLS, P. R. *Linear Free Energy Relationships* (Academic Press, 1968).

WIBERG, K. B. *Physical Organic Chemistry* (Wiley, 1964).

WILLIAMS, G. H. *Homolytic Aromatic Substitution* (Pergamon, 1960).

Review Periodicals

Advances in Physical Organic Chemistry. GOLD, V. (Ed.) (Academic Press, Vol. I, 1963–).

Organic Reaction Mechanisms. REES, C. W., PERKINS, M. J., and CAPON, B. (Eds.) (Interscience, Vol. I, 1966–).

Progress in Physical Organic Chemistry. STREITWIESER, A., and TAFT, R. W. (Eds.) (Interscience, Vol. I, 1963–).

INDEX

Acetaldehyde
carbanion from, 194
hydration, 181
Acetals, 183, 246
Acetate ion, delocalisation, 16
Acetoacetate, ethyl
acidity, 233
carbanion from, 175
tautomerism, 239, 241, 242
Acetone
bromination, 41, 249
from cumene, 118
hydration, 181
Acetonitrile, acidity, 233
Acetylacetone
acidity, 233, 235
tautomerism, 242
Acetylenes
acidity, 57, 233
addition to, 159
alkylation, 245
carbanion from, 193
carbonation, 244
hydrogenation, 167
Acid anhydrides, reactivity, 206
Acid azides
Curtius degradation, 112
formation, 112
Acid chlorides, reactivity, 205, 206
Acidity
and sp^1 carbons, 57, 223
and sp^2 carbons, 57
anomalous o-effect in, 61
constant, 53
effect of solvent, 53
hydrogen bonding in, 62
in C—H compounds, 53, 232, 242
kinetic control of, 232, 234, 241, 242
origin of in organic compounds, 54
Acid strength
alcohols, 55
aliphatic acids, 22, 56
alkanes, 54, 232
benzoic acid, 60
dicarboxylic acids, 62
imides, 66
phenol, 21, 55
substituted aliphatic acids, 58

Acid strength—*continued*
substituted benzoic acids, 60
substituted phenols, 59
Acrylonitrile, in cyanoethylation, 172
Activated complex, 39
Activation
energy, 40
enthalpy, 40
entropy, 40
free energy, 40
Acylium ions, 130, 208
Acyloins, formation, 191
Addition
electrophilic, 156
nucleophilic, 172, 174, 178
radical, 268
stereospecificity, 49, 157, 272
to benzene, 270
to C=C, 156, 268
to C=C—C=C, 168
to C=C—C=O, 174
to C≡C, 159
to C≡N, 209
to C=O, 181
Alcohols
addition to C≡N, 209
addition to C=O, 183
dehydration, 97, 162, 211, 214
protonation, 98
Aldol condensation, 194
formaldehyde as acceptor in, 195
reversibility, 194
Aldols, base-catalysed dehydrations, 196
reversal of, 175
Alkenes
polar addition to, 156–168
protonation, 97
purification, 229
radical addition to, 268–276
rearrangement, 106
relative stability, 26, 219, 223
Alkyl chlorosulphites, decomposition, 85
Alkyl hydrogen sulphates, hydrolysis, 163
Alkylonium salts, elimination, 218
Alkynes, addition to, 159

Allyl cation, 79, 99, 104, 169, 170
Allylic rearrangement, 103, 194
Aluminium isopropoxide in Meerwein-
      Ponndorf reduction, 187
Amides
      Hofmann reaction, 111
      hydrolysis, 205
      N-substituted from Beckmann, 113
      reduction, 187
Aminoazo compounds, 134
Anilinium cation, 68
      electrophilic attack on, 142
Anthracene
      delocalisation energy, 17
      photodimerisation, 287
      photo-oxidation, 282
Anti-oxidants, 282
Arndt-Eistert reaction, 110
Aromatic character, 12, 100, 235
Aromatic substitution
      electrophilic, 119–148
      homolytic, 283
      nucleophilic, 148–155
Arrows
      curly, 16
      double-headed, 16
Arylation, 283
      directive effects in, 284
      two-stage, 284
Aryl cations, 150
Atactic polymers, 275
Autocatalysis, 280
Autoxidation, 280
      benzaldehyde, 263, 281
      cumene, 118, 281
      ethers, 281
      hydrocarbons, 280
Azomethines
      tautomerism, 240
      thermal decomposition, 262

Baeyer-Villiger oxidation, 116
Basicity
      and multiply bonded N, 71
      anomalous *o*-effects, 69
      effect of solvent, 66
      hydrogen bonding in, 65
      origin of in organic compounds, 63
      steric effects in, 70
Basic strength
      aliphatic amines, 64
      amides, 66
      aniline, 67
      guanidine, 67
      heterocyclic bases, 71
      substituted anilines, 68
      tetraalkylammonium hydroxides, 66
Beckmann rearrangement, 113
      $H_2^{18}O$ in, 114
      polarity of solvent in, 115
      stereochemistry, 113

Benzene
      aromatic character, 12
      bond lengths, 14
      charge clouds, 14, 119
      delocalisation in, 14, 119
      electrophilic substitution, 119
      heat of hydrogenation, 15
      Kekulé structures, 13
      π orbitals, 13
      radical addition, 270
      stabilisation of, 14
Benzilic acid change, 201
Benzoin condensation, 200
Benzoyl cation, 79, 99
Benzoyl peroxide, fission, 262, 266
Benzyne intermediates, 154
      decomp. in mass spectrometer, 155
      dimerisation, 155
      trapping of, 154
Biphenylene, 155
Bisulphite, addition to C=O, 184
Bisulphonium ion, 127
Bond
      angles, 5
      breaking, 18
      energies, 8
      forming, 18
      heterolytic fission, 18, 256
      homolytic fission, 18, 256
Bond lengths, 6
      and hybridisation, 6
Bouveault-Blanc reduction, 192
Bredt's rule, 222
Bridged carbonium ions, 99, 109, 286
Bridged radicals, 273, 285
Bromoamides, in Hofmann reaction,
      112
1,2-Bromohydrins, pinacolones from,
      108
Bromonium ions, 158
N-Bromosuccinimide
      in allylic and benzylic bromina-
      tion, 279
      mechanism of action, 279
1-Bromotriptycene, 81
t-Butyl cation, 74, 76, 78, 96, 101

Cannizzaro reaction, 188
      crossed, 195
      intramolecular, 189
Canonical structures, 16
Carbamic acids, in Hofmann reaction,
      112
Carbanions, 19, 232–255
      addition to C=C, 173, 175
      addition to C=O, 192–203
      and acidity, 232
      displacement reactions, 245–249
      formation, 232–234
      from acetic anhydride, 197
      from acetylene, 193, 233

Carbanions—*continued*
  from aldehydes, 173
  from aliphatic nitro compounds, 175,
    196, 243
  from azomethines, 240
  from cyclooctatetraene, 236
  from cyclopentadiene, 18, 235
  from diethyl malonate, 175
  from ethyl acetate, 198
  from ethyl acetoacetate, 175, 198
  from ethyl cyanoacetate, 175
  from ketones, 194, 202
  from triphenylmethane, 232, 237
  in alkylation, 245
  in benzoin condensation, 200
  in decarboxylation, 253
  in elimination, 214, 218, 226, 228,
    230, 247
  in halogenation of ketones, 249
  in ion pairs, 234
  in Kolbe-Schmidt reaction, 249
  in Reimer-Tiemann reaction, 247
  in tautomerism, 238–244
  in Wurtz reaction, 246
  rearrangement, 254
  solvation, 237
  stability, 234–237, 251
  stereochemistry, 237
Carbenes, 49, 110, 229, 247
Carbinolamines
  acid-catalysed dehydration, 185
  as intermediates, 185
Carbonation, of organometallics, 244
Carbon, electron-deficient
  migration to, 103–111
Carbonium ions, 19, 96–111
  addition to C=C, 164
  allyl, 79, 99, 104, 169, 170
  benzyl, 79, 99
  bridged, 99 109
  t-butyl, 74, 76, 78, 96, 101
  cycloheptatrienyl, 100
  cyclopropenyl, 100
  elimination of proton, 101, 105, 213
  intermediates in S_N1, 75, 78
  methods of formation, 96
  phenonium, 99
  phenyl, 150
  β-phenylethyl, 99
  planar, 75, 78, 84
  propyl, 102
  reactions, 101–103
  rearrangement, 101, 103–111, 162
  shielding of, 84
  stabilisation, 78, 98, 101
  stereochemistry, 101
  tropylium, 100
Carbon–nitrogen bonds, 9
Carbon–oxygen bonds, 9
Carbonyl group, 178–209
  attack of Lewis acids, 97

Carbonyl group—*continued*
  C—C bond formation with, 192–
    203
  characterisation of, 185
  condensation reactions, 185–186
  hemi-acetal formation from, 181
  hydration, 181
  in carboxylic acid derivatives, 204–
    209
  nature of, 178
  nucleophilic addition to, 181–203
  protection of, 184
  protonation , 97, 178
  reduction, 186–192
  stereochemistry of addition, 203
  structure of, 21, 178
  substitution and reactivity, 179–181,
    204–209
Carboxylic acid derivatives
  acid hydrolysis, 206
  base hydrolysis, 204, 205
  esterification, 205
  reactivity, 204
Chain length, 275
Chain reactions, 256, 269, 271
Chain transfer agents, 275
Chloral
  aldehyde ammonia, 186
  hydrate, 182
Chloroform, hydrolysis, 49, 51, 229,
  247
Chugaev reaction, 231
Cinnamic acid, from Perkin reaction,
  197
Claisen ester condensation, 198
  mixed, 199
  reversibility, 200
Claisen-Schmidt condensation, 197
Configuration, 40
  and optical rotation, 82
  apparent retention, 86
  determination of relative, 82
  inversion in S_N2, 81
  oximes, 113
  racemisation in S_N1, 83,
  retention in S_Ni, 84
Conformation
  eclipsed, 6
  staggered, 6
Conjugate
  acids, 52
  bases, 52, 232
Conjugation, 10
Conjugative effect, 21
Coordination polymerisation, 276
Cope reaction, 231
Copolymerisation, 275
Cracking of petroleum, 106, 262
Cram's rule, 204
Cumene, phenol from, 118
Curtius reaction, 111

Cyanides
  addition of ethanol, 209
  hydrolysis, 210
  reduction, 210
Cyanoethylation, 172
Cyanohydrins, formation, 184
Cycloheptatrienyl cation, 18, 100
Cyclohexene, heat of hydrogenation, 15
Cyclooctatetraene dianion, 236
Cyclopentadiene
  acidity, 236
  carbanion from, 18, 235
Cyclopropanes
  from carbenes, 49, 229
  pyrolysis, 287
Cyclopropenyl cations, 18, 100

Debromination, 1,2-dibromides, 227
Debromodecarboxylation, 227
Decarbonylation
  acylium ions, 131
  aldehydes, 285
Decarboxylation, 253
Dehydration
  acid-catalysed, 97, 162, 186, 214
  base-catalysed, 196, 226
Delocalisation, 13
  conditions for, 16
  energy, 15
  in ions, 19, 21, 235, 236, 255
  in σ complexes, 121, 139, 140
  steric inhibition of, 26, 70, 152, 209
    238, 288
Deuteration, toluene, 120
Diamagnetism, 257
Diazoamino compounds, 133
Diazo coupling, 131
  effect of pH, 132
Diazo hydroxides, 132, 283
  anhydrides from, 283
α-Diazoketones, Wolff rearrangement,
  110
Diazomethane, 110
Diazonium cations, 26, 91, 131
  decomp. in mass spectrometer, 155
  displacement reactions, 150
Diazonium salts, 91
  and amines, 132
  and phenols, 132
  decomposition, 93, 98
  Sandmeyer reaction, 283
Diazotate anions, 132
Diazotisation, 91, 133
Dichlorocarbene, 49, 229, 247
Dieckmann reaction, 176, 199
Diels-Alder reaction, 154, 171
  stereochemistry, 171
Dienes
  addition to, 168
  conjugated, 10, 168
  isolated, 10

Dienes, conjugated, addition to, 168–
  172
  halogens, 169
  hydrogen halide, 170
  1,2- v. 1,4-, 170
  reduction, 169
Dienophiles, 171
β-Diketones
  fission, 200
  tautomerism, 239
Dimedone
  enolisation, 176
  formation, 175
Dinitrogen tetroxide, 91
2,4-Dinitrophenylhydrazones, formation, 185
1,2-Diols
  from alkenes, 164
  rearrangement, 107
Diphenyl disulphide, radicals from, 259
1,1-Diphenyl-2-picrylhydrazyl, 259
αω-Diphenylpolyenes, colour, 12
Dipole moments, 20, 137
Diradicals, 269, 282, 286, 287
Displacement, 32
Double bond, C=C, 7
  addition to
    polar, 156
    radical, 268
  conjugated, 10
  energy, 8
  isolated, 10
  length, 8
  migration of, 11
  restricted rotation about, 8
Double bond, C=O, 9
  addition to, 181
  conjugated, 174
  energy, 10
  length, 10
  structure and reactivity, 179

Electrolytic reduction, 263
Electromeric effect, 24, 141
Electron configuration, 9
Electron-deficient carbon, migration
  to, 104–111
Electron-deficient nitrogen, migration
  to, 111–116
Electron-deficient oxygen, migration
  to, 116–118
Electron density, 19
  factors influencing, 19–26
Electron-donating groups, 23, 135
Electronegativity, 20
Electron paramagnetic (spin) reso-
  nance, 264
Electron spin, paired, 2
Electron-withdrawing groups, 23, 135,
  172, 225, 233, 253, 267, 277
Electrophiles, 30, 119

Electrophilic addition to C=C, 156–168
  bromine, 156
  bromonium ions in, 158
  carbonium ion intermediates, 158, 160, 161, 162, 164
  effect of added nucleophiles, 157
  effect of Lewis acids, 157
  effect of substituents on rate, 159, 161
  hydration, 161, 162
  hydroboration, 163
  hydrogen halide, 160
  hydroxylation, 164
  hypochlorous acid, 162
  orientation of, 160
  osmium tetroxide, 164
  ozone, 167
  peracids, 165
  π complexes in, 160, 163
  stereospecificity in, 157, 165
  sulphuric acid, 163
Electrophilic addition to C=C—C=C, 168
Electrophilic addition to C≡C, 159
Electrophilic substitution, aromatic, 119–148
  acylation, 130
  alkylation, 127, 144
  as addition/elimination, 122
  condition of reaction, 141
  deuteration, 120
  diazo coupling, 131
  effect of substituents, 135
  electron density and rate, 136, 138
  energetics of, 121, 124
  halogenation, 125
  inductive effect in, 135
  isolation of intermediates, 124, 128
  kinetic control, 138, 144, 146
  kinetic v. thermodynamic control, 144, 146
  Lewis acids in, 120, 125, 127, 130
  mesomeric effect in, 136
  nitration, 122
  ortho/para ratios, 143
  π complexes, 119
  position of substitution, 135, 138
  σ complexes, 120, 139, 140
  sulphonation, 127
  thermodynamic control, 144, 146
  transition states, 124, 138
  v. addition, 120
α-Elimination, 229
  chloroform, 49, 229, 247
  2,2-diphenylvinyl bromide, 230
E1 elimination, 212–214
  effect of structure, 213, 221, 245
  ion pairs in, 212, 213
  orientation, 223
  steric effects, 213, 223
  v. S_N1, 213, 223

E2 elimination, 212, 214–219
  carbanions in, 214, 218, 226, 228, 230
  debromination, 227
  debromodecarboxylation, 227
  effect of activating groups, 225
  effect of structure, 216, 221
  orientation, 218
  rate-limiting step, 214
  size of base, 220
  size of leaving group, 220
  stereospecificity, 216, 228
  steric effects, 219, 220, 221, 222
  strength of base, 215
  v. S_N2, 224
Ei (cis) elimination, 230
  acetates, 230
  cyclic transition states, 230
  ion pairs in, 231
  tertiary amine oxides, 231
  xanthates, 231
Elimination v. substitution, 212, 223–225
  change of mechanism, 224
  concentration of base, 224
  steric effects, 223, 224
  temperature, 225
Energetics of reaction, 35
Energy profiles, 39
Enolisation, 238, 252
  dimedone, 176
Enthalpy
  and bond energies, 37
  of activation, 40
Entropy, 36
  of activation, 40
  translational, 37, 57
Epoxides
  from alkenes, 165
  hydrolysis, 166
  ring-opening, 86, 94, 165
Equilibrium
  constant, 36
  control of reaction, 44, 144, 171
Ester hydrolysis
  acid-catalysed, 206
  acyl-oxygen cleavage, 205, 206
  alkyl-oxygen cleavage, 207
  base-catalysed, 205
  ^{18}O in, 47, 205
  steric hindrance, 28, 205, 207
  transesterification, 205
Esterification
  mechanism, 206
  o-substituted benzoic acids, 28, 207
  reversibility, 206
  steric hindrance, 28, 207
  tertiary acids, 28
  transesterification, 205
Esters
  acyloin condensation, 190
  amination, 205

Index

Esters—*continued*
  Claisen condensation, 198
  Dieckmann reaction, 199
  reduction, 187, 191
Ethers
  cleavage, 93, 98
  protonation, 93, 98
Ethyleneimmonium salts, 88
Ethyl orthoformate, 184, 246
Excited state, carbon, 3
Exclusion principle, 2

Fenton's reagent, 262
Ferrocene, 236
Fluorination, alkanes, 278
Formaldehyde, hydration, 181
Free energy, Gibbs, 36
  change and equilibrium constant, 36
  of activation, 39
Free radicals, 18, 256
Friedel-Crafts reaction
  acylation, 130, 137
  acylium ions in, 130
  alkylation, 127, 144
  carbonium ions in, 102, 129
  dealkylation in, 129
  effect of solvent, 137
  formylation, 130
  Lewis acids in, 103, 127
  polarised complexes in, 128, 130
  polyalkylation, 130, 137
  rearrangements in, 103, 129
  reversibility, 129
  with alkenes, 129
  with cyclic anhydrides, 131

Gattermann-Koch reaction, 131
General acid catalysis, 182, 241
1,2-Glycols
  cis, formation, 165
  trans, formation, 166
  rearrangement, 107
Glyoxal
  Cannizzaro reaction, 189
  hydrate, 182
Gomberg-Bachmann reaction, 283
Grignard reagents
  addition to C=C, 172
  addition to C=O, 192
  as Lewis acids, 192
  carbonation, 244
  displacement reactions, 245
  from 1,2-dihalides, 228
  structure of, 192
Ground state, carbon, 3

Haloform
  hydrolysis, 229
  reaction, 251
Halogenation

Halogenation—*continued*
  electrophilic, 125
  ketones, 249
  radical, 268, 276
α-Halogenoethers, formation, 185
Heats of
  combustion, 10, 219
  hydrogenation, 10, 15, 169
Hemi-acetals, 183
Hexacene, 283
Hexachlorocyclohexane
  elimination, 217
  formation, 270
Hexaphenylethane, dissociation, 257
Hofmann
  elimination, 211, 218, 219
  reaction of amides, 48, 111
Hund's rule, 3
Hybridisation, 3
Hydration
  C=C, 162
  C=O, 181
Hydride transfer reactions, 186
Hydroboration, C=C, 163
Hydrogenation
  acetylenes, 167
  alkenes, 166
  heats of, 10, 15, 169
  stereospecificity, 166
Hydrogen bonding
  in amine cations, 65
  in carbonyl hydrates, 183
  in decarboxylation, 254
  in dicarboxylic acids, 62
  in keto-enol tautomerism, 243
  in salicylic acid, 62
Hydrogen bromide, addition to, C=C
  allyl bromide, 272
  cyclic alkenes, 273
  electrophilic, 160
  orientation, 160, 271
  propylene, 160, 245
  radical, 271
  rate, 161
  stereospecificity, 162, 272
  vinyl bromide, 161
Hydrogen peroxide
  Fenton's reagent, 262
  hydroxylation of C=C, 165
  oxidation of ketones, 117
Hydrolysis
  esters, 205
  halides, 40, 73
Hydroperoxides
  decomposition, 281
  formation, 280
  in ozonolysis, 168
  rearrangement, 118
Hydroperoxy radicals, 281
Hydroxamic acids, Lossen degradation, 112

1,2-Hydroxyamines, pinacolones from, 108
Hydroxylation of C=C, 164
Hydroxyl radical, 263
Hyperconjugation, 24, 78, 209, 219, 252, 260, 266

Imino-ethers, from nitriles, 209
Inductive effect, 19
Inductomeric effect, 24
Inhibitors, radical, 257, 265, 269, 271, 274
Initiators, radical, 257, 262, 265, 271, 274, 280
Intermediates
  bridged, 100, 109, 273, 285
  isolation of, 48, 112, 124, 128, 151
  spectroscopic detection, 48
  trapping, 49, 154, 253
Intermolecular rearrangements, 109 134
Intramolecular rearrangements, 109
Ionisation
  and dielectric constant, 53
  and $\Delta G°$, 56
Ion pairs, 18, 85
Ions, solvation of, 53, 75
Isocyanates, as intermediates, 112
Isomerisation, $cis \rightarrow trans$, 270
Isotactic polymers, 275
Isotope effects
  non-kinetic, 47
  primary kinetic, 46, 124, 127

K_a, 53
K_b, 63
Ketals, 184
Ketenes, in Wolff reaction, 110
β-Keto acids, decarboxylation, 253
Keto-enol tautomerism, 239, 242
β-Keto esters
  'acid decomposition', 200
  formation, 198
  tautomerism, 239
Ketones, halogenation, 249–253
  acid-catalysed, 252
  base-catalysed, 250
  position of substitution, 250, 252
Ketoximes, Beckmann rearrangement, 113
Ketyls, 190
Kinetic control
  in addition to dienes, 170
  in aromatic substitution, 138
  in naphthalene sulphonation, 146
Kinetic data, interpretation, 45
Kinetic isotope effects, 46, 124, 127
Kolbe electrolysis, 263
Kolbe-Schmidt reaction, 249

Lead alkyls
  as anti-knock, 261
  thermal decomposition, 261, 276
Leaving groups, 89
Lewis acids, 30, 268
  in Friedel-Crafts, 102, 127, 270
  in petroleum cracking, 106
  in polymerisation, 270
Lewis bases, 30
Lindlar catalyst, 193
Lithium aluminium hydride
  in reduction of amides, 187
  in reduction of esters, 187
  in reduction of ketones, 186
Lossen reaction, 111

Magnetic moment, electronic, 264
Malonate, diethyl
  acidity, 233
  carbanion from, 175
Mannich reaction, 202
Markownikov's rule, 161, 271
Meerwein-Ponndorf reduction, 187
Mercaptans, addition to C=O, 184
Mesomeric effect, 21, 240
Meta directing groups, 139
Metal alkyls
  carbonation, 244
  in Wurtz reaction, 246
  thermal decomposition, 261
Michael reaction, 175
Migratory aptitude, 109, 254
Molecularity of reaction, 74
Molozonides, 167
Mustard gas, hydrolysis, 88

Naphthalene
  delocalisation energy, 17
  sulphonation, 146
Neighbouring group participation, 86
Neopentyl rearrangement, 104
Ninhydrin, 182
Nitrating mixture, 122
Nitration, 122
  kinetics, 123
  no isotope effect, 46, 124
  with dilute nitric acid, 125
Nitrenes, 111
Nitriles
  addition of ethanol, 209
  hydrolysis, 210
  reduction, 210
Nitrobenzene, nucleophilic substitution, 149
Nitro compounds, aliphatic
  acidity, 225, 233
  carbanions from, 175, 196
  tautomerism, 239, 243
Nitrogen, electron-deficient
  migration to, 111–116
Nitronium ion, 122

Nitrosation
  amines, 90, 102
  phenol, 125
Nitroso-amines, 92
Nitrosonium ion, 91, 125
Nitrosotrialkylammonium cation, 92
Nitrosyl halides, 91
Nitrous acid, protonated, 91
Non-bonded interactions, 6
Nucleophiles, 30, 88
Nucleophilic addition to activated C=C
  carbanions, 173
  cyanoethylation, 173
  ethanol, 172
  Grignard reagents, 172
Nucleophilic addition to C=C—C=O
  carbanions, 175
  cyanide ion, 174
  Grignard reagents, 174
  hydrogen halide, 174
  hydroxyl ion, 175
  Michael reaction, 175
  1,2- *v.* 1,4-, 174
Nucleophilic addition to C=O, 178–209
  acetylide ion, 193
  acid-catalysis, 179
  alcohols, 183
  aldol condensation, 194
  amine derivatives, 179, 185
  bisulphite ion, 184
  benzilic acid change, 201
  benzoin condensation, 200
  carbanions, 193–203
  Claisen ester condensation, 198
  Claisen-Schmidt condensation, 197
  cyanide ion, 178, 184
  Dieckmann reaction, 199
  electrons from dissolving metals, 190
  Grignard reagents, 192
  halide ion, 185
  hydride ion, 186
  hydroxyl ion, 188, 201
  Mannich reaction, 202
  Perkin reaction, 197
  position of equilibrium, 181
  rate, 180
  stereochemistry, 203
  steric effects, 180
  thiols, 184
  water, 181
Nucleophilic addition to C=O in carboxylic acid derivs., 204–209
Nucleophilicity, 88
Nucleophilic substitution, aliphatic, 73–95
  alcohols, 85
  alkylation of reactive methylenes, 93, 245
  allyl halides, 79

Nucleophilic substitution—*continued*
  benzyl halides, 79
  bimolecular, 74, 81, 85
  1-bromotriptycene, 81
  t-butyl halides, 74, 77
  β-chloroamines, 88
  β-chlorohydrins, 86
  β-chlorosulphides, 87
  effect of entering group, 88
  effect of leaving group, 89
  effect of silver ion, 98
  effect of solvent, 75
  effect of structure, 76
  epoxides, 86, 94, 165
  α-halogenoacids, 87
  iodide ion as catalyst, 89
  kinetics, 73
  list of reactions, 92
  methyl bromide, 73
  neighbouring group participation, 86
  neopentyl halides, 80, 249
  nitrosation of amines, 90, 102
  proton as catalyst, 90
  S_N1 *v.* S_N2, 75, 77, 89
  stereochemistry, 81
  steric effects, 77, 78, 80, 81
  tosylates, 82, 90
  triphenylmethyl halides, 79, 81
  unimolecular, 75, 83
  v. elimination, 223
  vinyl halides, 80
Nucleophilic substitution, aromatic, 148–155
  activated phenyl halides, 151
  anionic intermediates, 149, 151, 152
  as elimination/addition, 154
  benzyne intermediates, 154
  p-chlorotoluene, 153
  diazonium salts, 150
  halogenopyridines, 152
  nitrobenzene, 148
  phenyl halides, 80, 153
  picryl chloride, 152
  pyridine, 150
  steric inhibition of delocalisation, 152

Orbitals
  anti-bonding, 5
  atomic, 1
  axial overlap of, 5
  bonding, 5
  cyclic delocalised, 17
  degenerate, 2
  delocalised, 11
  hybridisation of, 3
  lateral overlap of, 8
  localised, 5
  molecular, 4
  p, 2
  π, 7
  s, 2

Orbitals—*continued*
  σ, 5
  sp^1, 8
  sp^2, 7
  sp^3, 3
Order of reaction, 41, 74
Ortho/para directing groups, 140
Ortho/para ratios
  electronic effects, 143
  interaction with ortho substituent 144
  steric effects, 143
Osmic esters, cyclic, 165
Osmium tetroxide, addition to C=C, 164
Overlapping, orbitals, 5, 8
Oximes
  assignment of configuration, 113
  formation, 185
Oxygen
  as diradical, 269, 280, 281, 286
  migration to electron-deficient, 116–118
Ozonides, 167, 168
Ozonolysis, 167

Paramagnetism, 257, 264
Pentaerythritol, formation, 195
Peracids
  formation, 281
  in Baeyer-Villiger oxidation, 116
  in epoxide formation, 165
Perkin reaction, 197
Perlon, 116
Peroxide effect, 271
Peroxides
  as initiators, 256, 271, 273, 274
  as polymerisation catalysts, 274
  dimers, 168
  heterolytic fission, 118
  homolytic fission, 262, 266
  in HBr addition, 271
  rearrangement, 117, 118
  thermal decomposition, 262, 266
  transannular, 282
Peroxy radicals, 280
Peroxy zwitterions, 168
Petroleum cracking, 106, 262
Phenanthrene, delocalisation energy, 17
Phenol
  acidity, 21, 59
  diazo coupling, 132
  from chlorobenzene, 153
  from cumene, 118
  nitrosation, 125
  reaction with carbon dioxide, 249
  reaction with chloroform, 247
  reaction with formaldehyde, 143
Phenonium ion, 99
β-Phenylethyl cation, 99

Phenyl halides, nucleophilic substitution, 151–155
Phenylhydrazones, formation, 185
Phenylnitromethane, tautomerism, 243
Phenyl radicals, 283
Photochemical initiation, 268, 276
Photodimerisation, 287
Photo-oxidation, 282
π bonds, 7
π complexes, 119, 170
π-deficient heterocycles, 147
π-excessive heterocycles, 147
Pinacol/pinacolone rearrangement, 107
Pinacols, from ketones, 190
pK_a, 53
pK_b, 63
Polarisability, anions, 89
Polymerisation
  aliphatic aldehydes, 197
  alkenes, 164, 274
  coordination, 276
  electrophilic, 164
  oriented, 275
  radical, 262, 264, 274
  vinyl, 274
Propane, rearrangement, 103
Propylamine, nitrosation, 103
Propyl cation, 102
Proton
  acceptors, 30, 52
  donors, 30, 52
Prototropy, 239
Pseudo-acids, 239, 244
Pyridine
  basic strength, 71
  delocalisation energy, 18
  electrophilic substitution, 146
  nucleophilic substitution, 149
Pyrrole
  basic strength, 71
  electrophilic substitution, 147
  polymerisation, 148
  protonation, 72, 148

Quantum numbers, 1
Quasi-aromatic character
  cyclopentadienyl anion, 235
  cyclopropenyl cation, 100
  ferrocene, 236
  tropylium cation, 100
Quinuclidine
  basicity, 71
  complex with trimethylboron, 27

Racemisation, 83, 238, 250, 277
  in S_N1, 83
  in radical reactions, 277
Radical addition, 268
  halogens, 268
  hydrogen halide, 161, 271
  vinyl polymerisation, 274

Index

Radical anions
  from electrolysis of ketones, 263
  from reduction of aromatic ketones, 190
  from reduction of esters, 190
Radicals, 18, 31, 256–288
  addition reactions, 156, 161, 262, 268–276
  alkyl, 260, 261, 263, 264, 266
  allylic, 169, 260, 277
  apocamphyl, 265
  as electrophiles, 277
  benzoate, 262, 266
  benzoyl, 263, 281
  benzyl, 260, 270
  carboxyl, 263
  detection, 259, 264
  dimerisation, 246, 261, 274
  1,1-diphenyl-2-picrylhydrazyl, 259
  diradicals, 269, 282, 286, 287
  displacement reactions, 267, 276–285
  disproportionation, 246, 261, 274
  effect of solvent, 261, 267, 268, 277
  from electrolysis, 263
  from oxidation/reduction, 262, 281
  from photochemical fission, 260, 268, 276, 281
  from thermal fission, 261, 266, 276, 277
  half-life, 264
  hydroxyl, 263
  in arylation, 283
  in autoxidation, 263, 280
  in halogenation, 268, 276
  in polymerisation, 262, 274
  inhibitors, 257, 265 269, 271, 274
  initiators, 257, 262, 264, 265, 271 273, 274, 280
  long-lived, 257–260
  pentachloroethyl, 268
  perbenzoate, 281
  peroxy, 118, 280
  phenyl, 267, 283
  polar effects, 277
  rearrangements, 118, 285
  short-lived, 257, 260–288
  stabilisation, 258, 260, 265, 276, 285, 286, 287
  stereochemistry, 258, 265, 277
  stereospecificity in addition, 272
  triphenylmethyl, 257, 264, 265
Radical substitution, 276
  arylation, 283
  autoxidation, 280
  halogenation, 276
Raney nickel catalyst, 184
Rate constant, 41
Rate-determining step, 42
  and kinetic isotope effects, 47
Rate of reaction
  and catalysis, 43

Rate of reaction—*continued*
  measurement of, 41
Reaction
  energetics of, 35
  heat of, 36
  intermediates, 42, 47
  molecularity of, 74
  order of, 41, 74
  rate-determining step of, 42
  rate of, 38
  types of, 31
Reaction mechanism, investigation, 44
Reagents, classification, 29
Rearrangement, 33, 103–118
  alkanes, 103
  alkenes, 106
  allylic, 33, 103, 193
  Beckmann, 113
  benzilic acid, 201
  1,2-bromohydrins, 108
  carbanions, 254
  crossed, attempted, 109, 116
  Curtius, 111
  diazoamino compounds, 134
  α-diazoketones, 110
  Hofmann, 111
  1,2-hydroxyamines, 108
  in Friedel-Crafts, 129, 144
  in sulphonation of naphthalene, 146
  intermolecular, 109, 134
  intramolecular, 109, 113, 116, 189, 201
  keto-enol, 238
  Lossen, 111
  migratory aptitude in, 110, 117
  neopentyl, 104
  ozonides, 167
  peroxides, 117, 118
  phenylnitromethane, 243
  pinacol/pinacolon, 107
  radicals, 285
  steric effects, 110, 113, 117
  triphenylmethyl peroxide, 286
  Wagner-Meerwein, 105, 106, 129, 144
  Wolff, 110
Reimer-Tiemann reaction, 229, 247
Resonance
  energy, 15
  hybrids, 17

Sandmeyer reaction, 283
Sandwiches, molecular, 236
Saytzeff elimination, 218
Schiff bases, 186
Semicarbazones, formation, 185
σ bonds, 5
σ complexes, 120, 139, 140
Single bond, C—C, 5
  rotation frequency, 7
S_N1 mechanism, 75, 83, 150, 151
S_N2 mechanism, 74, 81, 152, 245, 247

300

S_N2 (*aromatic*) mechanism, 150, 152
S_N2' mechanism, 104
S_Ni mechanism, 84
Sodium borohydride reduction of aldehydes and ketones, 187
Solvation envelopes, 53, 57, 84
Solvolysis, 75
Spin, electronics, paired, 2
Statistical effect in radical halogenation, 276
Stereochemical criteria in mechanism, 49
Steric effects, 26, 203
  crowding, 6, 203
  hindrance, 27
  inhibition of delocalisation, 26, 70, 152, 209
  in rearrangements, 110
Substituent effects in aromatic substitution, 135
Substitution
  aromatic, 119–155
  electrophilic, 119–148
  nucleophilic, 73–95, 148–155
  radical, 276–285
Sulphonation, 127
  isotope effects in, 127
  of naphthalene, 145
  reversibility, 127
  sulphur trioxide in, 127
Sulphonium salts, 93
  cyclic, 87
Sulphur trioxide in sulphonation, 127
Sulphuryl chloride, chlorination with, 279
Super acids, 96

Tautomerism, 238–244
  acid-catalysed, 241, 252
  azomethines, 240
  base-catalysed, 241
  concerted *v.* stepwise mechanism, 239
  distinction from mesomerism, 240
  effect of solvent, 240, 243
  hydrogen bonding in, 243
  keto-enol, 239, 240, 242, 252
  phenylnitromethane, 243
  rate, 241
  structure and position of equilibrium, 242
Terminators, chain, 275
Tetraalkylammonium salts, 92
  elimination reactions, 218
Tetraarylhydrazines, radicals from, 259
Thermodynamic control of reaction
  in addition to conjugated dienes, 171
  in Friedel-Crafts alkylation, 144
  in phenylnitromethane tautomerism, 244
  in sulphonation of naphthalene, 146

Thermodynamics, second law of, 36
Thioacetals, 184
Thioketals, 184
  hydrogenolysis, 184
Thiols, addition to C=O, 184
Thiophenate radical, 259
Toluene
  deuteration, 120
  radical chlorination, 270
Transesterification, 205
Transition state, 39
  cyclic, 188, 192
  in aromatic substitution, 124
  in carbonyl addition, 181, 203
  three-membered, 255, 286
Tricyanomethane, acidity, 233
Triketohydrindene hydrate (ninhydrin), 182
Triphenylmethane
  acidity, 232
  carbanion from, 232, 237
  radical from, 257, 264, 265
Triphenylmethyl peroxide
  formation, 257
  rearrangement, 286
Triple bond, C≡C, 8
  addition to, 159
  energy, 9
  length, 9
Tropylium cation, 100
Tschitschibabin reaction, 150

Ultra-violet radiation, as catalyst, 261
Unexpected products, 50, 153
Unexpected rates, 51
$\alpha\beta$-Unsaturated acids from Perkin reaction, 197
$\alpha\beta$-Unsaturated carbonyl compounds, addition to, 174
Urethanes, in Curtius reaction, 113

Vinyl cyanide in cyanoethylation, 172
Vinyl polymerisation, 264, 274
  branching in, 275
  coordination, 276
  induction period in, 274
  oriented, 275
  termination, 274

Wagner-Meerwein rearrangements, 105, 106, 129
Water
  addition to C=C, 162
  addition to C=O, 181
Wolff rearrangement, 110
Wurtz reaction, 246

Xanthates, pyrolysis, 231

Zwitterions, peroxy, 168